Electronics Process Technology

Wilfried Sauer, Martin Oppermann, Gerald Weigert,
Sebastian Werner, Heinz Wohlrabe, Klaus-Jürgen Wolter
and Thomas Zerna

Electronics Process Technology

Production Modelling, Simulation and Optimisation

Translated by Anthony Rudd

With 281 Figures

 Springer

Wilfried Sauer, Prof. Dr.-Ing. habil.
Martin Oppermann, Dr.-Ing.
Sebastian Werner, Dipl.-Ing.
Heinz Wohlrabe, Dr.-Ing.
Thomas Zerna, Dr.-Ing.

Dresden University of Technology
Centre for Microtechnical
    Manufacturing
Helmholtzstraße 10
D-01069 Dresden, Germany

Gerald Weigert, Dr.-Ing.
Klaus-Jürgen Wolter, Prof. Dr.-Ing. habil.

Dresden University of Technology
Electronics Packaging Laboratory
Helmholtzstraße 10
D-01069 Dresden, Germany

British Library Cataloguing in Publication Data
A catalogue record for this book is available from the British Library

Library of Congress Control Number: 2006926887

ISBN-10: 1-84628-353-1   e-ISBN 1-84628-354-X   Printed on acid-free paper
ISBN-13: 978-1-84628-353-6

© Springer-Verlag London Limited 2006

Printed in Germany

9 8 7 6 5 4 3 2 1

Springer Science+Business Media
springer.com

Foreword

With the development of system theory, the requirement arose in the recent decades not only for its comprehensive use and a theoretically-based penetration into as many disciplines as possible, but also in engineering science. For a long time, production scientists, in particular, were greatly disadvantaged in their methodical and theoretical constructions compared with the natural sciences and there was only limited progress in building a scientific basis. However, with the introduction of powerful information, communications and computer technology, the accumulation and mastering of immense data flows and the continuously increasing interconnection made the requirement for a practicable theoretical basis for manufacturing processes ever more pressing. The strongly growing industrial branching point of electronics has attempted for many years to transfer or adapt models, methods, algorithms and programs from mechanical engineering. However, the specific characteristics of electronics production resulted in practical concepts for the provision of a specific theoretical basis with a more efficient use.

As early as 1970, "electronics process technology" was introduced as a university course and, within a few years, integrated into the new more comprehensive discipline of electronics technology still offered to this day as a successful study course. For the first time, this book attempts to present some of the methods developed in a closed form. Although from the theoretical and mathematical viewpoint they are not always new, the innovation lies in the application area and in the possibility of achieving practical solutions. The authors have gained many years of experience at the Dresden University of Technology, Germany, with these methods, not only in the education of students, but also in industrial use.

I would like to thank all the authors for the high motivation and competence they provided for the preparation of the manuscript. I would also like to especially thank Dr. Zerna who, with his extensive experience, had the time-consuming task of producing a consistent result. Finally, I would like to thank Anthony Rudd for his support in translating our German book into English and for all the support we have had from Springer.

Dresden, April 2006   The editor

Preface

Process Technology Scientific Discipline

Manufacturing of electronic products places ever-increasing demands on technological procedures and processes.

The increasing range of products, coupled with the reduction of the production interval, has led to the creation of flexible manufacturing processes. Furthermore, quality assurance assumes an ever-increasing importance in production, in particular, also because the functional scope of electronic products is increasing permanently. The integration level of electronic components will also further increase in the future, and finally, in the meantime, the miniaturisation of the components and modules has made the transition from the precision technology to microtechnology and, in a few years, to nanotechnology.

All these developments affect manufacturing and cause the manufacturing process itself to become a complex and often also fault-susceptible system that normally can only be maintained stable using process control. Consequently, for new products that have not yet achieved a fully perfected technology, the so-called process window proved to be very small, and, in many cases, the technological know-how has a high strategic significance for a company with regard to process mastering.

With regard to this development, the new discipline of electronics process technology was established in recent decades in research and education. In the meantime, many individual questionnaires and tasks can be combined to form a unit. Similarly, the range of methods for successful solutions has led to a theoretical basis that makes it desirable to present it in a closed form, especially since practical results have proved the usefulness of these methods and this discipline. Although this book pays particular attention to the specific characteristics of electronics manufacturing, transfer to other disciplines is possible.

A special characteristic of electronics manufacturing is the large number of inspection processes and test steps needed to determine faults in products immediately after their occurrence. This raises the question: should the faults be detected and corrected "immediately" after their occurrence or the faulty products separated out as reject, or should this decision be made "later", namely after one or more additional technological steps. This is a typical problem from electronics manufacturing. Figure 0.1 shows symbolically an SMT manufacturing line for the assembly of electronic modules with an inspection process and a repair process immediately after the solder paste printing.

A further problem situation concerns system theory. System theory has been used in electrotechnology for around 80 years. Operational calculus through to the so-called "symbolic methods" have proved themselves to be powerful tools for the calculation of electrical networks and circuits. It is now possible to show that the same methods can generally also be used for the analysis of manufacturing processes, where this produces a very interesting connection to reliability theory.

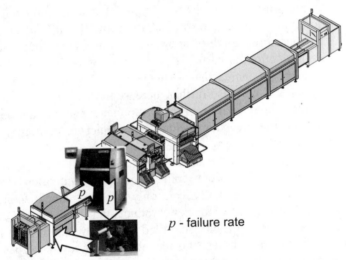

p - failure rate

Figure 0.1. Typical SMT assembly line with included inspection process

Structure analysis, the separation of systems into their elements, flow graph theory and operational calculus give process technology new procedures that can be used for analysis and synthesis. If must also be mentioned that most of the models not only require probability theory and the methods of mathematical statistics, but also can only be interpreted on the basis of the mathematical term of the random quantity and the resulting mathematical methods.

Contents of this Book

The first chapter discusses the modelling of manufacturing processes. Based on the system term and the definition of the process technology parameters, the process characteristic and the test characteristic are presented as generalised descriptions of the technological processes, with operational calculus methods developed as a basis for a theory of product-flow graphs. Some examples show how the presented methods can be used not only for the modelling of manufacturing processes but also for quantitative analysis and synthesis.

The second chapter prepares special methods of graphical theory for the electronics process technology, and presents examples that show their capabilities and limits. The transition from product design to the quantity balance of the manufacturing process and to its structure produces an interlacing graph. The simplest network planning model, CPM, the critical-path method, is described and this augmented and extended with flow graph theory. Similarly to the theory of the signal flow graphs, product-flow graphs developed in this way provide a tool that can be used to represent and compute manufacturing structures in closed form. Figure 0.2 outlines the product flow graph for the SMT manufacturing line shown before. Queue models and Petri nets complete this chapter.

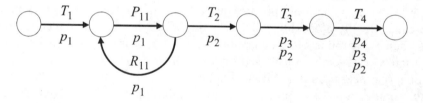

Legend: T - technology
P - inspection process
R - reject process (repair process)
p - failure rate

Figure 0.2. Product flow graph of the SMT assembly line

A chapter for the simulation of manufacturing processes follows. Simulation is often the only method that produces a solution when analytical solution procedures either do not exist or do not produce a result in an acceptable time. In recent years, simulation has been used in new application areas and has proved its efficiency, not only in planning and configuring, but also for direct manufacturing control under real-time conditions. For such tasks, in particular for process-accompanying execution simulation, new methods are shown, original tools presented and electronics manufacturing examples presented to show the effectiveness.

Optimisation is part of the fourth chapter. Starting with the classical extreme-value problems with examples from the yield optimisation and the problem area of the economical lot size, linear optimisation and dynamic optimisation is prepared for electronics process technology. The following section discusses new developments, such as heuristical procedures, and then makes the transition to simulation-based optimisation. Process simulation shows its full effectiveness only in conjunction with the optimisation of these processes. The optimisation always assumes a target function. Examples are provided to show that not only the costs, but also quality, quantity and time duration can be used as operational target parameters.

The fifth chapter, quality assurance, introduces the quality parameters on the basis of probability theory. Much importance is placed here on a clear and exact description, calculation and interpretation, because, especially in quality assurance, inadequate knowledge of the relationships can very easily lead to incorrect conclusions. As major tools of statistical process control (SPC), quality rule cards and sampling test schedules are described in detail, tailored to the requirements of electronics manufacturing, and using many examples. A section for the process and machine capability with new theoretical principles and many practical results completes the chapter.

The sixth chapter, process cost optimisation, illustrates the methods that lead to a low-cost creation of inspection processes during manufacturing. The actual manufacturing process that caused the fault is compared with a so-called quality process that detects and rectifies this fault, and then returns the fault-free products to the manufacturing process or alternatively declares the faulty products as rejects and separates them from the process. Because a quality process can be positioned either immediately after the cause of the fault or after some later process, an optimisation problem results that can be easily solved and leads to dynamic optimisation for large processes.

Statistical test designing has proved to be very effective for process analysis. The methods and examples introduced in Chapter 7 are directly tailored to electronics manufacturing. Because test designing produces regression equations for measurable characteristics, this method has also gained high importance for producing models from statistical data and thus forms a basis for process planning and optimisation.

Chapter 8 summarises the important reliability factors and models for electronics technological processes. The chapter describes the new procedures for determining test uncertainties and for the effect of the test characteristic on the quality factors and process equations. Concepts for a linear systems theory for manufacturing processes can be obtained by generalising the process equation.

The last chapter introduces the principles of a theory of assembly accuracy. The inaccuracies of the positioning action for electronic modules for the placement coupled with tolerance ranges leads to quality models that permit the calculation of the machine capability. Both rectangular and circular connection geometries of electronic components and circuit boards are significant. The theory proved to be usable for determining the operational limits for assembly robots while considering the achievable quality parameters.

The individual chapters introduce many matched methods for an optimum creation of electronics manufacturing processes that can be used both for the planning and preparation of the manufacturing process and also can lead to a high transparency and stabile mastering during the execution phase. The overall aim is always to achieve robust manufacturing processes that are resilient to fluctuations and malfunctions and that realise the optimum quantity, cost and quality for the manufactured products.

The figure summarises the major tasks of process-technology of electronics.

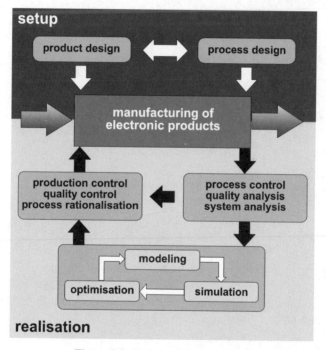

Figure 0.3. Process-technology tasks

Contents

List of Authors

Editor

Prof. Dr.-Ing.habil. Wilfried Sauer
Dresden University of Technology, Electronics Packaging Lab, Germany

Authors

Dr.-Ing. Martin Oppermann	Chapter 6
Prof. Dr.-Ing.habil. Wilfried Sauer	Chapters 1, 2, 4, 8
Dr.-Ing. Gerald Weigert	Chapters 1, 2, 4
Dipl.-Ing. Sebastian Werner	Chapter 3
Dr.-Ing. Heinz Wohlrabe	Chapters 5, 7, 8
Prof. Dr.-Ing.habil. Klaus-Jürgen Wolter	Chapter 8
Dr.-Ing. Thomas Zerna	Chapter 9

Dresden University of Technology, Electronics Packaging Lab, Germany

1 Modelling of the Manufacturing Processes

1.1 Manufacturing Systems in Electronics Production

In accordance with [10], a manufacturing system is considered to be "…the totality of all components (procedure and equipment) that take workpieces from one state into a subsequent state". This general definition suits both complete industrial plants and individual workplaces or manufacturing islands. As in an organism, manufacturing systems can only be understood in the mutually agreed interactions of many subsystems. Each of the subsystems has different specific tasks to perform that always produce two main streams:

- Material-flow system
- Information system.

The material-flow system contains everything required for the material part of the manufacturing process. This includes not only personnel, machines and equipment, the transport and handling systems and the warehouse, but also the supply with raw material, energy, water and other resources. In contrast, the information system covers the immaterial world of the data, and also the invisible control rules that organise the manufacturing processes. If one remains with the image of the organism, the information system is the "brain" and the material-flow system is the "body" of the manufacturing system. The main fields that the information system must cover can be understood with the terms *time*, *quality* and *costs*.

The aspect of time of a manufacturing process is usually considered completely separate from its technological orientation. The individual manufacturing steps are largely reduced to their duration. In this case, one also speaks about *manufacturing sequences*, meaning the time structure of the manufacturing process. The time structure also includes key times (*e.g.*, the earliest or latest completion time), a predetermined break or shift pattern, planned maintenance cycles, *etc*. The manufacturing sequences can be used to make statistical statements about average throughput times, maximum occupation of buffering plants, bottlenecks or the machine and personnel loading capacity. In contrast, quality is largely concerned with the suitability of the products. It is principally influenced by the physical principles in the individual process steps. Time and quality are the true technical engineering domains of the manufacturing process, whereas the costs strongly affect the economics. All three aspects play an important role in the control and optimisation of manufacturing processes and cannot be considered fully independent of each other.

The material-flow system and the information system are closely intertwined. Depending on the expansion level, the manufacturing system can be represented more or less well in the past, the present and the future. Nowadays, many companies conscientiously identify all significant production events so that information can be obtained even years later. However, the planning of manufacturing processes is much more difficult. Both the increasingly difficult to predict turbulences in the external environment and also deficits in understanding the natural laws that affect the internal environment cause these difficulties.

The scientific investigation of manufacturing systems assumes that one has at least concerned oneself with the actual production structures in the industrial practice in order to recognise not only a generally valid natural law but also industry-specific peculiarities. Generalisations can best be expressed as classifications. In accordance with [10], manufacturing systems can be grouped according to the type of the manufacturing organisation, *i.e.* based on the organisation of the material flow, as follows:

- Construction-site manufacturing
- Workshop manufacturing
- Group/line manufacturing
- Production-line manufacturing.

Line manufacturing certainly predominates in electronics manufacturing. In contrast to production-line manufacturing, the individual workstations of a line are not time dependent on each other. A typical example is the so-called SMT (surface mount technology) line for assembling circuit boards. The individual stations – solder paste printing, one or more placement machines and soldering oven – are arranged sequentially and normally coupled with each other using an automatic transport system. Larger production locations have several SMT lines that operate in parallel.

Even this simple example of an SMT line shows the fundamental differences in how the machines operate. Thus, an automatic placement machine can normally only accept and process a single circuit board at any one time. Only when the circuit board has been fully assembled can a new circuit board be started. If the assembling of a circuit board takes t_1 time units, a lot consisting of n circuit boards will require at least time $t = n \cdot t_1$. A continuous conveyor belt is used to move the circuit boards through the soldering oven. In contrast to the placement machine, the soldering oven is ready to receive again after time $\Delta t < t_1$ and so can process several circuit boards concurrently. A lot of size n under these conditions spends at least time $t = t_1 + (n - 1) \cdot \Delta t$ in the soldering process. At a test station operated by m persons, m circuit boards can also be tested concurrently. However, there is an important difference between the soldering oven and the test station in that the circuit boards at the test station do not affect each other. Thus, the minimum duration for testing the lot is $t \geq (n / m) \cdot t_1$. Whereas a so-called batch process describes the behaviour of the placement machine and of the test station, the soldering oven exhibits a continuous-process behaviour.

Models for describing manufacturing sequences must correctly reproduce the time behaviour of the individual process steps, when throughput times, delivery dates or other time-related parameters, including a comparison with the reality, are to be achieved. This also requires that the use of resources (*e.g.*, setup groups) or operating personnel must be correctly represented. Complex manufacturing processes also have a complicated relationship between the individual process steps specified by bills of materials or work plans.

An SMT line is a highly automated process that impedes (at a minimum) operator interventions. In semiconductor manufacturing, the conditions are further intensified by the cleanroom requirements. Under such conditions, the absence of buffer storage between individual workstations can lead to undesirable interactions that impede determination of exact deadlines. For example, automatic placement machines are often directly coupled and so not independent of each other.

In electronics production, the *commissioned production* predominates over the *make-to-inventory production*. This means that the manufacturing process is initiated by a customer order. This requires particularly powerful scheduling methods because the observance of agreed dates is an important criterion for customer satisfaction. In accordance with [6], in addition, significantly more components are processed for each module type as, for example, is usual in mechanical engineering. Most of these parts must be acquired, where the delivery times for some components are measured in months. Consequently, the importance of operational planning and controlling in electronics production cannot be judged highly enough.

It is often difficult for the user to find the suitable system for his company from the abundance of planning tools, control stations or ERP systems available on the market. Because the limits of the tool often become apparent only in practical use, it is certainly useful to consider in more detail the models that serve as a basis for the associated system. The following section attempts to classify such models and to show what the individual models can provide for describing the manufacturing processes.

1.2 Abstract Manufacturing Systems

Manufacturing systems can be described and modelled in various ways. The methods developed for such purposes range from network planning, include Petri nets and queues, through to computing-intensive simulation methods.

Each of the methods has both advantages and disadvantages. Although networks are a simple means of determining deadlines in the manufacturing process, networks cannot be used successfully everywhere. Similarly, although queuing theory has a proven mathematical basis, it is seldom used for controlling manufacturing processes.

This section discusses the similarities and, in particular, the differences between the individual concepts. Set-theory methods are used to describe the major components and structures of manufacturing systems. This basis then can be used to define a type of metamodel from which all models can be derived. These "abstract manufacturing systems" do not contain any machines, queues or manufacturing orders, but only sets and relations. Obviously one cannot "compute" with such models – they are not suitable for actual use in practical industrial applications. However, they can be used to classify and compare the various planning and control methods. Furthermore, the user is given an instrument that can be used to decide on one method using a scientific basis.

1.2.1 Principles and Terms

Date
One of the most important terms in manufacturing control is that of the *date*. Synonymous terms are *time stamp* or *point of time*. In the following section, the date t is defined as being an element of an abstract date set T that is not discussed in further detail: $t \in T$.

In practice, dates are assigned a value such as "$t_1 = 30.01.2003$ 12:30 p.m. (CET)" or "$t_2 = 05.02.2003$ 06:00 a.m. (CET)". The fact that t_1 and t_2 are recognised as being an earlier and a later date, respectively, means that an order relation is defined for the set T. The date set is indeed wellordered, namely, each subset of T contains just one first (smallest) element. The properties of the date set can be used to derive a first model classification.

Model classification I

Models with a countable date set are designated as *time-discrete*, whereas one with an uncountable date set is designated as *time-continuous*. If, in addition, the date set has a smallest and a largest date, it is a *finite* model, otherwise an *infinite* model.

The computer implementation of a model is always based on a finite and thus also countable date set. However, this also means that an earliest and a latest date must exist. The earliest date is set arbitrarily. This is normally the reference date "$t_0 = 01.01.1970\ 00:00$ a.m. (GMT)". The latest date results from the associated application program or from the used hardware. This means that computer models are always time-discrete finite models.

Time interval
Although the order relation in T allows statements such as "earlier" or "later" to be made, it is not possible to determine how much earlier or later a date t_i is in comparison to a date t_j. This requires another metric on T that specifies the distance between each two of its elements. This is defined with a function $d\,(t_i, t_j)$ on the Cartesian product $T \times T$, for which in general:

1. $d(t_i, t_j) = 0$ if and only if $t_i = t_j$
2. $d(t_i, t_j) = -d(t_j, t_i)$
3. $d(t_i, t_j) = d(t_i, t_0) + d(t_0, t_j)$

Case 3 is a special case of the familiar triangle inequality that becomes an equation on the one-dimensional time axis. The separation $d\,(t_i, t_j)$ is the time interval or the time duration between the two dates t_i and t_j and, for simplicity, often also designated as d or Δt. In contrast to the dates themselves, time intervals can be added and subtracted or multiplied by constant factors. Time intervals do not originate from the date set, but are elements of a specific set D. Time durations are measured in time units, such as second, hour or day.

The $T \times T \rightarrow D$ metric allows any ordered pair (t_i, t_j) of dates to be represented in a time interval $d_{i,j} \in D$. In contrast, the $\{t_0\} \times T \rightarrow D$ representation operates with a fixed reference date t_0, so that apparently each individual date t_i will be transformed into a duration $d_{0,i} = d_i$, as Figure 1.1 shows.

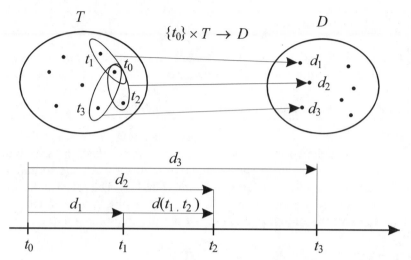

Figure 1.1. Transformation of dates in a time interval

However, one should not "calculate" carelessly with such time durations that result as transformation from dates. The following example with "$t_0 = 01.01.1970\ 00{:}00$ a.m. (GMT)" illustrates this:

i	t_i	d_i / s
1	30.01.2003 12:30 a.m. (CET)	1,043,926,200
2	05.02.2003 06:00 a.m. (CET)	1,044,421,200

The sum of d_1 and d_2 yields 2,088,347,400 s, which, although formally can be transformed in the date "05.03.2036 5:30 p.m. (CET)", this is scarcely sensible. In contrast, the difference yields 495,000 s, which corresponds to the time separation between the two dates. The example shows that one must carefully differentiate between dates and time intervals. However, one cannot always assume that dates are dimensionless quantities, in contrast, time intervals are measured in time units.

Model classification II

A model that can process dates is a *dates-based* model. In contrast, *time-based* models permit only the calculation of time intervals or relative dates.

State

Manufacturing systems are dynamic systems with an N-dimensional *state* [8]:

$$Z = Z_1 \times Z_2 \times \ldots \times Z_N$$

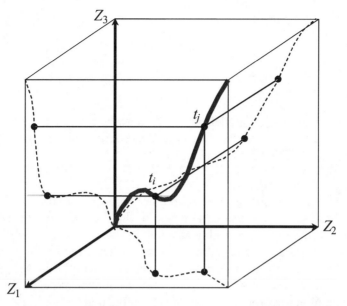

Figure 1.2. Trajectory of a manufacturing process in an assumed 3D state space

Models scarcely work with the complete state space, but only with projections of the state space on one or more subspaces Z^i. This allows simplification or the separation of significant items from nonsignificant items. It is not sensible in a model used to describe manufacturing processes to consider technological parameters, costs or energy consumption if these have only a limited (or no) influence on the time behaviour of the system. For this purpose, other state quantities are transformed to improve handling. If, for example, the machine state is represented as a traffic light (green = operating, amber = finished and red = defective), only the continuous "remaining processing time" is represented in a three-level discrete state quantity.

Model classification III

Manufacturing models are *state-discrete* if the Z set is countable. Otherwise it is a *state-continuous* model.

Simulation

The manufacturing process can be described with a time function in the state space $z = z(t)$. A transformation function is required to determine the system state $z(t+\Delta t)$:

$$z(t + \Delta t) = f(z(t), \Delta t)$$

For discrete systems it is:

$$z(t_j) = f(z(t_i))$$

Model classification IV

This is an *analytical* model, when f is defined explicitly for all Δt and for all i, j. If, however, f is defined explicitly only for $\Delta t \to 0$ and $j = i + 1$, it is called a *simulation model*.

In a simulation model, the computer processing time required to determine the state depends on the length of the time interval Δt. The length of the time interval does not play any role in the analytical model. If some arbitrary system state $z(t)$ is given, it can be used to calculate any other state without any significant affect of the interval length on the expected computer processing time. Obviously, any analytical model, *e.g.*, through the use of a numeric procedure, can also be simulated.

The difference between analytical methods and the simulation is illustrated in the state space for Figure 1.2. The analytical model allows the jump from t_i to t_j. In the simulation model, the complete trajectory, starting with t_i, is followed to reach t_j [13].

Event
The $T \to Z$ function assigns a state from Z to the date from T. The ordered pair

$$(t, z) \in E \subset T \times Z \tag{1.1}$$

is called an *event*. Events, like dates, do not have any chronological expansion. The set of all events E, the *manufacturing process*, can also be represented as a trajectory in the state space.

The order in the T date set is transferred to the set of events E. Consequently, finite systems have not only a first and a last date, but also a first and a last event. Furthermore, a decomposition $[E_1, E_2]$ can be defined in the set E, so that $E = E_1 \cup E_2$ or $\emptyset = E_1 \cap E_2$ applies and every event from E_1 is less than any event from E_2.

When, in addition, E_1 contains a last event and E_2 contains a first event, the decomposition forms a jump in the event set. When all jumps are marked in which a state change occurs, one obtains the generally more restricted event term (see, for example, [2, 10, 14]). However, for reasons that will only become clear in Section 1.2.3, a more general interpretation of the event term should be preferred here. Each point on the time axis has its own event, irrespective of whether a state change or another relevant parameter change is associated with the event.

Subspaces in the state space continue in the event space. For example, all events on a specific machine grouped in a dedicated event space E^i. Processes, however, cannot only be projected on machines and equipment, but also on products. This allows, depending on the requirement, an event-related or a machine-related view of technological processes to be derived. Each of this subspaces, like the date set, is well ordered. Consequently, concurrent events must lie in different event spaces.

Action

The *action* V can be defined from two decompositions $[E_1, E_2]$ and $[E_3, E_4]$ of the event set as follows:

$$V = E_3 \setminus E_1$$

Without any restriction on generality, it is assumed that $E_1 \subset E_3$ applies. The first event of the subset V – also called the start event – is then identical to the first event from E_1. The last event from E_3 is also the last event from V; it is called the final event.

Actions have a time expansion that results from the difference of the dates for the final event and the start event.

> **Model classification V**
>
> If the set of all actions contained in the model is countable, the model is called *action-discrete* or simply *discrete*. Otherwise it is an *action-continuous* model.

Only a countable event set can be constructed from a countable date set, which, in turn, permits only countable action sets. This means that time-discrete models are also action-discrete models. The reverse, however, does not apply. The state set does not have any affect whether or not a model is action-discrete or -continuous. Both model classes can be constructed from both countable and noncountable state sets.

Intensity

$N(t)$ is the number of those actions whose start event lies in the interval $[t_0, t)$. The average action intensity of a manufacturing process in the time interval Δt is then defined as follows:

$$\overline{I}(t) = \frac{N(t + \Delta t) - N(t)}{\Delta t} \tag{1.2}$$

In the case that the manufacturing process is stationary, the average action intensity for a sufficiently large time interval Δt is independent of the time.

$$\bar{I} = \frac{N(t+\Delta t) - N(t)}{\Delta t} \tag{1.3}$$

In the action-continuous system, the following applies:

$$\bar{I}(t) = \frac{\mathrm{d}N(t)}{\mathrm{d}t} \tag{1.4}$$

Control
The control of a model can be either time- or state-oriented. For time- or cycle-oriented procedures, the monitoring or control dates are specified and each associated system state is then determined. The exact date for a state change obviously cannot be determined more precisely than that associated time grid permits. Thus, to prevent errors, the time grid should be as fine as possible. This, however, leads to a relatively long computing time, which, in particular for systems with relatively few state changes, becomes more significant.

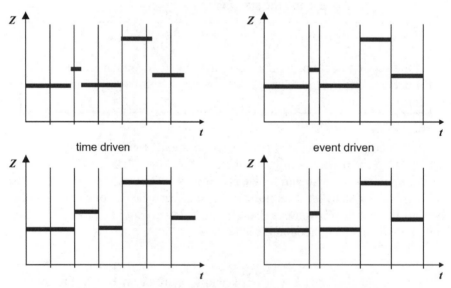

Figure 1.3. Comparison between time- and event-driven models

Event-oriented systems work much more effectively for those in which the date of the state changes is determined and not reversed. In general, event-oriented systems can resolve the time axis as finely as needed, namely they also work exactly as for time-continuous systems. Typical for event-driven models is that the time axis is represented distorted. Namely, the time interval in which few state changes occur will be shortened, whereas those with very many state changes will be lengthened. The significance of these models lies in the differentiated evaluation of the state trajectory [1, 7].

Model classification VI

Event-driven models are controlled by the system state. For *time-driven* models, the system state is determined at a specified time.

Workstep

A workstep is considered to be a single manufacturing step that cannot be further decomposed. The designations *operation*, *process step* or *activity* are also used as synonyms. The workstep is performed during the manufacturing process and thus converted into a concrete action. Each workstep can only be performed once. Thus, identical actions, which, for example, are performed on different products, also require different worksteps. The set of worksteps is the actual foundation on which the models for describing manufacturing systems are based. The following section assumes that the workstep set A is countable:

$$A = \{a_1, a_2, a_3, \ldots, a_n, \ldots, a_N\}$$

The structuring of the set A decides how the individual worksteps can be converted into actions. If A does not have any structure, each workstep a_n can be performed independently of all other worksteps. The number of permitted manufacturing processes will be reduced to the degree with which additional relations and functions are declared for A.

Order relations

The structure of the workstep set is principally determined by order relations that have their roots both in technological and in organisational conditions. The order relation $O \subset A \times A$ defines predecessor–successor relations between the individual worksteps. The workstep a_2 can be started only when workstep a_1 has been completed, when $(a_1, a_2) \in O$, namely $a_1 < a_2$ is true. Each workstep $a_n \in A$ divides the workstep set A into three disjoint subsets [5]:

$$\mapsto a_n = \{a \mid (a, a_n) \in O\}$$
$$a_n \mapsto\, = \{a \mid (a_n, a) \in O\}$$
$$\updownarrow a_n = \{a \mid (a_n, a) \notin O \wedge (a, a_n) \notin O\}$$

The workstep a_n can be started only when all actions from $\mapsto a_n$ have been completed. The worksteps from $a_n \mapsto$ can be started only when a_n has been completed. The worksteps from $\updownarrow a_n$ are concurrent to a_n.

Sequences
A workstep sequence F is a completely ordered subset of A.

$$F = \{a_{n-1}, a_n : a_{n-1} < a_n\} \subseteq A$$

Attributes
Like other objects of the manufacturing system, worksteps can also be given *attributes*. The attributes themselves are elements of a property set X. The primary attributes are normally assigned to the worksteps using the representation $f : A \rightarrow X$. The several property sets X^i or representations f^i often defined for a model describe the various aspects of the manufacturing process. The representation of the workstep set on the property sets is formed in A equivalence classes. All worksteps with the same $x \in X^i$ attribute value are equivalent with regard to the attribute X^i. The workstep set A is divided using the representation f^i into equivalence classes:

$$\tilde{a}_n^i = \{a : f^i(a) = f^i(a_n)\} \subset A$$

Typical workstep properties are processing times, workstations, cost centres, *etc.* Examples of equivalence classes are worksteps with the same start date (synchronised worksteps) or worksteps that use the same production unit (job, manufacturing order or manufacturing lot).

In addition to primary attributes, secondary attributes and higher-order attributes can be defined. For a workstep that can also be performed on different machines, the processing time, for example, is a secondary attribute of the form:

$$A \times X \rightarrow D$$

Where X is the set of machines and D is the set of time intervals.

Resources, capacity
Simple attributes have no affect on the structure of the representation or relation between their attribute set and the workstep set. Consequently, it is possible to link a determined attribute value (*e.g.*, processing duration, start date) with any number of worksteps. Attributes that can only be linked with a limited number of worksteps at any one time are called *resources*. The representation $Y \rightarrow N$ defined on the resource set $Y = \{y_1, y_2, \ldots, y_m, \ldots, y_M\}$ assigns the natural number K_m, the *capacity* of the

resource, to each resource y_m. In addition, a capacity requirement $a_n \in A$ is uniquely assigned to each workstep $k_n \in N$. The relation $R \subset A \times Y$ is completely described by the incidence matrix $G = (g_{n,m})$:

$$g_{n,m} = \begin{cases} 1 & \text{if} \quad (a_n, y_m) \in R \\ 0 & \text{else} \end{cases}$$

The loading k_m of the resource y_m then can be calculated as follows:

$$k_m = \sum_n g_{n,m} \cdot k_n \qquad (1.5)$$

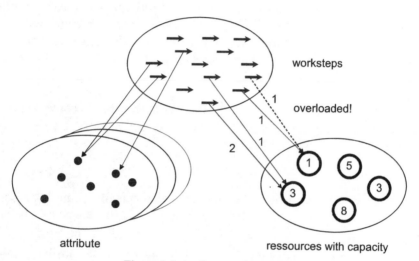

Figure 1.4. Attributes and resources

A relation $R \subset A \times Y$ is permitted when:

$$k_m \leq K_m \quad \forall y_m \in Y$$

namely, the workstep a_n can be linked with the resource y_m at a given time only when its capacity K_m permits this after taking into consideration the existing loading k_m. If this is the case, the associated resource is able to accept and the workstep can be performed at this time. Resources with infinite capacity can always accept, similarly, worksteps with a capacity requirement of zero can also be performed. In this regard, attributes can also be considered to be unlimited resources. Limited resources significantly affect the manufacturing process because they create indirect dependencies between the worksteps.

Although the loading of a resource is time dependent, it does not itself contain any time dimension. The loading curve of the resource y_m can be used to determine the performed work, the *work capacity* $K_{A,m}$, in the interval $[t_0, t]$ as follows:

$$K_{A,m}(t) = \int_{t_0}^{t} k_m(t_x) \cdot dt_x \qquad (1.6)$$

This allows the average capacity at time t, measured from time t_0, to be calculated:

$$\bar{k}_{m\,(t)} = \frac{K_{A,m}(t)}{t - t_0} \qquad (1.7)$$

In practice, often no difference is drawn between work capacity and capacity. Rather, the capacity term is used in conjunction with the processing duration or the availability duration of resources and so, corresponds better to the term work capacity used here.

Sequences

An arbitrary workstep sequence F in the workstep set A is converted with the representation $f^i : A \rightarrow X^i$ into a sequence $S(X^i, F)$ in the attribute set X^i.

$$S(X^i, F) = \langle x_1^i, x_1^i, \ldots, x_m^i, \ldots \rangle$$

In contrast to the sequence, the individual elements of a sequence are not necessarily different. Workstep sequences whose elements belong exclusively to an equivalence class $\tilde{a}_n^j$ with regard to the attribute set $X^j \neq X^i$ are specific sequences in X^i with regard to the attribute value $f^j(a_n)$.

Examples:

The *organisational sequence* is a specific sequence in the set of the manufacturing orders with regard to a selected workstation from the resource set.

The *technological sequence* is a specific sequence in the resource set with regard to a selected manufacturing order from the order set.

Model classification VII

In contrast to the *resource-related* models, only simple attributes and relations are defined in *resource-free* models .

Nondeterministic models

All previous discussions used unique relations, thus they describe deterministic models. In deterministic models, for example, just one attribute value from a selected attribute set X is assigned to a workstep. In non-deterministic models, several attribute values can also be assigned to the same workstep. A determined value is selected randomly. The associated representation is modified as follows:

$$f : A \to P(X) \backslash \emptyset$$

The set $P(X)$ designates the power set for X.

Special structures

Manufacturing systems are also differentiated using the organisation principle, namely according to which sequences are permitted. Manufacturing systems for which all sequences are permitted are also designated as a workshop or as a *job shop*. As in a workshop, a technological sequence is individually assigned to each product and according to the associated requirements. Similarly, the organisational sequences at the individual machines are independent of each other. Manufacturing systems in which neither technological nor organisational sequences are specified, are designated as an *open shop*. If all products exhibit the same technological sequence, these are called a manufacturing line or a *flow shop*. If, in addition, the products within a manufacturing line cannot overtake each other, one observes the same organisational sequence at all workstations [11].

The organisation principles represent, in effect, the corners of a definition space that can be used to represent real manufacturing systems. The pure form is seldom found. However, mixed and special forms occur frequently.

1.2.2 Scheduling

Scheduling is considered to be the chronological classification of the worksteps, namely, the calculation of the expected start and end dates. Obviously, the execution sequence must not contradict the model details. The accuracy with which the real manufacturing process matches the plan depends both on the structural agreement of the model with the reality and also on the expected malfunctions.

A model that already contains all the deadlines does not allow any planning freedom, it is the execution sequence itself. In general, the less information that a model contains, the easier is the planning task. A model that, for example, specifies only the processing duration per workstep allows practically any execution sequence to be developed without any conflicts. For example, all worksteps could start at the same time or, just as good, successively. The decision for one of the possible plans can be dependent on the optimisation criteria.

There are a large number of planning methods, and an even larger number of planning tools. On closer investigation, however, two main groups, each based on resource-free or resource-related models, can be differentiated. The latter can be further divided into models without interaction and those that take the interactions into consideration. Generally, it can be expected that resource-free models are the easiest to master, whereas interaction-related models are the most difficult to master. Although powerful analytical procedures exist for the former, simulation methods must be used when interactions need to be considered. In practice, however, such knowledge is still too little observed. The decision to use a specific planning method often depends on its availability, less from its general suitability for the actual problem. [4, 11, 18]

Infinite Scheduling

As the designation suggests, these planning methods are based on models that do not contain any resources. The planning uses only attributes and order relations. Because attributes behave just like resources with infinite capacity, this is also called planning with unlimited capacity.

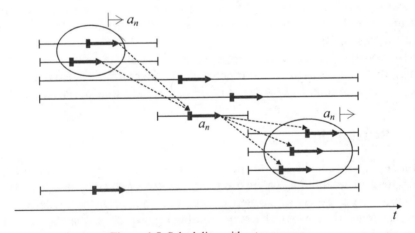

Figure 1.5. Scheduling without resources

The network planning and also the product-flow graphs belong to this category. In a permitted execution sequence, the following duration d_n applies to the start date $t_{n,A}$ and the end date $t_{n,E}$ of any workstep $a_n \in A$:

$$t_{n,A} + d_n \leq t_{n,E} \tag{1.8}$$

$$t_{n,A} \geq \underset{\mapsto a_n}{\text{Max}}\{t_E\} \tag{1.9}$$

$$t_{n,E} \leq \underset{a_n \mapsto}{\text{Min}}\{t_A\} \tag{1.10}$$

The predecessor and the successor of the workstep are determined only by the order relation. Worksteps that do not have any relation to each other are concurrent and so can be planned and executed independently of each other. The familiar algorithms of the network planning for the forwards and reverse scheduling are based on Equations (1.8)–(1.10).

Scheduling with Blocking-free Resources

In addition to the order relations, the resources, provided they are used in the model, influence the manufacturing process. Figure 1.6 shows a manufacturing system for which the worksteps are associated with resources whose capacity is assumed to be 1. Thus, worksteps associated with the same resource exclude each other. In the example, these are the worksteps a_2 and a_4, and a_3 and a_6, that compete for the shared resource S_3 or S_2. The mutual exclusion of worksteps can be achieved in various ways. One possibility involves also specifying the workstep sequence that is dependent on the critical resources (organisational sequences), namely, the order relation within the workstep set. In the network, this would make itself apparent though the inclusion of addition pseudoactivities.

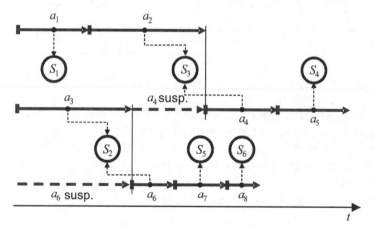

Figure 1.6. Execution sequence for limited resources

The cause of the suspension of worksteps, however, can be found in the resources themselves, which, in their effect, are comparable with the semaphores known from parallel programming theory. In Figure 1.6, the station S_3 acts as a semaphore set by the first workstep a_2 that arrives and then is released on completion. The workstep a_4 that impacts on the occupied station will be temporarily suspended. Which of the two worksteps is first served and which is suspended depends on many factors and cannot be determined just from the order structure. Consequently, the individual planning concepts do not fundamentally differ whether the resource binding was originally considered or not.

Scheduling with Blocking Resources

It is assumed for blocking-free systems that worksteps only use resources while they are being executed. In contrast, no resource binding occurs in the suspended state. Unfortunately, this initially plausible assumption does not always meet the reality. In general, not the suspended workstep itself, but rather its immediate predecessor, remains on the resources.

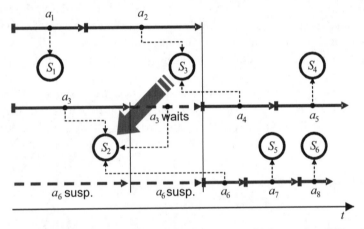

Figure 1.7. Execution sequence for limited resources with interaction of S_3 on S_2

As the example in Figure 1.7 shows, a_3 waits for resource S_2 while a_4, its immediate successor, remains suspended by S_3. Thus, the workstep a_2 causes indirectly also a suspension of the workstep a_6, although neither uses any shared resource. The interaction of S_3 on S_2 causes a back-up. These indirect relations between the worksteps are particularly difficult to calculate in advance and almost always can only be mastered using simulation methods.

It must be decided depending on the particular situation which model, under the specified practical conditions, is suitable for the operational planning. For high resource availability or low system load, a plan that assumes unlimited capacity can be sensible, in particular, because the associated algorithms are relatively simple. However, resource-free models must always be considered critically when they are to be used in manufacturing systems whose machines are either not, or only inadequately, uncoupled from each other using a buffer storage. In this case, the very complex relations that form between the individual worksteps can often only be solved using simulation methods.

1.2.3 Static Process Model

Complex Time Level and Process Tracks

As already mentioned in Section 1.2.1, a process consists of a sequence of events, where the event itself is defined as a link of system state z and date t. Events are transient and do not have any chronological expansion. In contrast, the lifetime of the process lasts from the date of its first event through to the date of its last event. The process is also transient, because it does not exist outside its lifetime and the time cannot be stopped or, in particular, reset. In the following section, a model will be developed that permits processes to be reproduced or predicted any number of times and at any time. The model operates with process copies whose events are linked with the original process using a virtual timestamp specified independently of the actual time t. As for a video, different playback speeds, and both forward and backward running processes, can be created.

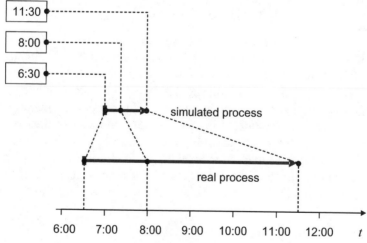

Figure 1.8. Parallelism for a real and simulated process

Figure 1.8 shows an example of a manufacturing process accompanied by a simulation running parallel in time. The manufacturing process begins at 6:30 a.m. and ends at 11:30 a.m.

The simulation is a faster-running film of the manufacturing process and has the goal of predicting the significant parameters of the manufacturing process. The simulation starts at 7:00 a.m. and completes at 8:00 a.m. The unusually long simulation duration is used here only for clarity, but does not change anything in general. The initial data for the simulation originate from the state of the manufacturing system at time 6:30 a.m., in contrast, the simulation result corresponds to the system state at 11:30 a.m.

Assuming that the simulation itself can be observed, initially only events from the past are visible. However, the simulation slowly approaches the present to provide a result that precedes the real manufacturing process by 4.5 h. The simulated process represents an accelerated manufacturing process whose events, depending on the point of observation t, are already in the past or provide an image of the future system state. The simulator provides all the events of the picture process with additional virtual dates t_v that correspond to the real dates of the manufacturing process. The virtual and real date agree only in the manufacturing process itself.

The real date t when the event actually occurs and the virtual date t_v when the simulated or calculated event corresponds to the reality are independent of each other. Thus, each event has a complex date v that consists of an ordered pair (t, t_v). The event now defined by the $T \times T \rightarrow Z$ function is an extension of the event term of Equation (1.1).

$$(t, t_v, z) \in E \subset T \times T \times Z \qquad (1.11)$$

Adding the virtual date to its own time axis produces the complex time level shown in Figure 1.9. The processes themselves are then directed lines in this time level, whose real and virtual lifetimes can be read from the projections on the associated time and date axis. Thus, the simulated process runs one hour in real time, which corresponds to an overall manufacturing time of 5 h. The intersection of the two straight lines marks the time when the verification of the past makes the transition to the prediction. It will be shown later that the complex time level has the additional advantage that the current execution speed of the simulation can also be represented exactly.

In Equation (1.11), the virtual date t_v can also be replaced by the age of the event:

$$\tau = t - t_v \qquad (1.12)$$

The event age is a time interval and, as Figure 1.10 shows, is represented as duration. The events of the real process then lie only on the t-axis and the simulation from Figure 1.8 is a downward-pointing arrow.

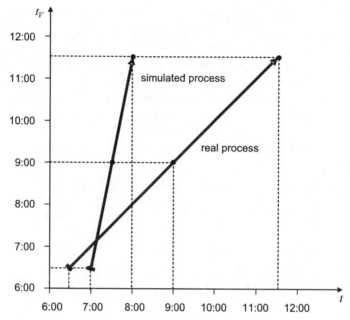

Figure 1.9. Representation of the virtual dates on their own time axis

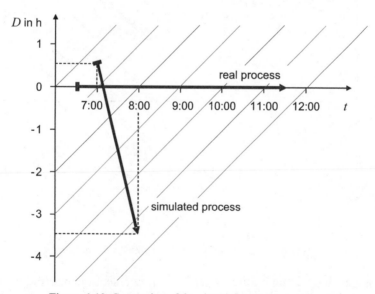

Figure 1.10. Conversion of the virtual dates to an event age

The simulator initially generates events from the past ($\tau > 0$) and future events ($\tau < 0$) for the continuing simulation duration. The system state, which, for example, is set to 8:00 a.m., is already simulated at 7:20 a.m., namely 40 min before its actual realisation in the manufacturing process.

The following section considers only time-based models, namely, time intervals are represented rather than dates. This does make any fundamental change, but some computation rules can be formulated. Figure 1.11 shows the time level represented by the time t and that is applied to the age τ. All events above the t-axis indicate a positive age and thus have already elapsed. In contrast, events below the t-axis will only be realised in the future because they have a negative age. Only those events on the t-axis itself are real, namely current events with the age zero. All other events are designated as virtual events because they are only representations of future or past events.

An event realised at time t_0 ages with continuing time:

$$\tau = \tau(t) = t - t_0$$

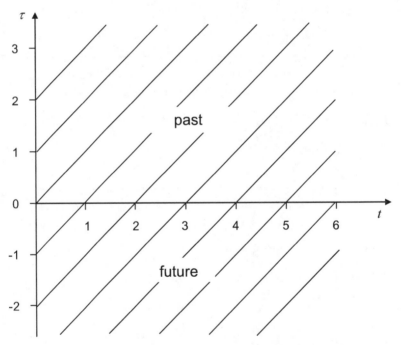

Figure 1.11. Complex time level with event lines

Virtual events that lie on the lines $t_0 = t - \tau$ represent the one and the same real event that was realised at time t_0 or will be realised (negative age). Thus, the line will also be designated as the event line in the following section.

The directed pair from time t and age τ can be interpreted as being a complex time v and formally also written as a complex number. In this case, the real part and the imaginary part of the complex time designate the observation time and the age of the event related to the real process, respectively.

$$v = t + j\tau$$

The function $z = z(v)$ assigns a unique system state z to each complex time v. This leads to a generalisation of the process term. Because the events are no longer transient, but are retained along their event line for any length of time in the past and in the future, this can also be called a *static process model*. When a track $\tau = \tau(t)$ is added to the complex time level, almost any process can be generated from the static process model. A virtual event created in the intersection point of the track with the event line $t_0 = t - \tau$ observed at time t and has age τ. Each track defines its own process instance.

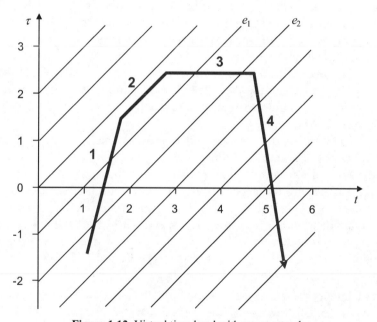

Figure 1.12. Virtual time level with process track

Figure 1.12 shows an example of a process track. In section 1, the events on the track apparently age faster as the real time progresses. A slow-motion recording of a future process can be observed in which the film is eventually overtaken by the reality, namely, past events are observed. In section 2, the age of the events does not change; this produces a stationary image of the process. In section 3, there is only a time displacement between the real process on the t-axis and the film. The observer sees the events running with the same speed as in reality. Section 4 represents a compressed time, because the events become more recent as time continues. After some time, the virtual process leaves the past and produces future events. The events e_1 and e_2 are created twice – once in section 1 and once in section 3.

In the simplest case, one can image the process model as being a storage medium, such as a video tape. The associated video recorder is an example of a process model that is only capable of storing past events for an unlimited time and replaying them at any time. In contrast, analytic models or simulation models also have the capability of predicting future events. This allows events such as sun or moon eclipses to be calculated for a long time in advance or in the past. The astronomy laws, however, are an ideal case. In most cases, the memory capability of a process model decreases with increasing distance from the time axis. A typical example here is the weather forecast. Not only are predictions over a longer period hardly possible, but also the ability to remember long-past weather events, at least specific events, decreases greatly. Process models can generally be evaluated according to the quality of their memory and whether they are only capable of returning just past events or whether they can also make predictions. The tracking of events in manufacturing processes varies greatly. The process data acquisition often applies to quality data, and only less frequently to execution data. The difficulty in predicting manufacturing processes is comparable to that of the weather: either suitable models or adequate data are lacking.

Each function $\tau = \tau(t)$ is a permitted process track. Namely, a process track generates at a specified time t a maximum one (virtual or real) event with the age τ, never, although more than one event.

Process tracks are not necessarily continuous, but, in general, can consist of a sequence of complex event times. Continuous sections of a process track create continuous processes. The speed with which the events pass by the observer is determined by the derivation of the process track $\tau(t)$ according to time, the ageing speed $\dot{\tau}(t)$, Figure 1.13 shows how the process acceleration can be calculated at any point of the process track.

The tangents to the process track form the angle φ with the time axis. Two events that follow each other in the real process in the separation dt_r have the chronological separation dt on the tangent. The acceleration a specifies by how much the time between two consecutive events in the virtual process appears reduced compared with the real process and can be written as the following differential quotient:

$$a = \frac{dt_r}{dt} \tag{1.13}$$

If the time interval on the track is longer than on the t-axis, the virtual process runs slower than the real process and the acceleration factor is less than 1. Figure 1.13 shows just this case.

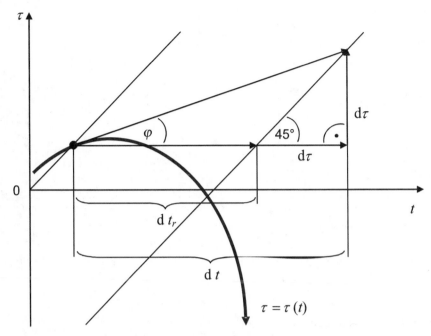

Figure 1.13. Calculation of the acceleration factor

This yields the following ageing speed on the track:

$$\dot{\tau} = \frac{d\tau}{dt} = \tan \varphi \tag{1.14}$$

This allows the acceleration factor a to be determined as in the following equation:

$$a = \frac{dt - d\tau}{dt} = 1 - \dot{\tau} = 1 - \tan \varphi \tag{1.15}$$

Thus, the process-track graph allows not only the determination of the virtual time of a process but also the execution speed. Figure 1.14 summarises the most important cases.

The process tracks allow the development not only of forward- and backward-running virtual processes, but also compressed-time and slow-speed representations. Which property the virtual process has at a given time can be simply read from the direction angle φ of the track tangent.

The static process model assumes a storage medium that behaves the same in the past and future, and in which the process acts. The ideal process medium exhibits an infinite memory in both directions, namely the observed system state may not change along an event line.

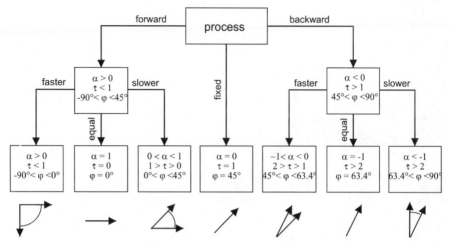

Figure 1.14. Relationship between the acceleration factor, ageing speed and gradient of the process track

Control of Processes

An external intervention in the process that changes the system state in some unforeseen manner is designated as a *control*. The following section assumes an event-related control at time t_0. The control itself obtains its information from the past of the process up to time t_0.

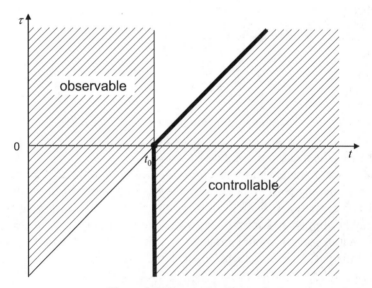

Figure 1.15. Event-related control

Figure 1.15 shows the principle. The thick line marks a fault in the event space caused by the control event. The controllable event space on the right contains all events that still can be changed by the control. The part of the event space that can be observed up to time t_0 is marked on the left.

The event-related control is an ideal case not achieved in practice. Before an event can be observed, it is already aged by the amount τ_o because the information must first be collected and prepared. The actual control intervention also requires a preparation time whose magnitude will not be less than τ_c. Thus, the time for reordering jobs or, in particular, for setting up machines, cannot be ignored.

Consequently, the control must be applied to a virtual event at time v_c during which the required information about the process course is obtained from the virtual event time v_o. The τ_o and τ_c quantities are characteristic parameters of the control system and designated in the following section as an observation or control horizon.

Figure 1.16 shows the control and observation space when a control and observation horizon is present. The control and observation space do not have any contact point, rather a time gap $(\tau_o + \tau_c)$ exists between the two during which the process is no longer being observed but the control has not yet become effective.

This "blind area" can only really be ignored when the process does not change significantly during this time or the process is stopped for the control. For obvious reasons, the latter is not used for manufacturing processes.

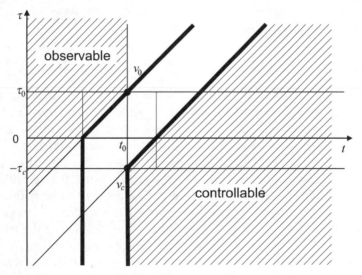

Figure 1.16. Delayed control

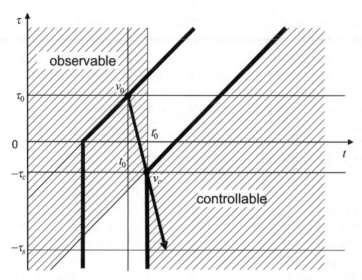

Figure 1.17. Predictive control with virtual extension of the observation space

Figure 1.17 shows how the observation space is expanded by the simulation of the manufacturing process so that the observation and control space can be combined again. The prerequisite is a simulation running in parallel with the actual manufacturing process (process-accompanying). In addition, the simulation must be significantly faster $(a \gg 1)$ than the manufacturing process and the simulation horizon τ_s must be larger than the control horizon τ_c. Under these conditions, the control of the manufacturing process can be assumed at time t_0' or later. This form of control is designated as predictive control.

Comparison of Processes

The predictive control of processes assumes the use of models that permit future events to be forecast with adequate reliability. The virtual expansion of the observation range can be used only when it also agrees with the subsequent reality. This raises the question of how simulation models can be compared with the reality and simulation models. A comparison should always take place at the process level, namely using the relevant state variables. To compare two system states z_1 and z_2 with one another, one normally defines a distance function $d(z_1, z_2)$ that yields the value 0 for equality. In the simplest case, if permitted, the difference of the two state values is used as the distance function. The time must also be considered for the comparison of two processes. It is decisive here that only states with the same virtual timestamp t_v can be compared with each other, irrespective of when each of the events are observed.

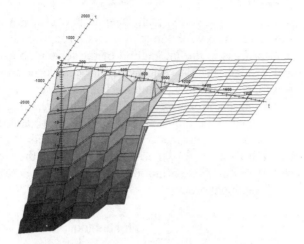

Figure 1.18. Distance function of an adapted simulation model

The static process model permits a time-independent comparison between the processes by the distance function being formed over the complex time level.

$$d(v) = d(z_1(v), z_2(v))$$

When a scalar distance function is used, it is even possible to calculate a "mountain range" over the time level that can be used to draw conclusions about the degree of agreement of a manufacturing process with its simulation model. The distance function can, as for the process itself, be determined along the process track. Furthermore, the static process model also has the advantage that the effect of synchronisation and adaptation methods can be shown clearly. Figure 1.18 shows the distance function $d(v)$ for a simulation model that was improved over the course of time. Whereas reliable forecasts are scarcely possible at the start of the simulation, the agreement between model and reality is almost perfect at the end. Chapter 3 describes in more detail how such a distance function can be created. [12, 15, 17, 16].

1.3 Characteristics of Technological Processes

1.3.1 The Technological Process Term

In Section 1.2.1, the workstep was used as the elementary term for the model formation using abstraction from the designations work action, process step and activity. The term "technological process" now introduced as a generalisation here always contains the workstep as its core.

A technological process is the interaction of the technological procedure and the technical product. The technological procedure is almost always implemented using technological equipment. The technical product, briefly: product, is generally an electronic product here. The technological process contains the three components: material, energy and information. Similar to a coin, the technological process has two sides: the procedure side and the product side. The unit formed from these two elements is the technological process (Figure 1.19). For the formation, the word *Process* should be assigned to the technological process.

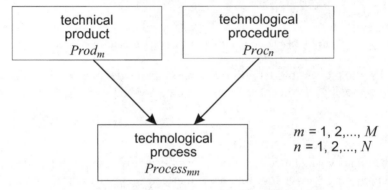

Figure 1.19. The two sides of the technological process

Technological processes can be further subdivided into individual processes. Conversely, several technological processes can be combined together. In this way, the technological process receives the property of elements connected using structures to form systems. The terms, complete process, elementary process, process element and others, used for the model formation are not further explained here, however, they are clearly defined for each actual case of modelling.

In accordance with the previous discussions, each technological process $Process_{mn}$ is uniquely determined by the technological procedure $Proc_n$ and the product $Prod_m$. Because each technological procedure with several products can form a technological process and, conversely, a specific product "flows" through several different processes, in the general case a technological process must be identified by *two* indices (m and n) (Figure 1.19): $m \hateq$ product, $n \hateq$ procedure.

The vector of the technical products and the vector of the technological procedures produce a matrix of technological processes that can be coupled using the appropriate flows (see Figure 1.20 and Figure 1.21).

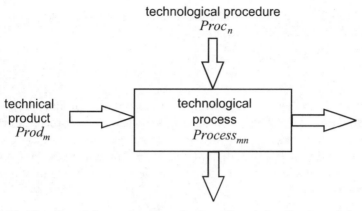

Figure 1.20. Direction of the product flow (horizontal) and the direction of the "virtual procedure flow" (vertical)

The duration d_{mn} (process duration) is an elementary characteristic value for each technological process $Process_{mn}$. Two different possibilities exist for the chronological sequence of several technological processes:

First possibility
As determined by the *technological sequence*, during the manufacturing process, a specific product $Prod_m$ passes successively through the individual processes $Process_{m1}$, $Process_{m2}$,...,$Process_{mn}$,...,$Process_{mN}$ (a total of N processes). These are the horizontal flows in Figure 1.21.

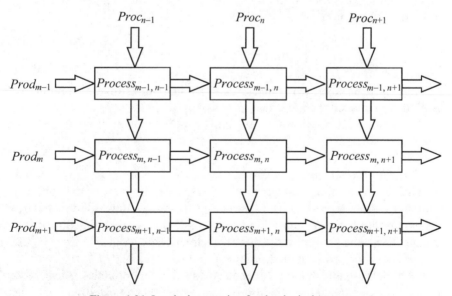

Figure 1.21. Interlacing matrix of technological processes

The complete process duration for this product is the sum of all individual durations.

$$d_{msum} = \sum_{n=1}^{N} d_{mn} \qquad (1.16)$$

This does *not* contain any waiting times of the product between the individual processes. This performs a process sequence, which, in principle, corresponds to a procedure sequence. Because each technological procedure (each piece of technological equipment) is present at a specific location (at a specific point at a "place"), the product performs a "movement" through the process chain.

Example:

A specific circuit board $Prod_m$ passes through the 4 processes, solder paste printing ($Process_{m1}$), placement ($Process_{m2}$), soldering ($Process_{m3}$) and test ($Process_{m4}$) in the specified sequence.

Second possibility

In accordance with the *organisational sequence*, a specific technological procedure $Proc_n$ is realised successively for several products $Prod_1$, $Prod_2, \dots, Prod_m, \dots, Prod_M$. The number of products is M units. These are the vertical flows in Figure 1.21. The technological procedure was, in effect, "split" into the various products to produce a *virtual flow of the technological procedure through the products*.

The complete process duration for this procedure is the sum of all individual durations

$$d_{sumn} = \sum_{m=1}^{M} d_{mn} \qquad (1.17)$$

This equation also does *not* contain any idle times of the technological equipment between the individual processes.

When practical applications are considered, three different cases must still be considered.

Case 1: All products are identical

This case occurs almost always, where M is determined by the lot size L (job size, series size, *etc.*). Because the process durations often differ from product to product, the duration must be handled as a random quantity in the model. The same applies to the quality characteristic values.

The sequence problem does not occur in this case.

Each individual technological process for identical products is essentially repeated under the same conditions. Consequently, repeated technological processes often exhibit a stochastic behaviour. If technological processes are described using a model, this always assumes repeated processes or the manufacturing of identical products.

The designation *repeated technological process* should be used only when there is no possibility of confusion with the individual technological process.

Case 2: All products are different
The technological procedures can be "parameterised" in many manu-facturing systems, namely a so-called *work point of the technological process* can be defined that can be set for the associated product. The various products to be used for the same technological procedure belong in general to a specific class. The specific technological process varies for each product. Varied technological processes are often called simply *technology*. Examples: placement technology, soldering technology.

Case 3: Several different lots
This is the typical case in electronics production. The number of identical products in the individual jobs, lots or items is different and the optimum organisational sequence can only be approximated (see Section 4.4).

1.3.2 Characteristics

Characteristics for the Product Set ("Process Set")

Process technology has many terms that concern the product set and that use the word *unit* as dimension. Examples are: series (or series size), batch, lot (or lot size, lot scope), job (also job volumes), requirement. These terms can be used in the same way as characteristic variables, *i.e.* they consist of a numeric value and a dimension that are linked multiplicatory. Each characteristic variable has a short designation (symbol), *e.g.*, the following notation can be used for a lot size of 200 units: $l = 200$ units.

As previously introduced in Section 1.2 about abstract manufacturing systems, this notation is used in the following section. Consequently, the terms sample (sample size, sample scope) and basic total are also treated as characteristic variables for technological processes. Another term frequently used in practice is the *unit count*, often also used together with attributes such as *produced, tested, transported, etc.* It must be noted here that although the term unit count is always understood, it must be appreciated that it is obsolete and should not be used as a characteristic variable.

Finally, another important term for product sets should be discussed, the *inventory stock*, or simply inventory. A store with an inventory $B > 0$ must be provided between neighbouring technological processes. Obviously this becomes a "pseudostore" when a rigid chaining of both processes exists because the inventory is always zero. On the other hand, an element for the store should also be provided for the process modelling using graphic methods (if the technological processes are modelled as arrows, the stores form the nodes (see Section 2.2). The store inventory level is a very important characteristic variable for modelling in process technology. Depending on the model type, it is described as a determined time function, as a random variable or as a time-dependent random vector.

The Process Duration

The technological process T is characterised by its duration d (process duration). The duration can fluctuate for a repeated run. It is then desirable to consider the duration as being a random variable with the probability density $f(t)$ (see Figure 1.22). A suitable modelling of practical relationships occurs when the shortest duration a and the longest duration b are used in the distribution. The most probable duration (the so-called modal value m) lies between a and b, and normally deviates from the average duration μ.

The average duration in probability theory is called the *expected value*.

$$\mu = \int_a^b t \cdot f(t)\, dt \tag{1.18}$$

It is the theoretical average value determined experimentally (namely estimated) using the empirical average value $\bar{t}$. Note that when the sampling method is used, the possible preference for longer durations must be taken into consideration for the estimation (corresponding to a law for the renewal theory).

The fluctuation of the duration is taken into consideration with the *variance σ^2*.

$$\sigma^2 = \int_a^b (t - \mu)^2 \cdot f(t)\, dt \quad = \int_a^b t^2 \cdot f(t)\, dt - \mu^2 \tag{1.19}$$

If the deviation is relatively small, the duration is considered as being determined by $d = \mu$.

$$v = \frac{\sigma}{\mu} \tag{1.20}$$

The variation coefficient v can be used as a criterion here. The numeric value α used as a criterion for the decision and, provided $v \leq \alpha$, will be calculated using a determined process. Typical values for α are $\alpha = 0.05$ or $\alpha = 0.01$ depending on the "decision sharpness".

Some models use a specific probability distribution. An example here is the network planning, PERT (program evaluation and review technique) that uses the Beta distribution. Where

$$f(t) = \begin{cases} \dfrac{(b-a)^{1-p-q}}{B(p,q)}(x-a)^{p-1}\cdot(b-x)^{q-1} & \text{for } a < x < b \\ 0 & \text{otherwise} \end{cases} \tag{1.21}$$

with $p = 3 + \sqrt{2}$, $q = 3 - \sqrt{2}$.

The Beta function is

$$B(p,q) = \int_0^1 x^{p-1}\cdot(1-x)^{q-1}\,dx \;\;=\; \frac{\Gamma(p)\cdot\Gamma(q)}{\Gamma(p+q)}$$

with the Gamma function

$$\Gamma(p) = \int_0^\infty t^{p-1}e^{-t}\,dt \quad p > 0 \tag{1.22}$$

Although these theoretical relationships are relatively complicated, they produce very simple equations for the expected value and the dispersion, *i.e.*

$$\mu = \frac{a + 4m + b}{6} \tag{1.23}$$

$$\sigma^2 = \left(\frac{b-a}{6}\right)^2 \tag{1.24}$$

Thus, the estimated values a, m and b are used in the above equations to calculate the expected value and standard deviation for the duration that are then used in subsequent network calculations (Figure 1.22).

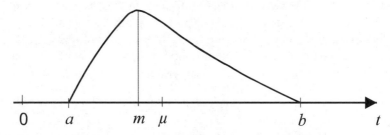

Figure 1.22. Beta distribution for the duration

Example:

For a repair process, the estimated or experimental values are used to calculate:

shortest duration	$a = 2$
most probable duration	$m = 5$
longest duration	$b = 14$

Which produces: $\mu = 6$, $\sigma = 2$

Intensity Characteristics for Technological Processes

The intensity I of a technological process is defined as being the quotient between the number ΔM of products flowing through the technological process in the time interval $\Delta t = t_2 - t_1$

$$I = \frac{\Delta M}{\Delta t} \tag{1.25}$$

The usual dimension used here is *units per hour* or *units per minute*, less frequently *units per second*.

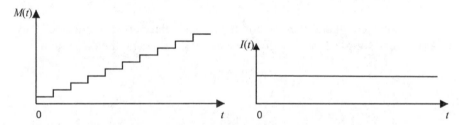

Figure 1.23. Quantity and intensity graph for constant duration

Other terms for the intensity are: throughput, yield, productivity; terms such as performance, effectiveness are also used. The lack of a standard notation, means, analogous to the electrical current, we suggest the use of the symbol I here, designated as intensity, where ΔM would be equivalent to a quantity of electrical charge. Because discrete processes are involved here (often called unit processes), Δt cannot be chosen arbitrarily small.

If the product quantity flowing through the technological process is considered as being a function of time, this produces a step function $M(t)$ (Figure 1.23) that also produces a constant intensity for a constant d:

$$I = \frac{1}{d} \tag{1.26}$$

I is time dependent for a fluctuating duration, *i.e.* $I(t)$ also fluctuates (Figure 1.24).

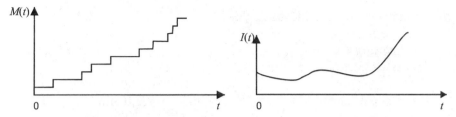

Figure 1.24. Quantity and intensity graph for a nonconstant duration

The exact curve $I(t)$ depends on the size of the time interval Δt. If a fluctuating duration is considered as being a random value with the density $f(t)$, the intensity can also be considered as being a random value with a density $f(I)$. The calculation of the distribution $f(I)$ from the distribution $f(t)$ is generally very difficult.

For expected values, however,

$$\mu_I = \frac{1}{\mu_d} \tag{1.27}$$

i.e. the average duration of technological processes and the average intensity are mutually reciprocal.

For practical calculations, it is desirable to always use the time range, *i.e.* $f(t)$ is used to calculate the relationships.

For further considerations, it often suffices to use the intensity equations for constant duration or for the expected value.

A very important intensity characteristic value is the maximal possible intensity I^0 for a technological process, the so-called *limit intensity*. This limit intensity can under no circumstances be exceeded, not even briefly. The actual intensity can always be only less than or equal to this limit intensity. A reduction of the intensity occurs when the technological process is not fully loaded. Either a certain time is not used for the production or, because of faults, the duration lasted longer than technologically necessary, *i.e.*

$$0 \le I(t) \le I^0 \quad \text{or} \quad \text{Max}\{I(t)\} = I^0 \tag{1.28}$$

If the individual technological processes in a process chain have different limit intensities, the limit intensity of the complete process is determined by the minimum of the limit intensities of the individual processes.

$$\text{Max } I_{sum} = \underset{(n)}{\text{Min}} \{I^0_n\} = I^0_{sum} \tag{1.29}$$

Thus, the usual min-max law applies to the process chain. This states that the maximum loading is the same as the minimum of the loadability.

The following *example* illustrates this law:

For the process chain

Process$_1$ solder paste printing	with	$d_1 = 1.5$ min
	gives	$I^0_1 = 40$ units/h
Process$_2$ placement	with	$d_2 = 3.0$ min
	gives	$I^0_2 = 20$ units/h
Process$_3$ soldering	with	$d_3 = 1.0$ min
	gives	$I^0_3 = 60$ units/h
Process$_4$ testing	with	$d_4 = 2.0$ min
	gives	$I^0_4 = 30$ units/h

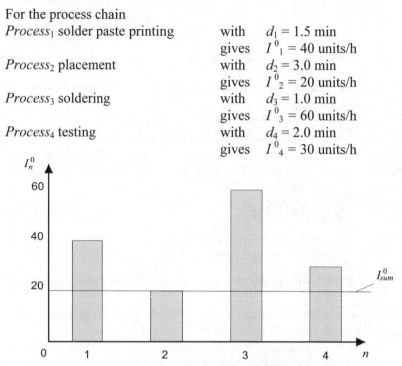

Figure 1.25. Limit intensities of technological processes (example)

Thus, the limit intensity of the complete processes is 20 units/h. The bottleneck process is the placement. All other processes have process reserves.

$$\Delta I^0_n = I^0_n - I^0_{sum} \qquad (1.30)$$

These reserves in technological processes allow a rationalisation strategy to be achieved, as shown in the following *example*:

Initially the costs of the complete process (without component and material costs) are analysed. The analysis result gives $C'_{sum} = €\,80$ per hour. C_{sum} are the costs of the complete process per hour. This produces a value for the process cost $c = €\,4$ for each module as manufacturing cost.

To increase the limit intensity of the complete process, the limit intensity of the bottleneck, *i.e.* of the placement process, must be increased. If this is achieved, for example, by the acquisition of a second placement machine, this doubles the limit intensity of the placement process and the inspection process becomes the new bottleneck.

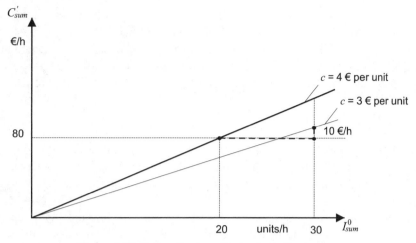

Figure 1.26. Rationalisation of technological processes

As a consequence, the total limit intensity in the example does not double, as possibly hoped, but rather increases to 30 units/h. If it is assumed that the rationalised process for the module assembly causes an additional cost increase of €10/h, then this rationalised process example not only increases the intensity but also reduces the unit cost to € 3. The limit of the true rationalisation is reached only when the additional placement machine increases the cost by €40/h.

Quality Characteristics for Technological Processes

Explanation of terms

The quality of a product is the satisfying of the condition with regard to its suitability, defined and assumed requirements (so-called quality requirements) [3].

A significant basic term used to determine the product quality is the defect (*the nonfulfilment of a requirement*). It is possible to manufacture defective products in almost all electronic technological processes. To determine whether or not a defect is present requires that value is placed on a very conscientious defect definition. This affects the selection of the product characteristics that determine the quality in this process and the assumed defect criterion used to make the decision whether or not a defect is present. Sometimes several product characteristics with their own defect criteria must be used for the quality evaluation of a technological process.

The *example* of solder paste printing is used to explain these discussions.

The requirements for quality-conform circuit boards printed with solder affect: 1) the volume and 2) the position deviations (the *offset*) of the solder paste deposit. The solder paste volume (V) must be neither too small nor too large, *i.e.* a two-sided tolerance range exists. The position deviations (L) must not exceed a maximum value (the maximum offset). To evaluate the quality, the numerical values of both characteristics must be determined and compared numerically with the tolerance ranges. Thus, a further process is required, generally designated as an inspection process in the following section.

The technological process (T) and the inspection process (P) are responsible for *creating the quality* and for *determining the quality*, respectively. Obviously, further technological processes can also be present between T and P. In this case, the faulty products are not determined *immediately* but rather *later*, which sometimes can be more cost effective.

Yield

The yield y is an important term in process technology and has a fundamental importance in the manufacturing of electronic components, in particular, for integrated circuits. The yield y is defined as the number of defect-free (good) products n_g divided by the total number of considered (tested) products n_b.

$$y = \frac{n_g}{n_b} \leq 1 \tag{1.31}$$

Obviously, the yield in technological processes should be high, *i.e.* lie near to 1 or 100%. This is counteracted not only by cost considerations, but also technological limits. For process chains, the overall yield is calculated from the product of the individual yields.

$$y_{sum} = y_1 \cdot y_2 \cdots y_n \cdots y_N = \prod_{n=1}^{N} y_n \tag{1.32}$$

In microelectronics, the number of individual processes N has increased continuously in recent years and currently is as large as 500 (and sometimes even more).

Assuming that the total individual yields have the same values, this gives

$$y_{sum} = y^N \tag{1.33}$$

For many applications, the yield is bound to a fixed product quantity and determined regularly in the running manufacturing process.

A typical example here is the determination of the yield per wafer in microelectronics. Because the number of defective chips per wafer fluctuates, some models also use probability theory here. Thus, one assumes that the yield is a random value Y with a probability distribution $f(y)$ (with the expected value μ_y and variances σ_y^2) and characterises the quality of the technological process. $f(y)$ also means quality distribution.

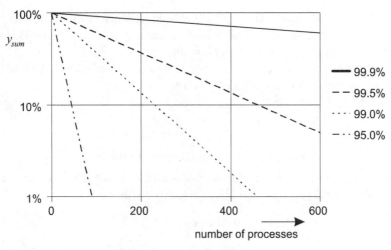

Figure 1.27. Yield multiplication law

The usual statistical methods can also be used here, namely, for example, the experimentally determined empirical average value $\bar{y}$ for several wafer yields can also be considered as an estimate for the expected value at the same time (point estimate).

$$\bar{y} = \hat{\mu}_y \qquad (1.34)$$

A corresponding point estimate is also possible for σ_y^2 and, furthermore, interval estimates can also be performed from the experimental data (see Section 5.6).

The yield y, however, cannot just be interpreted as a *random variable*. Its definition as a quotient that can assume values between 0 and 1, and, actually also corresponds to a frequency, means that the yield can also be interpreted as *probability*, *i.e.* more accurately: the yield y is the probability that the product does not have any defect or that the associated numerical value of the product characteristic lies within the permitted tolerance range $[T_l, T_u]$.

$$y = \begin{cases} P\{X = 0\} & X = \text{number of product defects} \\ & \text{for discrete characteristics} \\ P\{T_l \leq X \leq T_u\} & X = \text{measurable product characteristic} \\ & \text{with lower and upper tolerance limit} \end{cases} \qquad (1.35)$$

Considering the yield as probability makes possible the use of probability theory models. This is now illustrated with an example.

Example: soldering circuit boards

A populated circuit board with n solder points (n very large) is subjected to the reflow soldering technological process. The soldering defect rate p is small (*i.e.* it lies in the DPM range; DPM – defects per million = number of defects per million possible defects). How large is the yield? The Poisson distribution is appropriate as a model for this case.

The probability that the circuit board has exactly k soldering defects can be calculated using the Poisson distribution

$$p_k = \frac{\lambda^k}{k!} \cdot e^{-\lambda} \qquad (1.36)$$

with $\lambda = n \cdot p =$ mean value of soldering defects on the PCB.

The yield, namely the probability that a circuit board does not have any fault can be calculated from the above equation with $k = 0$ is

$$p_0 = y = e^{-\lambda} = e^{-n \cdot p} \qquad (1.37)$$

A *numerical example:* $n = 4000$ solder points, $p = 100$ DPM, $\lambda = 0.4$; *i.e.* the average number of soldering defects on a circuit board is 0.4.

Yield $y = e^{-0.4} \approx 0.67 = 67\%$

The yield is also called the *FPY* (first-pass yield) for the manufacturing of reparable products.

Defect rate
The defect rate p can be calculated from the yield y:

$$p = 1 - y \qquad (1.38)$$

However, in particular in the manufacturing of electronic modules and devices, the term yield in contrast to the defect rate is used only very rarely. The defect rate is also the basic characteristic of the quality assurance and statistic process control (SPC). Analogous to the yield definition, the defect rate can also be directly defined as a quotient. The defect rate p is defined as the quotient from the number of defective products n and the total number of the considered (tested) products n_b, *i.e.*

$$p = \frac{n_d}{n_b} \qquad (1.39)$$

As for the yield, the considerations for the statistical and the probability-theory-based interpretation also apply to the defect rate. The interpretation of the defect rate p as the probability that the product is defective allows it to be also used very easily for individual elements. A typical example here is the solder joint. The soldering defect rate is then the probability that the solder joint is defective.

An exact analysis shows the following: if one takes the product as the consideration unit, then the observation whether or not it is defective is certainly clearly defined – even when there are various defect possibilities and thus defect types. The total number of considered products means that the value for the defect rate can also be calculated uniquely. This is not the case for elements such as the solder joint. There are also various defects here: solder bridges, solder beads, solder droplets, cold solder joint, bad solder joint, tombstone effect, *etc*. To make the calculation, one requires as denominator, the number of considered (or possible for the defect characteristic) solder joints and, for example, those definitions whether one or two defective solder joints is assumed for a solder bridge.

Test blurring, 1st and 2nd type
In real manufacturing systems, the defect rate p caused by a technological process can only be determined by a subsequent inspection process. The inspection process is the interaction between the test procedure and the product, where the term workstep receives the same importance as the technological process. Compared with the technological process, an inspection process is characterised by two features:

1. The product state is not changed in the inspection process, but rather information is obtained.
2. Each inspection process has *two* exits, *i.e.* the input flow of the products becomes the flow of nonfaulty products and the flow of defective products.

Thus, in general, the inspection process makes the defect rate p "visible" and is brought from being hidden for the subsequent manufacturing process. The stream of defective products should then pass through a different technological process (*e.g.*, through a defect-correction or repair process) from the stream of nondefective products. If one analyses the inspection process with regard to its characteristic variables, this yields the following: not only technological processes, but also information can be *faulty*, for example, a defect-free technical product can be declared as being faulty as a result of the inspection process.

Thus, a similar situation is present here as for a technological process *Process* with the defect rate p. The inspection process with regard to defect-free products (with the share y on the product stream) has an information defect rate that should be designated as *1st type test blurring α*. *α* is a characteristic variable of the inspection process. The probability theory again gives a possible interpretation: α is the probability that the testing of a technical product results in it being declared as being defective although the product leaves the technological process as defect free. Or stated concisely: α is the probability of declaring a defect-free product as being defective.

As with the yield and the defect rate, the 1st type test blurring α can also be defined as a quotient (in particular for the experimental determination) and finally also as a random value. Analogous to the 1st type test blurring, the *2nd type defect rate β* behaves as a probability; the declaration of a technical product in the result of the inspection process as defect free although it leaves the result of the technological process as defective. Or stated concisely: *β* is the probability of declaring a defective product as being defect free.

Both test blurrings cause a change in the true yield or defect rate. To make a simple differentiation between the two terms, the yield and the defect rate determined by the test are designated as *acceptance rate a* and as *rejection rate r*, respectively.

Acceptance and rejection rate

An inspection process with the test blurring α and β changes the yield y and the defect rate p as follows:

$$a = y \cdot (1-\alpha) + p \cdot \beta \tag{1.40}$$

$$r = p \cdot (1-\beta) + y \cdot \alpha \tag{1.41}$$

Stated in words: The true yield (process yield) y will be reduced by the value $\alpha \cdot y$, because this proportion is "unintentionally" classified as being defective. Conversely, a proportion $\beta \cdot p$ of the defective products is "unintentionally" added to the yield.

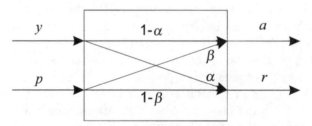

Figure 1.28. Test blurring, 1st and 2nd type

A *numerical example* demonstrates the above relationships. $N = 1000$ products are tested, of which 600 are defect free and 400 defective. The inspection process has the test blurring $\alpha = 0.08$ and $\beta = 0.05$. Then

$$N \cdot a = 600 \cdot 0.92 + 400 \cdot 0.05$$
$$= 552 + 20 = 572$$

$$N \cdot r = 400 \cdot 0.95 + 600 \cdot 0.08$$
$$= 380 + 48 = 428$$

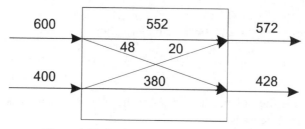

Figure 1.29. Real inspection process (example)

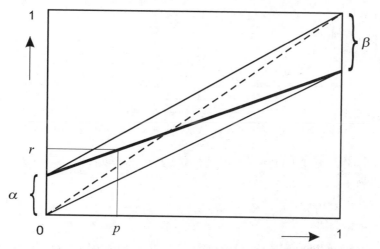

Figure 1.30. Test diagram of the real inspection process

The inspection process causes the true yield to sink from $y = 60\%$ to the acceptance rate or the tested yield $a = 57.2\%$.

The special case when α and β are independent of the defect rate p allows the relationships to be represented clearly (see Figure 1.30).

In practice, however, y and p are not known, but rather a and r as a result of the inspection process. Unfortunately, one wants to know how large is the true yield or the true defect rate when the inspection process has blurring.

Acceptance and rejection average outgoing quantity

The relative proportion of the defective products in the products accepted after the test also represents a defect rate that should be designated as an *acceptance average outgoing quantity* D_a

$$D_a = \frac{p \cdot \beta}{a} \tag{1.42}$$

Thus, the inspection process reduces the defect rate from the original value p before the inspection to the value D_a after the inspection, however, this also causes the scope to be reduced from 100% to $a\%$.

In the numerical example:

The defect rate in the accepted items, *i.e.* after the inspection is

$$D_a = \frac{0.04 \times 0.05}{0.572} = 0.035$$

or 3.5%. This is the defect rate that after the inspection process is passed to the subsequent processes.

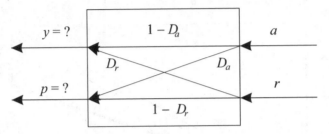

Figure 1.31. Acceptance and rejection average outgoing quantity

Similarly, a rejection average outgoing quantity should also be introduced:

$$D_r = \frac{y \cdot \alpha}{r} \tag{1.43}$$

that specifies the relative proportion of the defect-free products in the rejection stream.

Numerical example:

$$D_r = \frac{0.6 \times 0.8}{428} = 11.2\%$$

i.e. the repair process (or possibly the rejects) contains 11.2% fault-free products.

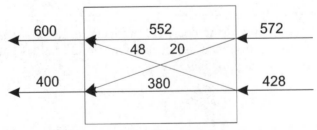

Figure 1.32. Acceptance and rejection average outgoing quantity (numerical example)

If D_a and D_r are known, a and r can be used to calculate the true yield y or the true defect rate p):

$$N \cdot y = a \cdot (1 - D_a) + r \cdot D_r \tag{1.44}$$

$$N \cdot p = r \cdot (1 - D_{r)} + a \cdot D_a \tag{1.45}$$

(see Figure 1.31)

The 1st and 2nd type test blurring can be derived from the above equations:

$$\alpha = \frac{r \cdot D_r}{y} \tag{1.46}$$

$$\beta = \frac{a \cdot D_a}{p} \tag{1.47}$$

This duality for inspection processes should not be forgotten, where the knowledge of the characteristic variables D_a and D_r or α and β is important for the process technology.

1.4 Process Characteristic

1.4.1 Explanation of the Process Characteristic Term

Because the density $f(x)$ of the characteristic value X for the product quality characterises the technological process, it should be called the process characteristic.

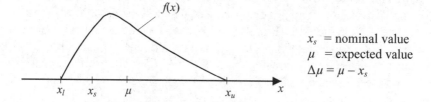

x_s = nominal value
μ = expected value
$\Delta\mu = \mu - x_s$

Figure 1.33. $f(x)$ process characteristic

The procedure parameters and product properties are both integrated into this process characteristic.

The process characteristic can be augmented with additional details, which, in particular, affect the technological procedure, because the precise course of the process characteristic always also depends on the procedure parameters.

The process characteristic is multidimensional $f(x)$ if the product quality is determined by several characteristic variables (characteristic variable vector x). Initially, the one-dimensional process characteristic should always be handled in the associated section.

If possible, known probability distributions, such as the normal distribution or the Weibull distribution for continuous quality characteristic values and, for example, the binomial distribution and the Poisson distribution for discrete quality characteristic values, should be used for the mathematical description of the process characteristic.

Examples:

1. A manufacturing process for thick-film resistors with a nominal value of $x_s = 1,000\ \Omega$ can be characterised using a normal distribution with $\mu = 1,010\ \Omega$ and $\sigma = 25\ \Omega$.
2. A soldering process for modules can be characterised with a Poisson distribution with the number of soldering points n and the soldering defect rate p, $n = 1,500$, $p = 200$ DPM.

The distributions selected in the examples are very simple and serve only to illustrate the principle.

Note: The designation $f(x)$ for the process characteristic can also be used for discrete distributions. In this case, the Dirac function $\delta(x-k)$ is used and written as

$$f(x) = \sum_{k=0}^{K} p_k \cdot \delta(x-k) \qquad (1.48)$$

This is a finite Dirac series. However, $K \to \infty$ must be selected for the Poisson distribution. Thus, the subsequent discussions always write $f(x)$ for the process characteristic. Finally, the nominal value x_s and the tolerance range with the lower tolerance limit T_l and the upper tolerance limit T_u also belong to the process characteristic. The two tolerance limits are used to calculate the tolerance middle T_m

$$T_m = \frac{T_u + T_l}{2} \qquad (1.49)$$

and the (absolute) tolerance width ΔT (= half-"tolerance")

$$\Delta T = \frac{T_u - T_l}{2} \tag{1.50}$$

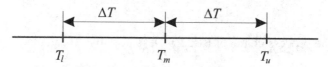

Figure 1.34. Tolerance range (two-sided)

The relative tolerance width δT is also defined as the quotient formed from the absolute tolerance width and tolerance average

$$\delta T = \frac{\Delta T}{T_m} \tag{1.51}$$

and the specific tolerance width τ as the quotient formed from the absolute tolerance width and the variation width

$$\tau = \frac{\Delta T}{b} = \frac{\Delta T}{\sigma} \tag{1.52}$$

All three tolerance widths are defined as being positive.

Continuation of the examples: manufacture of resistors
The resistors should lie in the tolerance range $[T_l = 950\ \Omega;\ T_u = 1{,}050\ \Omega]$. Thus $T_m = T_s = 1{,}000\ \Omega$, $\Delta T = 50\ \Omega$, $\delta T = 5\%$, $\tau = 2$.

Soldering process:
The nominal value is $k_s = 0$ defects. Only $k = 0$, *i.e.* zero defects, is also accepted as the tolerance for the soldering process.

One-sided tolerance ranges often arise. There are two possibilities here:

- One-sided tolerance range with lower tolerance limit T_l
- One-sided tolerance range with upper limit T_u.

The terms tolerance middle and tolerance width are not used for one-sided tolerance ranges.

The tolerance ranges and the process fluctuations represent the two sides of a technological process (duality)

- technological procedure: μ, σ
- technical product: $T_m, \Delta T$

1.4.2 Operating Point for Technological Processes

The expected value μ is called the operating point x_A of the technological process:

$$x_A = \mu \tag{1.53}$$

Using the process characteristic $f(x)$, it is possible to calculate

$$\mu = \int_{x_l}^{x_u} x \cdot f(x) \, dx \tag{1.54}$$

With reference to the examples in Section 1.4.1, this gives for the resistor production: $\mu = 1{,}010 \ \Omega$ and for the soldering process $\mu = 0.3$ defects per module.

In practice, the operating point is estimated using the average value $\overline{x}$ (point estimation):

$$\hat{x}_A = \overline{x} \tag{1.55}$$

The $\overline{x}$ card has proved useful for monitoring the operating point as an element of the statistical process control (SPC). An interval estimate for the operating point is used to determine the intervention limits and for the quality assurance. The following properties characterise the operating point of technological processes:

1. The operating point of technological processes *can be set*. This means changes to the procedure parameters can also be used to change (control) the operating point.
2. The operating point of technological processes is *timedependent* $x_A(t)$; *i.e.* various actions such as wear, ageing, fatigue, change of the bath composition, smearing of the print masks, *etc.*, cause the operating point to "wander" during the course time from the set value and so exhibit a drift or a trend. Chapter 8 discusses this in more detail.

1.4.3 Adjustment Point and Eccentricity

The operating point $x_A (t = 0)$ set at the process begin (*i.e.* $t = 0$) is called the adjustment point x_E.

$$x_E = x_A(t = 0) \tag{1.56}$$

By making an appropriate selection of the technological parameters, one attempts explicitly "to hit" a determined point. For a symmetric process characteristic, this is the tolerance centre (in general the same as the setpoint) because the yield is largest there, *i.e.* the tolerance centre is the optimum adjustment point x_E^* for this case.

$$x_E^* = T_m \qquad (1.57)$$

For an unsymmetrical process characteristic, one receives the optimum adjustment point x_E^* (*i.e.* the maximum yield) with the consideration shown in Figure 1.35.

The process characteristic $f(x)$ with the expected value μ and the standard deviation σ is normalised using the transformation z

$$\frac{x - \mu}{\sigma} = z$$

i.e. one receives

$$f(x) = f(z \cdot \sigma + \mu) = f_0(z) \qquad (1.58)$$

$f_0(z)$ is the "normalised" process characteristic.

One now calculates the yield

$$y = \int_{T_l}^{T_u} f(x)\,dx = \int_{\frac{T_l - \mu}{\sigma}}^{\frac{T_u - \mu}{\sigma}} f_0(z)\,dz = F_0\left(\frac{T_u - \mu}{\sigma}\right) - F_0\left(\frac{T_l - \mu}{\sigma}\right) \qquad (1.59)$$

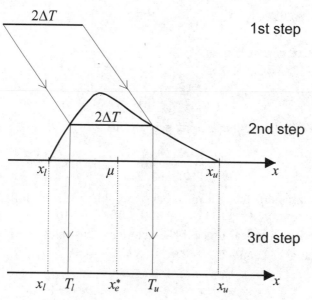

Figure 1.35. Normalised process characteristic

The expected value is identical to the operating point, which now should be "set", *i.e.*

$$\mu = x_A = x_E$$

This provides for the yield depending on x_E

$$y(x_E) = F_0\left(\frac{T_u - x_E}{\sigma}\right) - F_0\left(\frac{T_l - x_E}{\sigma}\right) \qquad (1.60)$$

The maximum yield is obtained when the differential quotient of the yield for $x_E = x_E^*$ produces the value zero,

$$\left.\frac{dy}{dx_E}\right|_{x_E = x_E^*} = 0$$

which gives

$$f_0\left(\frac{T_u - x_E^*}{\sigma}\right) = f_0\left(\frac{T_l - x_E^*}{\sigma}\right) \qquad (1.61)$$

Or in other words: the optimum adjustment point x_E^* has for the form of a technological process the significance of a nominal value for the operating point to be set. However, the true value of the set operating point normally deviates. This deviation should be designated the absolute *eccentricity e*. A symmetrical process characteristic with $x_E^* = T_m$ gives:

$$e = \Delta\mu = \mu - T_m \qquad (1.62)$$

The relative eccentricity is

$$\delta e = \frac{e}{T_m} = \frac{\Delta\mu}{T_m} \qquad (1.63)$$

The same applies to the specific eccentricity

$$\varepsilon = \frac{e}{\sigma} = \frac{\Delta\mu}{\sigma} \qquad (1.64)$$

In the *example* of the resistor production, this gives the following values for the eccentricity: $e = +10\ \Omega$, $\delta e = 0.01 = 1\%$, $\varepsilon = +0.4$.

For a nonsymmetrical process characteristic, the value x_E^* is used, rather than the tolerance centre T_m.

$$e = \Delta\mu = \mu - x_E^* \qquad\qquad \delta e = \frac{\Delta\mu}{x_E^*}$$

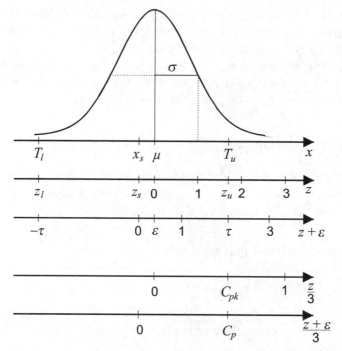

Figure 1.36. Optimum adjustment point x^*_E (maximum yield)

Because $\Delta\mu$ can assume positive or negative values, all coefficients for the eccentricity can also be positive or negative, depending on whether the operating point is greater or less than the tolerance middle.

1.4.4 Variation Width of Technological Processes

The standard deviation σ is called the absolute variation width b of the technological process.

$$b = \sigma$$

The variance coefficient v is also designated as the relative variation width b_{rel}.

$$b_{rel} = \frac{\sigma}{\mu}$$

Example:
For the resistor production, $b = 25\ \Omega$ and $b_{rel} = 2.5\%$. For the soldering process, $b = \sqrt{\lambda} = 0.55$ defects per product and $b_{rel} = 1.8$.

The absolute dispersion width is always a measure of the quality of the technological equipment; it cannot normally be influenced in the manufacturing process. Precision machines have a small dispersion width and are often more expensive than those with a large dispersion width. However, it must be taken into consideration that the product properties and characteristic values can also affect the dispersion width.

1.4.5 Process Capability

The process capability specifies the degree of suitability of a technological process for the manufacturing of quality-conform products. The process capability is calculated using the following coefficients.

1. Usual process-capability coefficient C_p

$$C_p = \frac{1}{3}\tau \qquad (1.65)$$

where τ is the specific tolerance width.

2. Critical process-capability coefficient C_{pk}

$$C_{pk} = \frac{1}{3}\left(\tau - |\varepsilon|\right) \qquad (1.66)$$

Although this definition applies initially to the normal distribution, it can also be used for other distributions. Section 5.8 discusses process-capability coefficients in more detail.

1.4.6 Normal-distributed Process Characteristic

Many processes can be described using a normal distribution

$$f(x) = \frac{1}{\sqrt{2\pi}\cdot\sigma}\cdot e^{-\frac{1}{2}\left(\frac{x-\mu}{\sigma}\right)^2} \qquad (1.67)$$

with the tolerance limits $[T_l; T_u]$.

This easily allows the yield y to be calculated:

$$y = \Phi(\tau + \varepsilon) + \Phi(\tau - \varepsilon) - 1 \qquad (1.68)$$

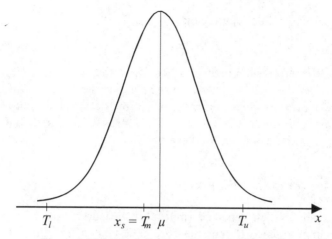

Figure 1.37. Normal-distributed process characteristic

1.5 Measurement Characteristic

1.5.1 Measurement Process – Principles and Terms

The importance of the inspection process in electronics production for determining the product quality has also been discussed in Section 1.3. If the value of the quality characteristic value is now determined using a measurement process, which is true in most cases, one observes the following:

The measurement changes the true value "created" by the technological process. However, the change is often insignificant. The difference between the measured and the true value is called the *deviation of measurement*. The true value is also called the exact value, correct value.

Obviously, in individual cases, the deviation of measurement can also assume the value zero. In this case, the measured value also agrees with the true value. However, the true value is often not known. If a theoretical model exists, it can sometimes be calculated. For some discussions in the next sections, the true value is assumed as being known.

The following symbols are used for the unique designation of the various values:

x = true value
x' = deviation of measurement
x^* = measured value

This gives the simple relationship

$$x^* = x + x'$$ (1.69)

Because the error of dimension "falsifies" the true value, it was often designated as the *error of measurement* in the past. However, this designation can be used when a criterion for the definition of the error exists (*e.g.*, tolerance limits for not permitted deviations) and the deviation of the measurement satisfies the fault criterion.

1.5.2 Measurement and Chance

Experience shows that repeated (multiple) measurement of the same true value x can produce different measurement results, *i.e.* the measurement results vary. Consequently, the random value X^* with a probability density $g(x^*|x)$ is introduced for the measured value. This means that this density function exists only when the true value *does not* change and thus the fluctuations only affect the measurement process (Figure 1.38).

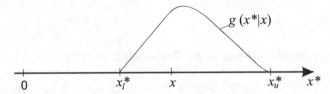

Figure 1.38. Measured process characteristic for a single measured value x

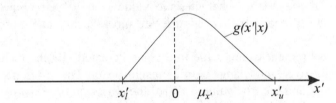

Figure 1.39. Measurement characteristic

If x is known, it can be used to derive a density $g(x')$ for the random variable

$$X' = X^* - x$$ (1.70)

i.e. for the "true" deviation of measurement, which, in effect, represents the measurement process without measuring the true value and is designated as the *measurement characteristic*. More exactly, one should also write here: measurement characteristic $g(x'|x)$, *i.e.* the "form" of the measurement characteristic can also depend on which true value should be measured. This dependency should, however, be ignored for the subsequent discussions because the values of a process characteristic to be measured fluctuate in a relatively small range and a possible effect on the form of the measurement characteristic is not significant.

1.5.3 Measured Process Characteristic

If one wants to determine a process characteristic by measurement, one must differentiate between the true process characteristic $f(x)$ and the measured process characteristic $h(x)$. The deviation of measurement in the measurement process causes a changed measured value to be assigned to each true value. Adding the random values

$$X^* = X + X'$$

gives a convolution for the densities [9]

$$h(x) = f(x) * g(x) \tag{1.71}$$

As for addition, subtraction, multiplication and division, the convolution, often designated as the 5th arithmetical operation, can be used on two functions to produce a result function. The symbol for the convolution operation is an asterisk (*). For the convolution of the densities, the variables for the true value, the error of dimension and the measured value must have a uniform designation; x has been chosen here.

The convolution of functions is defined by an integral:

$$h(x) = \int_{-\infty}^{+\infty} f(x-z) \cdot g(z)\,dz = \int_{-\infty}^{+\infty} f(z) \cdot g(x-z)\,dz \tag{1.72}$$

(z is the integration variable, but it can be replaced by any other character.)

If one assumes that each characteristic lies only in a range between the smallest and the largest value, $i.e.$

- process characteristic $x_l \leq x \leq x_u$
- measurement characteristic $x_l' \leq x' \leq x_u'$
- measured process characteristic $x_l^* \leq x^* \leq x_u^*$,

this produces for the integration limits:

$$h(x) = \int_{x_{l'}}^{x_{u'}} f(x-z) \cdot g(z)\,dz = \int_{x_l}^{x_u} f(z) \cdot g(x-z)\,dz \tag{1.73}$$

and

$$x_u^* = x_u + x_u'$$

$$x_l^* = x_l + x_l' \qquad \text{with } x_l' < 0 \text{ (Figure 1.39)} \qquad (1.74)$$

This also provides the following relationships for the expected value and the variance

$$\mu_{x^*} = \mu_x + \mu_{x'} \qquad (1.75)$$

$$\sigma_{x^*}^2 = \sigma_x^2 + \sigma_{x'}^2 \qquad (1.76)$$

The expected value

$$\mu_{x'} = \int x' \cdot g(x') \, dx' \qquad (1.77)$$

is also called the *systematic error of measurement* (previously known as the systematic measurement error). This systematic error of measurement should be applied as a correction to each measured value. This correction is always possible, even when the measurement characteristic $g(x'|x)$ and, thus possibly $\mu_{x'}$, depends on the measured value x. A measurement process with $\mu_{x'} = 0$ is called unbiased.

It behaves differently with the deviation

$$\sigma_{x'}^2 = \int (x' - \mu)^2 \cdot g(x') \, dx' \qquad (1.78)$$

$\sigma_{x'}$ is called the *stochastical error of measurement* (previously the stochastical measurement error). It cannot be eliminated and "propagates" the process characteristic, whereas the systematic error of measurement "displaces" the process characteristic.

The common special case of the symmetrical measurement characteristic in which the probabilities for positive and negative deviations are equally large, *i.e.*

$$g(-x') = g(x') \qquad (1.79)$$

has immediate practical significance. These values are always unbiased, *i.e.* they do not have any systematic error of measurement.

1.5.4 True and Measured Yield

Because the true yield y with

$$y = \int_{T_l}^{T_u} f(x) \, dx \qquad (1.80)$$

is not known in manufacturing processes, it must be determined by measurement, *i.e.* the acceptance rate (measured yield) results

$$a = \int_{T_l}^{T_u} h(x)\, dx = \int_{T_l}^{T_u} \left[f(x) * g(x) \right] dx \qquad (1.81)$$

that deviates from the true yield (it is normally smaller) (Figure 1.40).

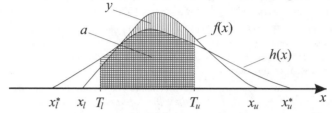

Figure 1.40. True and measured yield

The error of measurement means that type 1 and 2 test blurring can occur in the vicinity of the tolerance limits T_l and T_u (see Sections 1.3 and 8.7).

References

[1]   BOSSEL, H.: *Simulation dynamischer Systeme*. Vieweg, Braunschweig/Wiesbaden 1992

[2]   DANGELMAIER, W.; WARNECKE, H.-J.: *Fertigungslenkung – Planung und Steuerung des Ablaufs der diskreten Fertigung*. Springer-Verlag, Berlin 1997

[3]   DIN 55350: Technische Begriffe der Qualitätssicherung und Statistik. Teil 11 Grundbegriffe der Qualitätssicherung. Hrsg. Deutsches Institut für Normung. Berlin, Cologne: Beuth-Verlag 1995

[4]   DOMSCHKE, W.; SCHOLL, A.; VOSS, S.: *Produktionsplanung – Ablauforganisatorische Aspekte*. Springer-Verlag, Berlin, Heidelberg, New York, Tokyo 1997

[5]   DRESZER, J.: *Mathematik Handbuch – für Technik und Naturwissenschaft*. Fachbuchverlag Leipzig, Leipzig 1975

[6]   HAMPEL, D.: *Simulationsgestützte Optimierung von Fertigungsabläufen in der Elektronikproduktion*. Verlag Dr. Markus A. Detert, Templin 2002

[7]   KIENCKE, U.: *Ereignisdiskrete Systeme - Modellierung und Steuerung verteilter Systeme*. R. Oldenbourg Verlag, Munich, Vienna, 1997

[8]   LOCKE, M.: *Grundlagen einer Theorie allgemeiner dynamischer Systeme*. Akademie-Verlag, Berlin 1984

[9]   MAIBAUM, G.: *Wahrscheinlichkeitstheorie und mathematische Statistik*, VEB Deutscher Verlag der Wissenschaften, Berlin 1980

[10]   MEINBERG, U.; TOPOLOWSKI, F.: *Lexikon der Fertigungsleittechnik – Begriffe, Erläuterungen, Beispiele*. Springer-Verlag, Berlin 1995

[11]   PINEDO, M.: *Scheduling – Theory, Algorithms, and Systems*. Prentice Hall, Englewood Cliffs, New Jersey 1995

[12]   SAUER, W.; WEIGERT,G.; GOERIGK, P.: *Real Time Optimization of Manufacturing Processes by Synchronized Simulation*. In: International Journal of Flexible Automation and Integrated Manufacturing 4 (1996) No.1, p. 15–27

[13]   SCHMIDT, B.: *Systemanalyse und Modellaufbau – Grundlagen der Simulationstechnik*. Springer-Verlag, Berlin Heidelberg New York Tokyo 1985

[14]   VDI-RICHTLINIE 3633: *Simulation von Logistik-, Materialfluss- und Produktionssystemen. Begriffsdefinitionen*. VDI-Verlag, Dusseldorf, 2000

[15]   WEIGERT, G; WERNER, S.; KELLNER, M.: *Fertigungsplanung durch prozessbegleitende Simulation*. In: Frontiers in Simulation. Anwendungen der Simulationstechnik in Produktion und Logistik. Conference Proceedings for the 10th ASIM Conference. SCS, Gent, 2002, p. 42–51

[16]   WERNER, S.; WEIGERT, G.: *Process Accompanying Simulation - A General Approach for the Continuous Optimization of Manufacturing Schedules in Electronics Production*. In: 2002 Winter Simulation Conference, San Diego, California USA, December 2002, Proceedings p. 1903–1908

[17]   WEIGERT, G.; WERNER, S.; SAUER, W.: *Adaptive Simulation Systems for Reducing Uncertainty of Predictions*. In: 9th International Conference FAIM'99, Flexible Automation & Intelligent Manufacturing, Tilburg, The Netherlands, June 1999, Proceedings p. 691–701

[18]   ZWEBEN, M.; FOX, M. S.: *Intelligent Scheduling*. Morgan Kaufmann Publishers, San Francisco 1994

2 Structures, Graphs and Networks

2.1 Directed Graphs

An important graph in process technology is the directed graph (Gozinto graph) that specifies the interconnection of components, assemblies and modules within a plant or a device. An example demonstrates the procedure.

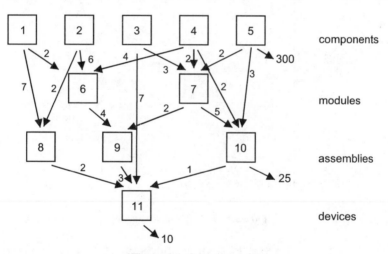

Figure 2.1. Gozinto graph

The nodes are numbered in increasing sequence and define a cost coefficient a_{ij} that specifies which set of products E_i is required to produce just *one* item of the product E_j.

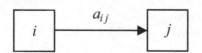

Figure 2.2. Coefficient of costs

When all the coefficients in a matrix are noted, this produces the costs matrix.

$$A = (a_{ij})$$ (2.1)

Because no closed-arrow sequences are possible in directed graphs, a numbering of the nodes can always be achieved in which each arrow points from a node with a smaller number to a node with a larger number. This means coefficients of costs are only possible outside the leading diagonals. For example, $a_{5;7} = 2$ in Figure 2.1 means that exactly 2 components of type E_5 are required to manufacture 1 item of the E_7 module.

The costs matrix obviously always has the value zero in the leading diagonals. A zero must be entered in all the empty fields.

However, for clarity, this is not done. For the example, the costs matrix is shown in Figure 2.3.

$$A = \begin{pmatrix} 0 & & & & & 2 & 7 & & & \\ & 0 & & & & 6 & 2 & & & \\ & & 0 & & & 3 & & & 7 & \\ & & & 0 & & 4 & 2 & & 2 & \\ & & & & 0 & 2 & & & 3 & \\ & & & & & 0 & 4 & & & \\ & & & & & & 0 & 2 & 5 & \\ & & & & & & & 0 & & 2 \\ & & & & & & & & 0 & 3 \\ & & & & & & & & & 0 & 1 \\ & & & & & & & & & & 0 \end{pmatrix}$$

Figure 2.3. Costs matrix (example)

The calculation of the sets z_i of products E_i ($i = 1, \ldots, N$) is now performed as follows.

Add all set requests made to node i:

$$z_i = a_{ij} \cdot z_j + a_{ik} \cdot z_k + a_{il} \cdot z_l + \cdots + a_{iN} \cdot z_N + y_i$$ (2.2)

This equation assumes that an arrow is also possible to all subsequent nodes, otherwise the corresponding cost coefficient is zero. The set y_i is the set that leaves the system.

Example: i = 5

$$z_5 = 2 \times z_7 + 3 \times z_{10} + 300$$

where $a_{5;6} = 0$, $a_{5;7} = 2$, $a_{5;8} = 0$, $a_{5;9} = 0$, $a_{5;10} = 7$, $a_{5;11} = 0$ and $y_5 = 300$.

When Equation (2.2) is written for all nodes and generalised, this produces the interconnection equation

$$\mathbf{z} = \mathbf{A} \cdot \mathbf{z} + \mathbf{y} \tag{2.3}$$

with the production vector z

$$\mathbf{z} = \begin{pmatrix} z_1 \\ z_2 \\ \cdot \\ \cdot \\ \cdot \\ z_N \end{pmatrix} \tag{2.4}$$

and the delivery vector (output vector) y.

$$\mathbf{y} = \begin{pmatrix} y_1 \\ y_2 \\ \cdot \\ \cdot \\ \cdot \\ y_N \end{pmatrix} \tag{2.5}$$

The task now involves calculating the production vector z for a given matrix A and for a given output vector y.

Equation (2.3) produces

$$\mathbf{z} = \mathbf{B} \cdot \mathbf{y} \tag{2.6}$$

where the demand matrix B

$$\mathbf{B} = (\mathbf{E} - \mathbf{A})^{-1} \tag{2.7}$$

with the identity matrix E.

However, it is not necessary to solve this task using a matrix inversion but with an algorithm designated as a mirror algorithm. To do this, the costs matrix is used to create a table. The leading diagonals represent a mirror. The algorithm runs as follows: start in row 11 and set the value for $y_{11} = 10$ in the column z, *i.e.* $z_{11} = 10$.

Continue with row 10: search for $z_{11} = 10$ and mirror this number to the cost coefficient $a_{10;11} = 1$, form the product and add the value $y_{10} = 25$, *i.e.*

$$z_{10} = z_{11} \times a_{10;11} + y_{10}$$
$$= 10 \times 1 + 25 = 35$$

	1	2	3	4	5	6	7	8	9	10	11	z	y	x
1						2		7				380		380
2						6		2				760		760
3							3				7	775		775
4						4	2			2		1,020		1,020
5							2			3		875	300	875
6									4			120		
7									2	5		235		
8											2	20		
9											3	30		
10											1	35	25	
11												10	10	

Figure 2.4. Mirror algorithm

Similarly with row 9

$$z_9 = z_{11} \times a_{9;11} + z_{10} \times a_{9;10} + y_9$$
$$= 10 \times 3 + 35 \times 0 + 0 = 30$$

etc.

For completion: row 5
All z_6 to z_{11} are known, then it follows:

$$z_5 = 35 \times 3 + 235 \times 2 + 300 = 875$$

The vector x represents the purchased parts and agrees with the corresponding product sets of the production vector z.

2.2 Network Planning

2.2.1 Principles

Technological complete processes can be represented as networks. Thus, the network is a model for the manufacturing process that permits calculations based on practice, clear graphical representations and structure analyses. Although there are many types of networks, all have the process duration d as the most important characteristic value. It is usual to represent the product flow through a network as a model. The oldest network method is known by the name CPM (critical path method). The process durations are considered as being determined, where the individual processes are represented as arrows and are generally designated in the network as *activities* or actions. The nodes between the arrows are called *events* in the network.

This network is thus a directed graph with a source and a sink. Thus, only arrows begin in the source, only arrows end in the sink. The duration of an activity is also called the *length* of an arrow. The sequence of arrows forms a *path*. The length of the path is the sum of the lengths of all arrows along the path. The critical path is the longest of all possible paths from the source to the sink.

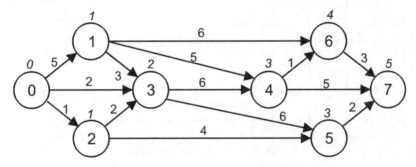

Figure 2.5. A network as a graph

A CPM network is a very understandable and simple model for a complete process. Although it does not demand any particular mathematical requirements, a good structure analysis of the product flow is necessary.

2.2.2 CPM Network and Equivalent Representations

Network as graph

The network consists of $N + 1$ nodes and at least N arrows, where a closed arrow series is not possible. This allows the nodes to always be numbered so that each arrow points from a smaller to a larger number. The source and the sink receive the number 0 and N, respectively. When all arrows that leave the source are removed, this produces (at least) one new source. This receives the number 1 (rank 1). When there are several new sources, they all receive the rank 1. Then continue and assign rank 2, *etc.* The numbering then continues with increasing rank (algorithm from Ford). The rank r_i of a node (i) is equal to the maximum number of arrows from the source 0 to the node (i). In other words: the rank r_i is the length of the critical path when each arrow has the length "1".

Figure 2.5 shows an example of a network. It has 8 nodes and 14 arrows. 17 possible paths lead from the source to the sink: the longest path is the *critical path* with length 19. The rank is shown above the nodes. It can be seen on the network whether activities can be moved or whether events have buffer times.

Network as matrix
Form a square matrix with $N + 1$ rows and write the durations between the nodes at the appropriate locations.

$t_f(i)$	i \ j	0	1	2	3	4	5	6	7
0	0	-	5	1	2				
5	1		-		3	5		6	
1	2			-	2		4		
8	3				-	6	6		
14	4					-		1	5
14	5						-		2
15	6							-	3
19	7								-
$t_s(j)$		0	5	6	8	14	17	16	19

Figure 2.6. A network as amatrix

There are *no* entries in the leading diagonal and below the leading diagonal. If one entered a zero there, this would correspond to an arrow with the duration $d = 0$, which would be equivalent to a "pseudoactivity". Pseudoactivities are necessary when chronological and technological dependencies require them. Similarly, it is also not permitted to enter zeros at free locations above the leading diagonals. Thus, although no matrix exists in the mathematical sense, this term will be used in the discussed case.

Section 2.2.3 discusses the specified column with the designation $t_f(i)$ and row with the designation $t_s(j)$ in Figure 2.6.

The structure and other properties of the graphs can also be read from the matrix. For example, it is easy to see from a column (j) from which nodes the arrows end in the nodes (j). Similarly, take a row (i) to see where the leaving arrows lead.

If most fields above the leading diagonal are set, a chain structure is in effect present. If all fields are occupied, this is called a complete graph. Complete graphs seldom occur in process technology.

The network can be uniquely obtained from the matrix. Both representations are equivalent.

Network as GANTT diagram

The GANTT diagram is the representation of all activities on the time axis. Two rules are prescribed:

1. All activities $(i; j)$ are represented proportional to their duration $d(i; j)$ *as early as possible* (*i.e.* as far left in the diagram as possible).
2. All activities are drawn in reciprocal lexicographical sequence. In other words: first the activity with $j = 1$, then the activities with $j = 2$ with increasing i, then with $j = 3$, *etc.*

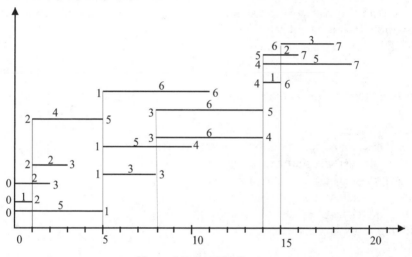

Figure 2.7. GANTT diagram

Figure 2.7 shows the GANTT diagram for the example network. The critical path is easy to see from the diagram. It consists of the 4 activities (0;1), (1;3), (3;4) and (4;7).

The GANTT diagram is also equivalent to the other two network representations. Each of the 3 representation forms can be derived from each of the other forms and also transformed into each other form.

2.2.3 Determining Dates in the Network

Event dates

Each event can be characterised by 2 dates.

$t_e(i)$ – earliest date of the event (i)
$t_l(i)$ – latest date of the event (i).

Both dates are calculated recursively (by choosing the indices appropriately).

$$t_e(0) = 0 \qquad\qquad\qquad (2.8)$$

$$t_e(j) = \underset{i}{\text{Max}}\left\{ t_e(i) + d(i; j) \right\}$$

$$t_l(N) = t_e(N)$$

$$t_l(i) = \underset{j}{\text{Min}}\left\{ t_l(j) - d(i; j) \right\}$$

These calculations can be performed very simply with the matrix (see Figure 2.6). To calculate t_e of the row i go to the column i, add the values in this column to each of previously calculated values t_e and determine the maximum. $t_l(j)$ is calculated similarly.

For critical events (i_k)

$$t_l(i_k) = t_e(i_k). \qquad\qquad\qquad (2.9)$$

The critical events in the example network are (0), (1), (3), (4) and (7).

Activity dates

Four dates can be specified for each activity $(i; j)$

$t_{es}(i; j)$ – earliest start date of the activity $(i; j)$

$t_{ee}(i; j)$ – earliest end date of the activity $(i; j)$

$t_{ls}(i; j)$ – latest start date of the activity $(i; j)$

$t_{le}(i; j)$ – latest end date of the activity $(i; j)$

with the following computing rules

$$t_{es}(i; j) = t_f(i) \qquad\qquad\qquad (2.10)$$

$$t_{ee}(i; j) = t_f(i) + d(i; j)$$

$$t_{ls}(i; j) = t_s(j) - d(i; j)$$

$$t_{le}(i; j) = t_s(j)$$

Figure 2.8 shows the dates for the activity (2, 5):

$t_e(2) = 1$	$t_{es}(2; 5) = 1$
$t_l(2) = 6$	$t_{ee}(2; 5) = 5$
$t_e(5) = 14$	$t_{ls}(2; 5) = 13$
$t_l(5) = 17$	$t_{le}(2; 5) = 17$

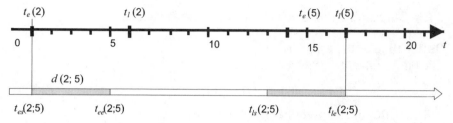

Figure 2.8. Date calculations for the activity (2;5)

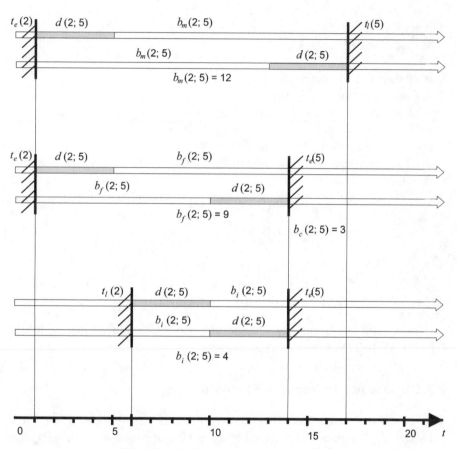

Figure 2.9. Buffer times for the activity (2; 5)

2.2.4 Calculation of Buffer Times in the Network

Buffer time of the event (i)
The buffer time $b(i)$ of the event (i) is calculated as

$$b(i) = t_l(i) - t_e(i) \tag{2.11}$$

The buffer time is zero for critical events i_k.

$$b(i_k) = 0 \tag{2.12}$$

The actual event date $t(i)$ always lies between $t_e(i)$ and $t_l(i)$. This means that critical events do not have any freedom for date displacements. Buffer times are also called slack times.

Buffer times of the activities ($i; j$)
There are 4 possible buffer times:
$b_m(i; j)$ – maximum buffer time of the activity ($i; j$)
$b_c(i; j)$ – conditional buffer time of the activity ($i; j$)
$b_f(i; j)$ – free buffer time of the activity ($i; j$)
$b_i(i;j)$ – independent buffer time of the activity ($i; j$)
The calculation equations are

$$b_m(i, j) = t_l(j) - t_e(i) - d(i; j) \tag{2.13}$$

$$b_c(i, j) = t_l(j) - t_e(j) = b(j)$$

$$b_f(i; j) = t_e(j) - t_e(i) - d(i; j)$$

$$b_i(i; j) = \mathrm{Max}\{0; t_e(j) - t_l(i) - d(i; j)\}$$

The activity (2; 5) is again used as an example for calculating the buffer times (Figure 2.9).

2.2.5 Strategies in Network Planning

The critical path determines the duration of a complete process. The critical activities $(i, j)_k$ cannot be displayed and must begin exactly at their start date

$$t_s(i; j)_k = t_{es}(i; j)_k = t_{ls}(i; j)_k \tag{2.14}$$

Each network has (at least) one critical path. The example has 4 critical activities: (0; 1), (1; 3), (3; 4), (4; 7). The other 10 activities are not critical. The actual start date can be freely chosen between the latest (latest permitted) and the earliest (earliest possible) start date. The free choice of the start date of a critical activity, however, is subject to restrictions:

Choice of the maximum buffer time (Figure 2.9, top)
If the activity $(i; j)$ is started at time $t_e(i)$, all activities that end at (i) will be forced to be completed by time $t_e(i)$. Thus you do not have the possibility, which you actually have, to end at $t_l(i)$. Consequently, you must intervene in the "freedom" of the preceding activities. The same thing occurs if you allow an activity to end at time $t_l(j)$. The subsequent activities that begin at (j) must wait and cannot, which would be possible, begin already at $t_e(j)$. Thus, if an activity (i, j) uses the maximum possible buffer time and occupies the complete interval $t_e(i)$ to $t_l(j)$, both the preceding activities and the subsequent activities will be hindered.

Thus, laying claim to the maximum buffer time is the *most inconsiderate behaviour* of an activity in the network.

Choice of the free buffer time (Figure 2.9, centre)
In this case, the activity $(i; j)$ claims the interval between $t_e(i)$ and $t_e(j)$. This takes account of all subsequent activities that are fully free in deciding to start as early as possible. This is a *desirable* behaviour in the network, but, however, requires all previous activities to complete at the earliest possible date.

Choice of the independent buffer time (Figure 2.9, bottom)
The activity $(i; j)$ occupies only the interval $t_l(i)$ to $t_e(j)$. This allows each subsequent activity to freely decide on the start and end, and will not be hindered. This is the absolutely *most considerate* behaviour of the activity $(i; j)$ in the network.

Choice of independent and free buffer time
Figure 2.9 shows that the activity $(i; j)$ in this case occupies the interval between $t_l(i)$ and $t_e(j)$. This strategy allows all preceding activities to freely decide but, if necessary, be complete at the latest permitted date. This behaviour is *conditionally sensible.*

Thus, the choice of the actual start dates $t_s(i; j)$ for an activity $(i; j)$ is always associated with a strategy. In the case of the independent buffer time, it is, however, possible that such a strategy is not available because the computed value yields

$$t_e(j) - t_l(i) - d(i; j) < 0$$

namely, it is negative. Then, the definition states $b_i = 0$, because the maximum formed between 0 and a negative number gives the value zero. Thus, in this case, the "possible" strategy cannot be achieved.

Thus, one should always perform an exact network analysis and control the true process execution using a good strategy.

2.2.6 CPM Cost

Previously, the duration $d(i; j)$ was always considered to be an invariant quantity, only the start of the activity $t_s(i; j)$ was considered to be variable within certain limits. To better conform to the practical conditions, the costs are also included in the CPM network planning as follows.

The duration is considered to be a quantity that possibly may be shortened, although such a shortening of the duration causes higher costs. When one creates a cost–duration diagram, this produces the general behaviour shown in Figure 2.10.

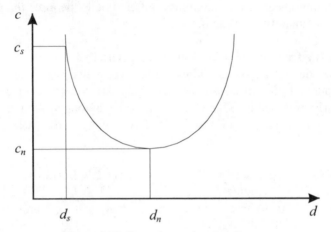

Figure 2.10. General cost–duration diagram

The duration of an activity can in general only be reduced to d_s, even if the costs further increase. Conversely, an arbitrary reduction of the costs, despite the lengthening of the duration, also cannot be achieved. Quite the contrary, the costs will further increase for a longer duration. The actual "stable" costs–duration characteristic curve lies between d_s and d_n. If this part is replaced with a straight line, this produces the linear cost–duration diagram shown in Figure 2.11.

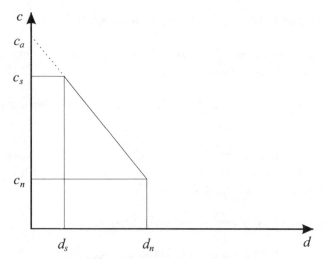

Figure 2.11. Linear cost–duration diagram

The complete duration D of the network and the complete costs C lie in an area (Figure 2.12).

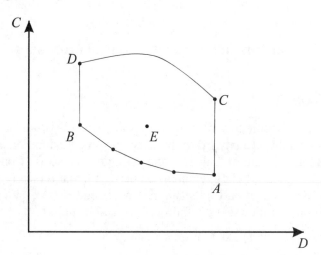

Figure 2.12. Cost–duration area of the complete network

Operating point A: *cheapest* and *longest* network. All activities $(i; j)$ have the duration $d_n(i; j)$, *i.e.* each with the longest duration and the least cost.

Curve A–B: the reduction of the critical path starting at A. Initially starting at *that* activity that when shorted shows the smallest cost increase, *etc*. The path is formed from those straight lines that always show the largest increase.

Operating point B: the *cheapest* and *shortest* network.

Section B–D: reduction of the non–critical activities starting at B; this makes the network more expensive but not shorter.

Operating point D: all activities for d_s and c_s; the *shortest* and *most expensive* network.

Section A–D: starting at A. Whereas non–critical activities will be shortened, critical ones will not be shortened. The critical path remains, but all non–critical activities are at d_s and c_s.

Operating point C: the *longest* and *most expensive* network.

Operating point E: the normal, not optimised, network.

2.3 Product-flow Graphs

2.3.1 Weighting Function for Technological Processes

Introduction

The duration of technological processes is described by a random variable with the probability density $f(t)$. In practice, the process duration always lies only between a smallest value a and a largest value b. Consequently, Section 1.3.2 specifies the integration limits with a and b (see Figure 1.22). Obviously the integration limits can also be extended $(-\infty, +\infty)$ when one takes into consideration that the following equation applies

$$f(t) = 0 \quad \text{for } t < a \quad \text{and } t > b \tag{2.15}$$

For example, using Equation (2.15), it can also be written as:

$$\int_a^b f(t)\,dt = \int_{-\infty}^{+\infty} f(t)\,dt = 1 \tag{2.16}$$

The following considerations do not assume a specific distribution. Other than the process duration, a probability w is assigned to each technological process as an additional characteristic value. w should be called the realisation probability and specifies the probability with which a technological process occurs.

$$0 < w \leq 1 \tag{2.17}$$

If, for example, defect products with a defect rate p are determined in an inspection process and these products are subsequently repaired, the repair process then has the realisation probability $w = p$.

The product of w and $f(t)$ is called the *weighting function* $g(t)$.

$$g(t) = w \cdot f(t) \tag{2.18}$$

This allows many relationships from the linear system theory [7] to be used for technological processes. Several examples are used to demonstrate this procedure.

Weighting Function for Constant Process Duration

The technological process is not stochastic, bur rather determined with the constant duration d. In this case, the density $f(t)$ becomes the *Dirac function* at position d, and the weighting function reduces to

$$g(t) = w \cdot \delta(t - d) \tag{2.19}$$

With regard to the distribution theory, the Dirac function is also a probability density here with the area 1, although it assumes the zero value and for $t = d$ the infinite value for $t \neq d$. It is an infinitely narrow and an infinitely high *pulse* and thus a special or boundary case for ordinary functions.

One can illustrate the occurrence of the Dirac function as follows: for constant area and constant expected value μ, the ordinary distribution $f(t)$ is made ever narrower (and thus ever higher) until finally the Dirac pulse occurs in the boundary case. The weighting function (2.19) also characterises the technological process so that a product with the probability w (in practical terms: the *proportion w* of a *product set*) flows through the technological process and requires the duration d from start to end, *i.e.* appears at the output after duration d. The system theory assigns an input x_1 and an output x_2 to the technological process. The technological process takes place between this input and output. This can be interpreted as a delay system for a constant process duration.

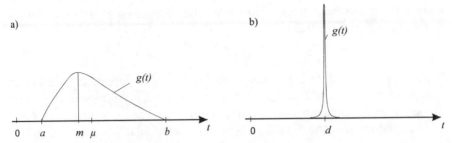

Figure 2.13. Weighting function for technological processes – (**a**) stochastic, (**b**) determined

Weighting Function and Pulse Response

If a Dirac pulse $\delta(t)$ is applied at the input $x_1(t)$ of a technological process at time $t = 0$, the so-called pulse response appears at the output $x_2(t)$ as a response (see Figure 2.14). The system theory proves that the pulse response is always identical to the weighting function $g(t)$, *i.e.*

$$x_1(t) = \delta(t) \quad \Leftrightarrow \quad x_2(t) = g(t) \tag{2.20}$$

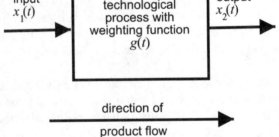

Figure 2.14. System representation of the technological process

Weighting Function and System Equation

The system equation describes the general relationship between $x_1(t)$, $g(t)$, $x_2(t)$, where $x_1(t)$ and $x_2(t)$ are also probability densities. The system equation is a convolution equation defined by the following integral:

$$x_2(t) = g(t) * x_1(t) \tag{2.21}$$

$$= \int g(u) \cdot x_1(t-u)\,\mathrm{d}u$$

$$= \int g(t-u) \cdot x_1(u)\,\mathrm{d}u$$

Calculation of the Process Parameters Using the Weighting Function

Parameters, such as expected value, deviation, standard deviation and variation coefficient, process duration, but also the realisation probability of technological processes, have major significance for practical investigations and for configuring. We now show here how some of these parameters can be calculated using the weighting function $g(t)$.

1. Calculation of the realisation probability
Integrating Equation (2.18) and using Equation (2.17) it is easy to show:

$$w = \int g(t)\,\mathrm{d}t \tag{2.22}$$

To simplify the description, we omit the specification of the integration limits a and b or $-\infty$ and $+\infty$ here and in the following section, *i.e.* if *no* limits are specified, the integral should always be performed over the complete domain.

2. Calculation of the expected value μ of the process duration
Equation (1.18) with Equation (2.18) yields

$$\mu = \frac{1}{w} \cdot \int t \cdot g(t)\,\mathrm{d}t \tag{2.23}$$

3. Calculation of the variance σ^2 of the process duration
Equation (1.19) with Equation (2.18) yields

$$\sigma^2 = \frac{1}{w} \cdot \int t^2 \cdot g(t)\,\mathrm{d}t - \mu^2 \tag{2.24}$$

2.3.2 Operator Technological Processes

Introduction

Using the Laplace transformation on the weighting function $g(t)$ produces the transfer function $G(s)$, also called the *technological operator*.

$$G(s) = \int_{-\infty}^{+\infty} g(t) \cdot e^{-st} dt = L\{g(t)\} \tag{2.25}$$

The symbol L is used as a short form for the multiplication of the function to be transformed with the so-called kernel e^{-st} and subsequent integration. This makes the character of a representation more apparent, *i.e.* the original domain (or t-domain, time domain) the function $f(t)$ is "represented" in the range (or s-domain, operator domain), *i.e.* transformed and called $G(s)$ there.

The Laplace transformation is a very powerful and popular method not only for solving differential equations but also for calculating time activities in systems used in the automatic control theory, in system theory and in electrical engineering.

The Laplace reverse transformation can be used to obtain the weighting function from any operator. This means $g(t)$ and $G(s)$ are fully equivalent.

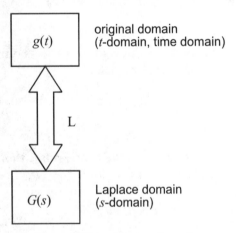

original domain
(t-domain, time domain)

L

Laplace domain
(s-domain)

Figure 2.15. Laplace transformation

Consequently, a function in the range can be assigned to any function in the original domain and *vice versa*.

Operator for Constant Process Duration

For determined processes with a constant duration d, the operator can be calculated using Equation (2.16) in Equation (2.25).

$$G(s) = w \cdot e^{-sd} \qquad (2.26)$$

This is the transform of the Dirac function (Equation (2.19)) in the range (using a realisation probability w and duration d). For the limit case $w = 1$ and $d = 0$, this yields $G(s) = 1$ and $g(t) = \delta(t)$.

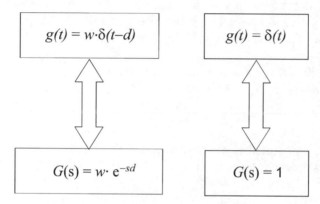

Figure 2.16. Operator for determined technological processes

System Equation in the Range

The system Equation (2.18) can be fully transformed in the range

$$X_2(s) = G(s) \cdot X_1(s) \qquad (2.27)$$

where $X_1(s)$ is the Laplace transform of $x_1(t)$; $X_2(s)$ is formed similarly. Consequently, the convolution in the t-domain (2.21) corresponds to multiplication in the s-domain. This allows $G(s)$ to be determined very easily by forming the quotient

$$G(s) = \frac{X_2(s)}{X_1(s)}$$

The technological process is represented as an arrow, where $X_1(s)$ and $X_2(s)$ are interpreted as the node potential and $G(s)$ is interpreted as the transfer function.

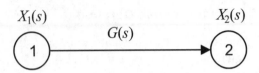

Figure 2.17. Element of a product-flow graph

Calculating Parameters from the Technological Operator

The realisation probability w, the expected value μ and the variance σ^2 of the process duration cannot only be calculated from the weighting function $g(t)$, but also from the technological operator $G(s)$.

1. Calculation of the realisation probability w
Using Equation (2.25) with Equation (2.18) yields

$$G(s) = w \cdot \int f(t) \cdot e^{-st} dt \qquad (2.28)$$

Setting $s = 0$ (mathematically, one should actually form a limit value: $s \rightarrow 0$), this produces from Equation (2.26) with Equation (2.17)

$$w = G(s)\big|_{s=0} = G(0) \qquad (2.29)$$

If $G(s)$ is known, then it is very easy to calculate the value $G(0)$.

2. Calculation of the expected value
Differentiating with s on Equation (2.28) yields

$$\frac{dG(s)}{ds} = G'(s) = -w \cdot \int t \cdot f(t) \cdot e^{-st} dt \qquad (2.30)$$

Setting $s = 0$, Equations (1.18) and (2.29) yield

$$\mu = -\frac{G'(s)}{G(s)}\bigg|_{s=0} = -\frac{G'(0)}{G(0)} \qquad (2.31)$$

This expression can be written as

$$\mu = -\frac{d}{ds}\{\ln G(s)\}\bigg|_{s=0} = -(\ln G(s))'\big|_{s=0} \qquad (2.32)$$

3. Calculation of the variance σ^2

Further differentiation of Equation (2.30) with Equation (1.19) yields

$$\frac{d^2G(s)}{ds^2} = G''(s) \quad = w \cdot \int t^2 f(t) \cdot e^{-st} dt \qquad (2.33)$$

$$= w \cdot (\sigma^2 + \mu^2)$$

and then with Equations (2.29) and (2.30)

$$\sigma^2 = \left[\frac{G''(s)}{G(s)} - \left(\frac{G'(s)}{G(s)} \right)^2 \right]_{s=0} \qquad (2.34)$$

$$= \left[\frac{G'(s)}{G(s)} \right]'_{s=0} = \left[\frac{G'(0)}{G(0)} \right]'$$

$$= \left[\ln G(s) \right]'' \Big|_{s=0}$$

2.3.3 Basic Structures of Technological Processes

We demonstrate the procedure using the three basic structures: chain structure, parallel structure and feedback structure.

Chain Structure

Two technological processes T_1 and T_2 with operators $G_1(s)$ and $G_2(s)$ form a chain structure

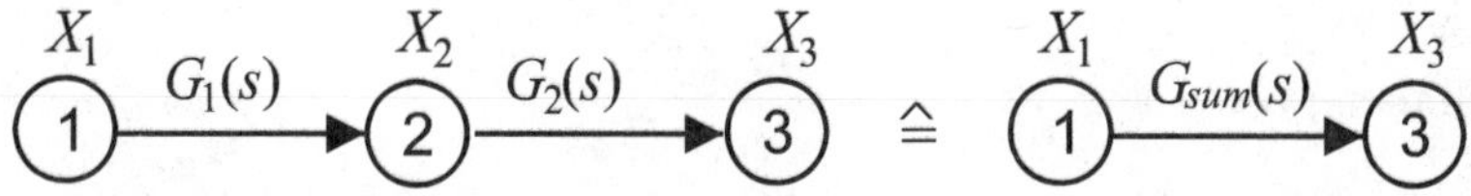

Figure 2.18. Chain structure consisting of 2 technological processes

This gives for the process

$$X_2(s) = G_1(s) \cdot X_1(s) \qquad (2.35)$$
$$X_3(s) = G_2(s) \cdot X_2(s)$$

with

$$G_{sum} = \frac{X_3(s)}{X_1(s)} \qquad (2.36)$$

this then yields

$$G_{sum}(s) = G_1(s) \cdot G_2(s) \qquad (2.37)$$

Using Equations (2.29), (2.32) and (2.34) yields the following calculation equations

$$w_{sum} = w_1 \cdot w_2 \qquad (2.38)$$

$$\mu_{sum} = \mu_1 + \mu_2$$

$$\sigma_{sum}^{\ 2} = \sigma_1^{\ 2} + \sigma_2^{\ 2}$$

It is easy to extend to more than 2 process elements using the above equations. The chain structure of technological processes is the simplest and most common form used in real manufacturing systems.

Parallel Structure

Two technological processes T_1 and T_2 with operators $G_1(s)$ and $G_2(s)$ form a parallel structure.

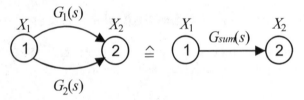

Figure 2.19. Parallel structure formed from 2 technological processes

This gives for the complete process

$$G_{sum} = G_1(s) + G_2(s) \qquad (2.39)$$

which then gives

$$w_{sum} = w_1 + w_2$$

$$w_{sum} \cdot \mu_{sum} = w_1 \cdot \mu_1 + w_2 \cdot \mu_2 \qquad (2.40)$$

$$w_{sum} \cdot \left(\sigma_{sum}^{\ 2} + \mu_{sum}^{\ 2}\right) = w_1 \cdot \left(\sigma_1^{\ 2} + \mu_1^{\ 2}\right) + w_2 \cdot \left(\sigma_2^{\ 2} + \mu_2^{\ 2}\right)$$

It is relatively easy to extend to more than 2 process elements results using Equations (2.40). The parallel structure of technological processes always occurs in real manufacturing processes when the products at the end of an evaluation process are split into two (or more) product flows, which then pass through different processes and are recombined to form a single flow.

Feedback Structure

To derive the overall operator, a structure is formed from 4 individual processes.

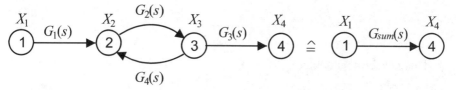

Figure 2.20. Feedback structure

The node equations are:

$$X_2(s) = G_1(s) \cdot X_1(s) + G_4(s) \cdot X_3(s) \tag{2.41}$$
$$X_3(s) = G_2(s) \cdot X_2(s)$$
$$X_4(s) = G_3(s) \cdot X_3(s)$$

The overall operator G_{sum} is calculated using:

$$G_{sum}(s) = \frac{X_4(s)}{X_1(s)} = \frac{G_1(s) \cdot G_2(s) \cdot G_3(s)}{1 - G_2(s) \cdot G_4(s)} \tag{2.42}$$

The feedback in real manufacturing systems is often caused by $G_2(s)$ and $G_4(s)$ representing an inspection process and a reworking or rejection process, respectively. This is sometimes also known as a "repair loop". Such structures occur frequently in electronics production. From the viewpoint of process modelling, the considered feedback structure can be interpreted as being a "pure chain" with 4 elements.

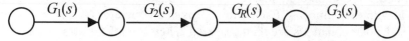

Figure 2.21. Modelling of the feedback structure as a chain structure

$$G_{sum}(s) = G_1(s) \cdot G_2(s) \cdot G_R(s) \cdot G_3(s) \tag{2.43}$$

Using the equations from Section 1.3.1, this gives

$$w_{sum} = w_1 \cdot w_2 \cdot w_R \cdot w_3 \tag{2.44}$$
$$\mu_{sum} = \mu_1 + \mu_2 + \mu_R + \mu_3$$
$$\sigma_{sum}^2 = \sigma_1^2 + \sigma_2^2 + \sigma_R^2 + \sigma_3^2$$

If one designates the operator for the chain in the feedback loop as

$$G_0(s) = G_2(s) \cdot G_4(s) \tag{2.45}$$

this produces the following operator for the feedback process:

$$G_R(s) = \frac{1}{1 - G_0(s)} \tag{2.46}$$

This is used to calculate

$$w_R = \frac{1}{1 - w_0} \qquad \mu_R = \frac{w_0}{1 - w_0} \cdot \mu_0 \tag{2.47}$$

$$\sigma_R{}^2 = \frac{w_0}{1 - w_0}\left(\sigma_0{}^2 + \frac{1}{1 - w_0} \cdot \mu_0\right)$$

with

$$w_0 = w_2 \cdot w_4 \tag{2.48}$$
$$\mu_0 = \mu_2 + \mu_4$$
$$\sigma_0{}^2 = \sigma_2{}^2 + \sigma_4{}^2$$

2.3.4 General Structures, Regularities, Relationships

Operator Equation

To calculate the overall operator of a manufacturing process with K nodes, the nodes are numbered successively. A potential $X_k(s)$ is assigned to each node with the number k. Each arrow is given an operator $G_n(s)$.

$K-1$ node equations are then prepared. They use all the arrows leading to the associated node and add the transferred potential proportions.

The overall operator is obtained by eliminating the potentials of the intermediate nodes. The following example demonstrates the procedure.

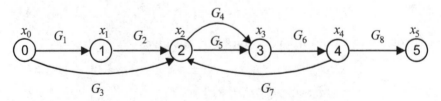

Figure 2.22. Product-flow graph for a manufacturing system

The 5 node equations are

$$X_1 = G_1 \cdot X_0 \tag{2.49}$$
$$X_2 = G_2 \cdot X_1 + G_3 \cdot X_0 + G_7 \cdot X_4$$
$$X_3 = G_4 \cdot X_2 + G_5 \cdot X_2$$
$$X_4 = G_6 \cdot X_3$$
$$X_5 = G_8 \cdot X_4$$

We use

$$X_5 = G_{sum} \cdot X_0$$

to obtain the overall operator

$$G_{sum} = \frac{(G_1 \cdot G_2 + G_3) \cdot (G_4 + G_5) \cdot G_6 \cdot G_8}{1 - (G_4 + G_5) \cdot G_6 \cdot G_7} \tag{2.50}$$

A symbolic method also allows G_{sum} to be specified when the arithmetic rules and the hierarchical system form are used for the chain, parallel and feedback structure:

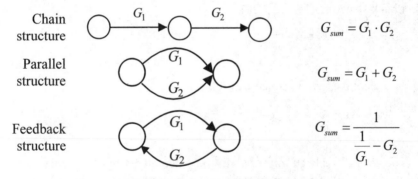

Chain structure	$G_{sum} = G_1 \cdot G_2$
Parallel structure	$G_{sum} = G_1 + G_2$
Feedback structure	$G_{sum} = \dfrac{1}{\dfrac{1}{G_1} - G_2}$

Figure 2.23. Basic structures and their operator equations

Realisation Probability

The sum of the realisation probabilities for all arrows *leaving* a node is always 1.

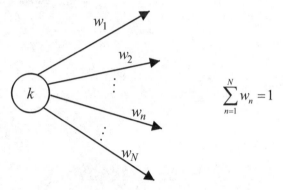

Figure 2.24. Law of conservation

This means that $w = 1$ only when *one* single arrow leaves a node. Every structure with K nodes always has $K-1$ conservation equations for the realisation probabilities that directly correspond with the structure.

As a first example, the conservation equations are specified for the feedback structure (Figure 2.20):

$$w_1 = 1 \qquad w_3 + w_4 = 1 \qquad (2.51)$$
$$w_2 = 1$$

The second example concerns Figure 2.22

$$w_1 + w_3 = 1 \qquad w_4 + w_5 = 1 \qquad w_7 + w_8 = 1 \qquad (2.52)$$
$$w_2 = 1 \qquad w_6 = 1$$

The overall realisation probability is calculated from the structure using Equation (2.29). This is done most easily by replacing $G_n(s)$ or G_n by w_n for all n in the operator equation.

This is demonstrated using the feedback structure as an example. Equation (2.42) then yields

$$w_{sum} = \frac{w_1 \cdot w_2 \cdot w_3}{1 - w_2 \cdot w_4} \qquad (2.53)$$

Set the conservation Equations (2.51) in Equation (2.53) to obtain $w_{sum} = 1$. This relationship obviously results for all systems provided they have "completed", *i.e.* they must have a node as source (only leaving arrows) and a node as sink (only entering arrows).

The relationships are *independent* of the associated operators, *i.e.* they are also independent of μ, σ^2 and other duration parameters!

They characterise only the structure of the manufacturing system.

Realisation probabilities characterise the quality behaviour of technological processes.

Expected Value of the Duration

The relationships of the process durations and the calculation of the complete duration are very important in electronic process technology. The method of product-flow graphs can now use the basis of manufacturing structures to derive the general equations that can be used to perform the dimensioning in an actual case for real manufacturing systems.

For a manufacturing system with N separate technological processes, the expected value of the complete duration μ_{sum} is always a sum of all expected values of the individual technological processes, where an intensity coefficient c_n must be considered for each process T_n

$$\mu_{sum} = \sum_{n=1}^{N} c_n \cdot \mu_n \tag{2.54}$$

The feedback structure (Figure 2.20) is used as an example to explain this relationship.

Setting Equations (2.47), (2.48) and (2.51) into Equation (2.44) yields

$$\mu_{sum} = \mu_1 + \frac{1}{1 - w_4} \cdot \mu_2 + \mu_3 + \frac{w_4}{1 - w_4} \cdot \mu_4 \tag{2.55}$$

The intensity coefficients depend only on the realisation probabilities. This is true for all manufacturing systems.

In particular

$$c_1 = 1 \qquad\qquad c_3 = 1 \tag{2.56}$$

$$c_2 = \frac{1}{1 - w_4} \qquad c_4 = \frac{w_4}{1 - w_4}$$

The model is explained using a numerical example.

Assuming a repair probability of $p = w_4 = 0.2$, one obtains $c_2 = 1.25$ and $c_4 = 0.25$. This means: the process T_1 has a defect rate of 20% and a yield $FPY = 0.8$ (first-pass yield). The repair process T_4 is loaded with 25% of the manufacturing flow because the repair loop is continually "repeated" in the model!

In other words: 20% of the originally manufactured products are repaired and retested; also afterwards 20% pass through the repair loop a second time, *i.e.* 4% of the originally manufactured products have two repair loops; 0.8% three repair loops, *etc.* This means the inspection process is loaded in total with 125%.

The intensity coefficients c_n can be calculated from the process equation $G_{sum}(s)$, also with total differentiation:

From

$$G_{sum}(s) = G_{sum}\left(G_1(s), G_2(s), \cdots, G_n(s), \cdots, G_N(s)\right) \tag{2.57}$$

yields

$$G'_{sum}(s) = \sum_{n=1}^{N} \frac{\partial G_{sum}}{\partial G_n} \cdot G'_n(s) \tag{2.58}$$

and

$$\mu_{sum} = \sum_{n=1}^{N} \frac{\partial G_{sum}}{\partial G_n} \cdot \left. \frac{G_n(s)}{G_{sum}(s)} \right|_{s=0} \cdot \mu_n \tag{2.59}$$

thus

$$c_n = \frac{w_n}{w_{sum}} \cdot \frac{\partial w_{sum}}{\partial w_n} \tag{2.60}$$

The expected value of the duration of the complete process thus depends neither on the deviations nor on the associated density change, but only on the individual expected values and their intensity coefficients. Consequently, the expected value can also be calculated when the process model is assumed to be deterministic (see Section 1.3.2), *i.e.*

$$G_n(s) = w_n \cdot e^{-\mu_n \cdot s} \tag{2.61}$$

Duration Deviation

The further differentiation for s on Equation (2.58) yields

$$G''_{sum}(s) = \sum_{n=1}^{N} \frac{\partial G_{sum}}{\partial G_n} \cdot G''_n(s) + \sum_{m=1}^{N}\sum_{n=1}^{N} \frac{\partial^2 G_{sum}}{\partial G_m \partial G_n} \cdot G'_m(s) \cdot G'_n(s) \tag{2.62}$$

Dividing this $G_{sum}(s)$ with Equations (2.29), (2.31) and (2.34)

$$\sigma^2_{sum} + \mu^2_{sum} = \sum_{n=1}^{N} c_n \cdot \left(\sigma^2_n + \mu^2_n\right) + \sum_{m=1}^{N}\sum_{n=1}^{N} d_{mn} \cdot \mu_m \cdot \mu_n \tag{2.63}$$

with c_n according to Equation (2.60) and

$$d_{mn} = \frac{w_m \cdot w_n}{w_{sum}} \cdot \frac{\partial^2 w_{sum}}{\partial w_m \cdot \partial w_n}$$

(2.64)

In practice, however, the deviation equation is almost never used.

2.4 Queue Models

Although queuing theory is generally suitable for the analytical modelling of manufacturing processes, its use also requires good knowledge of mathematics, in particular, probability theory. Unfortunately, the high abstraction level and structural reasons mean narrow limits are set on its practical use. Thus, complex systems with complicated controls can, at best, only be approximated. Queue models are infinite time-continuous models controlled stochastically by events. Furthermore, the models are both event and state-discrete, and consider not only the finiteness of the resources but also their interactions.

Queuing systems consist essentially of three basic elements

- service stations
- queues
- event or demand flows.

The stations and queues are connected with each other with a network consisting of directed edges. The demand flows run along these edges; the demand flows can be separated or combined at special nodes of the graphs (Figure 2.25).

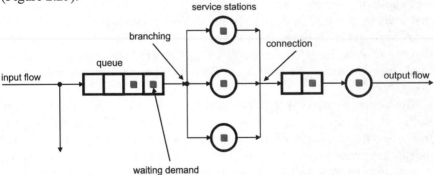

Figure 2.25. Example of a queuing system

The demand represents a very complex entity in queuing theory that is not always easy to interpret and that can best be compared with a product or produce in the real manufacturing. However, it is also possible to consider other tasks, such as the repair of a machine, as a demand. Generally, the demand contains a whole complex of tasks to be performed at the various service stations. This work schedule is however not explicitly linked with the demand object itself, but rather contained implicitly in the structure graphs of the queuing system.

A specific demand passes through the queuing system along the previously defined flow edges. When the demand arrives at a service station, it remains for the duration of the servicing in the station. The servicing duration is a property of the station and is generally considered to be a stochastic quantity. Service stations also have a limit capacity that specifies how many demands can be served concurrently. The following section assumes that this limit capacity always has the value 1. There are two possibilities for demands that arrive at an occupied station:

1. The demand waits in front of the station in a specially provided queue.
2. The demand will be rejected and diverted to a so-called loss flow when the acceptance capacity of the queue is reached.

Systems for which loss flows can occur are also designated as loss systems. Systems that do not accept or issue any external demands are called closed queuing systems. The number of demands is constant in closed systems. Such systems can be used for such things as the modelling of the manufacturing control.

2.4.1 Demand Flows

Because the demand flows form the backbone of the queuing theory, we will consider them in more detail here. We will make the basic assumption that a demand e_k arrives at the system at some random time (date) T_k. The dates must generally obey the relation: $T_k \leq T_{k+1}$. Because the arrival of a demand is normally equivalent to an event, the terms demand and event are used synonymously. The term event flows is often used instead of demand flows.

The system analysis is greatly simplified when simultaneous events are excluded, namely $T_k < T_{k+1}$ is assumed. This is no significant restriction from the practical point of view, because the refinement of the time grid means simultaneous events can always be transformed into sequential events. Event flows with this property are also called ordinary event flows. As for the event times themselves, the demands are linearly ordered and form the sequence $\{e_k\}$.

The dates can be interpreted as being a physical time when they relate to a common start date. The date T_0 of the first demand e_0 is a natural candidate here. Thus, the subsequent discussion will not differentiate between the random event date T_k and the random physical event time $T_k - T_0$.

The event flow is characterised by the so-called intermediate arrival time $\Delta T_k = T_k - T_{k-1}$ that results from the time difference between two consecutive events. The intermediate arrival time is a random quantity whose distribution function largely determines the properties of the event flow.

As Figure 2.26 shows, there are several equivalent representations for event flows:

1. sequence of event times (dates) $\{T_k\} = T_0 < T_1 < \ldots < T_{k-1} < T_k < \ldots$
2. sequence of intermediate arrival times $\Delta T_k = T_k - T_{k-1}$ with $k \geq 1$
3. continuously increasing counting process $N(t) = k$ for $T_k \leq t < T_{k+1}$

The intensity λ can be defined for a stationary event flow that specifies the average number of arriving demands until time t.

$$\lambda = \frac{E(N(t))}{t} \tag{2.65}$$

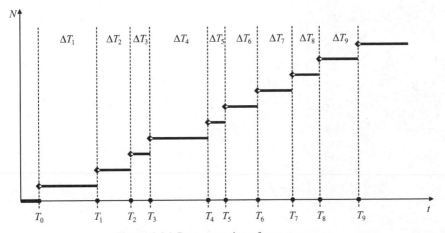

Figure 2.26. Representation of event currents

Event currents whose intermediate arrival times ΔT_k are independent exponentially distributed random values play a special role in queuing theory. In this case: $P(\Delta T_k > \Delta t) = e^{-\alpha \Delta t}$. The growths $\Delta N(t) = N(t+\Delta t) - N(t)$ are then also independent random quantities. However, because they obey the Poisson distribution, these flows are also designated as Poisson flows:

$$P(\Delta N(t) = k) = \frac{(\alpha \cdot \Delta t)^k \cdot e^{-\alpha \cdot \Delta t}}{k!} \qquad (2.66)$$

The Poisson flows have properties that significantly simplify the mathematical analysis. To understand this, we must first consider the exponential distribution and its special properties.

A continuous random value X has an exponential distribution when the following equation governs its distribution density:

$$f(x) = \begin{cases} 0 & \text{for } x < 0 \\ \alpha \cdot e^{-\alpha x} & \text{for } x \geq 0 \end{cases} \qquad (2.67)$$

Figure 2.27 shows the associated distribution function $F(x)$, for example, for three parameters: $\alpha = 2$, $\alpha = 1$ and $\alpha = 0.5$. Exponentially distributed random values produce small values particularly frequently. The larger an observed value is, the more seldom it occurs. Negative values are not created. If all values of an observation series are assigned to interval classes with width 0.2, this produces for $\alpha = 1$ the histogram shown in Figure 2.28.

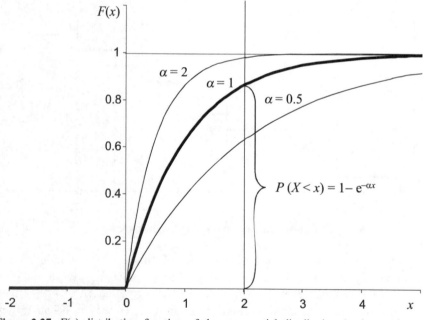

Figure 2.27. $F(x)$ distribution function of the exponential distribution (various values of parameter α)

The probability that a value x lies in the interval [0; 0.2) is approximately 0.18. However, in the interval [1.8; 2.0) it has already fallen to less than

0.03. The following equation governs the expected value and the deviation of the exponential distribution:

$$E(X) = \sqrt{D^2(X)} = \frac{1}{\alpha} \tag{2.68}$$

The intensity of the Poisson event flow can be determined from the counting process:

$$E(N(t)) = \sum_{k=1}^{\infty} k \cdot P(N(t) = k) \tag{2.69}$$

The intensity $\lambda = \alpha$ is obtained from Equation (1.36). Namely, the intensity of the Poisson event flow is identical to the reciprocated expected value of the intermediate arrival times.

However, another property, in particular, exclusively characterises the exponential distribution and thus the Poisson event flows and makes it important for queuing theory. When one queries the probability that $X \geq y + x$ subject to the condition that $X \geq y$ applies, this produces the following equation:

$$P(X \geq y + x \mid X \geq y) = P(X \geq x) \tag{2.70}$$

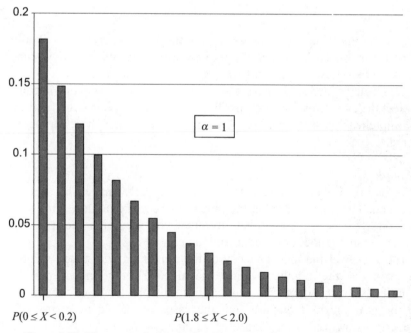

Figure 2.28. Histogram of an exponentially distributed random value for $\alpha = 1$

This surprising property is easy to prove:

$$P(X \geq y + x \mid X \geq y) = \frac{P((X \geq y + x) \wedge (X \geq y))}{P(X \geq y)} = \frac{P(X \geq y + x)}{P(X \geq y)} \quad (2.71)$$

When $P(X \geq x) = e^{-\alpha x}$, then:

$$P(X \geq y + x \mid X \geq y) = \frac{e^{-\alpha(y+x)}}{e^{-\alpha x}} = e^{-\alpha x} = P(X \geq x) \quad (2.72)$$

Transferred to the exponentially distributed intermediate arrival times of a Poisson event flow, this means: it is assumed that an event e_{k-1} has arrived by time T_{k-1}. Although the flow since this time was then observed constantly during the duration Δt_{k1}, the subsequent event e_k has still not yet arrived. The probability that the remaining wait time is greater than or equal to Δt_{k2} does not change, irrespective of how far the previous event occurred in the past. This independency of the remaining observation time of the previously elapsed observation time often conflicts with our experience.

Example 1:
The failure of an incandescent lamp becomes more probable with increasing operation duration.

Example 2:
The remaining time until the arrival of the long-expected bus should be shorter, the longer that the passenger has already waited at the bus stop. No Poisson event flow is apparently present in all these cases.

This can be deduced using a simple consideration without needing to subject the event flow and the distribution of its intermediate arrival times to a comprehensive statistical test. A counterexample, however, can be easily found.

Example 3:
It is assumed that a lottery participant regularly makes his bet every weekend. The probability that the time until he makes a large win is longer than one year is P_1. This probability is usually very high and it is easy to appreciate that it does not change after a year.

The cause for the independency of the remaining wait time (waiting for the win) is the number-picking machine. The individual draws are completely independent. The drawing from the previous week has no effect on the new draw. If this was not the case, you could make use of this property to increase your chance of winning, which certainly is not in the interest of the lottery company.

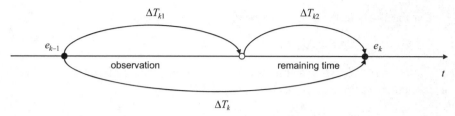

Figure 2.29. Observation duration and the remaining observation duration

The examples from daily life show how simple considerations can often be used to test whether or not Poisson event flows are involved. These considerations also clarify the nature of Poisson flows. Not only are the intermediate arrival times independent random variables, but the individual events arrive independently and are not coupled with each other with any internal structure other than their membership of the same event flow.

The exponential distribution has in the queuing theory a comparable role as the normal distribution has in mathematical statistics. In both cases, the reasons lie in the independency of the events or effects and in the possible simplification of the associated equation apparatus.

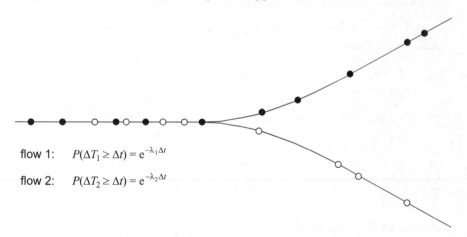

flow 1: $P(\Delta T_1 \geq \Delta t) = e^{-\lambda_1 \Delta t}$

flow 2: $P(\Delta T_2 \geq \Delta t) = e^{-\lambda_2 \Delta t}$

Figure 2.30. Overlaying of Poisson event flows

Figure 2.30 shows two event flows that are separated or combined. The following general equations apply to the complete flow

$$P(\Delta T \geq \Delta t) = P((\Delta T_1 \geq \Delta t) \wedge (\Delta T_2 \geq \Delta t)) \qquad (2.73)$$

or for independent flows

$$P(\Delta T \geq \Delta t) = P(\Delta T_1 \geq \Delta t) \cdot P(\Delta T_2 \geq \Delta t) \qquad (2.74)$$

For the Poisson event flows case, in particular:

$$P(\Delta T \geq \Delta t) = e^{-\lambda_1 \cdot \Delta t} \cdot e^{-\lambda_2 \cdot \Delta t} = e^{-(\lambda_1 + \lambda_2) \cdot \Delta t} \qquad (2.75)$$

Equation (2.75) shows that separated or combined Poisson event flows are themselves Poisson flows [6].

2.4.2 Classification of Queuing Systems

The designation key introduced by the English mathematician D. G. Kendall is generally used for the characterisation of queuing systems [2]. The key contains the character of the event flow, the property of the service station, the number of service stations and the length of the queues; the individual items are usually separated with a slash. These designations are found very often in modified form in the literature (see Table 2.1).

Systems with general distribution, multiple service stations m and limited queue length n: $G/G/m/n$ represent some of the largest challenges to the queuing theory. However, only the simpler systems with exponential distribution are investigated in the following section.

Table 2.1. Classification of queuing systems

	Explanation
M	Markov – exponentially distributed intermediate arrival times or serving times (Poisson event flow)
D	Deterministic – deterministic intermediate arrival times or serving times
GI	General Independent – The intermediate arrival times or the serving times are independent of each other. In addition, no demands are placed on the distribution function.
G	General – as GI, the independency between the random variables is however not demanded.

2.4.3 Loss System M/M/1/0

The M/M/1/0 system is the simplest possible loss system. It consists of just a single service station (machine) and does not have any queue. If the service station is free, the arriving demand will be processed immediately, otherwise it will be rejected. Rejected demands will be lost and form the loss flow. Without doubt, the best-known loss system is the telephone. If the dialled number is busy, one must hang up and redial later. However, even manufacturing systems contain substructures that can be described by a simple loss system.

Example:
A machine group contains a preferred machine that processes all incoming jobs. If the machine is occupied, incoming jobs will be diverted to the other machines. The problem situations could be described as: What is the probability that a job will not be processed on the preferred machine? How high is the loading on the preferred machine?

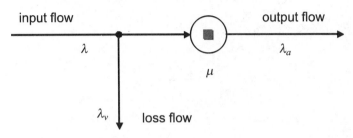

Figure 2.31. M/M/1/0 loss system

The following section assumes a stationary Poisson input flow with intensity λ. The intermediate arrival time of two successive events is T_A. Simultaneous events are ignored or excluded. The service station can always process just one demand for which it requires time T_B. Both the servicing time and the intermediate arrival time have an exponential distribution with servicing intensity μ.

$$P(T_A \geq \Delta t) = e^{-\lambda \cdot \Delta t} \tag{2.76}$$

$$P(T_B \geq \Delta t) = e^{-\mu \cdot \Delta t}$$

The loss system knows two discrete states. The task involves determining the state probabilities for an arbitrary time t.

State	Probability	
Z_0	$P(Z_0, t) = p_0(t)$	System is empty
Z_1	$P(Z_1, t) = p_1(t)$	System is busy

The probability that the system within the time interval Δt switches from state Z_i to state Z_j is the transition probability $p_{i,j}(\Delta t)$.

$$p_{i,j}(\Delta t) = P((Z_j, t + \Delta t) \,|\, (Z_i, t)) \tag{2.77}$$

This allows the state probabilities $p_i(t + \Delta t)$ to be calculated depending on the state probabilities at time t as follows.

$$p_0(t + \Delta t) = p_0(t) \cdot p_{0,0}(\Delta t) + p_1(t) \cdot p_{1,0}(\Delta t) \tag{2.78}$$

$$p_1(t + \Delta t) = p_0(t) \cdot p_{0,1}(\Delta t) + p_1(t) \cdot p_{1,1}(\Delta t)$$

The state graph is shown in Figure 2.32. Systems for which the following state is determined exclusively by the previous state are also called Markov systems.

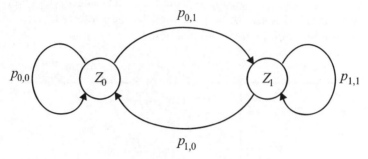

Figure 2.32. State graph of the M/M/1/0 loss system

This yields for the transition probabilities in a time interval Δt:

$$p_{0,0}(\Delta t) = P(T_A \geq \Delta t) = e^{-\lambda \cdot \Delta t} \qquad (2.79)$$

$$p_{1,0}(\Delta t) = P(T_B < \Delta t) = 1 - e^{-\mu \cdot \Delta t}$$

$$p_{0,1}(\Delta t) = P(T_A < \Delta t) = 1 - e^{-\lambda \cdot \Delta t}$$

$$p_{1,1}(\Delta t) = P(T_B \geq \Delta t) = e^{-\mu \cdot \Delta t}$$

One can now generate the individual functions in Taylor series and so obtain the probability polynomials in Δt for the transition. All elements with powers greater than 1 have each been combined in a remainder element $o(\Delta t)$ in Equation (2.80).

$$p_{0,0}(\Delta t) = 1 - \lambda \cdot \Delta t + o(\Delta t) \qquad (2.80)$$

$$p_{1,0}(\Delta t) = \mu \cdot \Delta t + o(\Delta t)$$

$$p_{0,1}(\Delta t) = \lambda \cdot \Delta t + o(\Delta t)$$

$$p_{1,1}(\Delta t) = 1 - \mu \cdot \Delta t + o(\Delta t)$$

Now setting the transition probabilities in Equation (2.78) yields the differential equation (2.81) for the state probabilities.

$$\frac{p_0(t + \Delta t) - p_0(t)}{\Delta t} = -\lambda \cdot p_0(t) + \mu \cdot p_1(t) + \frac{o(\Delta t)}{\Delta t} \qquad (2.81)$$

$$\frac{p_1(t + \Delta t) - p_1(t)}{\Delta t} = \lambda \cdot p_0(t) - \mu \cdot p_1(t) + \frac{o(\Delta t)}{\Delta t}$$

Because the term $o(\Delta t)$ contains only powers of Δt^2 and higher, the quotient $o(\Delta t)/\Delta t$ vanishes for the limit transition $\Delta t \rightarrow 0$ in both equations. The result is the differential equations for the state probabilities.

$$\dot{p}_0(t) = -\lambda \cdot p_0(t) + \mu \cdot p_1(t) \qquad (2.82)$$
$$\dot{p}_1(t) = \lambda \cdot p_0(t) - \mu \cdot p_1(t)$$

Because the loss system assumes just one of the two states p_0 or p_1 at any time:

$$p_0(t) + p_1(t) = 1 \qquad (2.83)$$

This allows the system of differential equations (2.82) to be separated into two independent differential equations for p_0 and p_1.

$$\dot{p}_0(t) = -(\lambda + \mu) \cdot p_0(t) + \mu \qquad (2.84)$$
$$\dot{p}_1(t) = -(\lambda + \mu) \cdot p_1(t) + \lambda$$

Equation (2.84) shows linear differential equations of first degree with constant coefficients that can be solved without difficulty. The result depends on the initial conditions. For the case that the loss system is initially empty, namely $p_0(0) = 1$, this yields the following state equations.

$$p_0(t) = \frac{\mu}{\lambda + \mu} + \frac{\lambda}{\lambda + \mu} \cdot e^{-(\lambda + \mu)t} \qquad (2.85)$$

$$p_1(t) = \frac{\lambda}{\lambda + \mu} - \frac{\lambda}{\lambda + \mu} \cdot e^{-(\lambda + \mu)t}$$

Figure 2.33 shows the transient response for $\lambda/\mu = 2$. The quotient from the input intensity λ and servicing intensity μ is also designated as the *traffic value* for which the Greek letter ρ is used,

$$\rho = \frac{\lambda}{\mu} \qquad (2.86)$$

The traffic value is a measure for the service-station loading. A high traffic value also means a high relative loading and consequently more demands enter the loss flow. The loss system quickly stabilises itself to the stationary end state. This can be calculated with the traffic value as follows:

$$p_0(t = \infty) = \frac{1}{1 + \rho} \qquad (2.87)$$

$$p_1(t = \infty) = \frac{\rho}{1 + \rho}$$

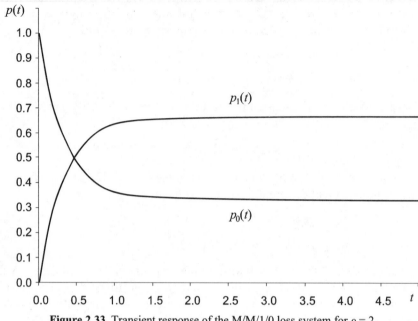

Figure 2.33. Transient response of the M/M/1/0 loss system for $\rho = 2$

Although the loss system is comparatively simple, it provides interesting and generally usable results for practical use.

Let us assume that a testing position is occupied by a single person and the duration required for the inspection of a product has an exponential distribution with testing intensity μ. The products to be inspected form a Poisson input flow with intensity λ. A product will be tested only when the testing person has currently nothing to do, otherwise the product will be forwarded. For this type of sample test, one is now interested in the number of tested products compared with the number of untested products.

The testing position separates the product flow into two flows: the flow of tested products with intensity λ_a and the flow of untested products with intensity λ_v. Obviously intensity λ_a cannot be larger than the testing intensity μ. In practice, a capacity consideration is often used for the calculation, for which the output intensity λ_a remains equal to the input intensity λ while $\lambda < \mu$. Afterwards, the output intensity is equal to the testing intensity μ. Figure 2.34 shows the relationship in its general form, in which both the input intensity and the output intensity are shown relative to μ. Formally, in addition to the traffic value ρ at the entry to the system, a traffic value $\rho_a = \lambda_a / \mu$ at the exit is also obtained. The representation is thus normalised and applies independently of the absolute value of the intensities.

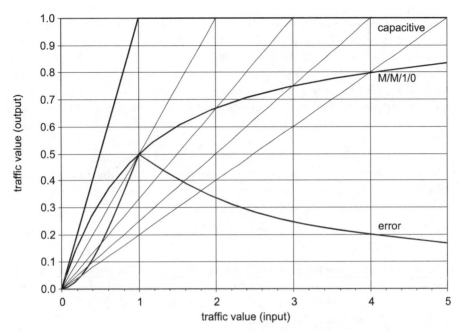

Figure 2.34. Loss system – capacitive and stochastical analysis

Actually, the M/M/1/0 loss system behaves quite differently. Between the simplified capacity consideration and the more accurate calculation using queuing theory there is an error that attains its maximum for the traffic value $\rho = 1$.

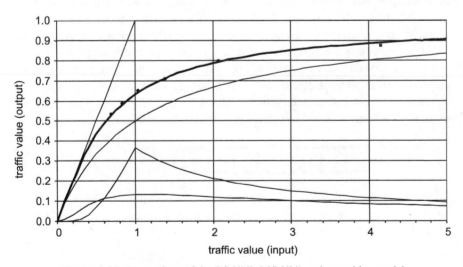

Figure 2.35. Comparison of the D/M/1/0, M/M/0/1 and capacitive model

The error arises from the fact that the simple capacity models do not take into consideration that phases of inactivity also occur for overloading, so that the theoretical intensity limit value can never be achieved. A rough estimate of the capacity is actually only suitable in two cases: for chronic underloading or for extreme overloading of the system. Otherwise, it can be shown that the M/M/1/0 loss system marks a limit curve. The system behaviour of an arbitrary queue system is characterised by a characteristic curve that lies in the area between the curve of the simple loss system as lower limit and the curve of the capacitive model as upper limit.

We use the D/M/1/0 system as an example in which the Poisson input flow has been replaced with a deterministic input flow. The traffic value at the system exit can be calculated as follows:

$$\rho_a = \rho \cdot (1 - e^{-1/\rho}) \tag{2.88}$$

The function $\rho_a = \rho(\rho)$ lies in the expected range. The marked points are the result of a simulation experiment and confirm the calculation. As the error curves below show, both the capacitive model and the M/M/1/0 loss system yield significant deviations.

2.4.4 Systems with Queue

The M/M/1/0 loss system is surely the simplest of all server systems and, in the true sense of the word, not really a queue system. Of more interest for manufacturing processes are systems with queue in which no demands or manufacturing jobs can get "lost". Systems whose queues have infinite capacity are easier to calculate than those with limited capacity. The M/M/1/∞ queue system is considered as a typical representative here.

Figure 2.36 shows a simple server system with an infinite queue and a service station. The intensity of the input flow is λ and the servicing intensity is μ. Both the intermediate arrival times and the serving times are mutually independent and have an exponential distribution.

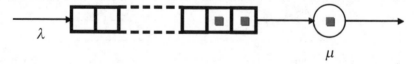

Figure 2.36. Server system with infinite queue

The system can, depending on the maximum permitted queue length n, assume exactly $n+1$ discrete states, which are defined as follows:

State	Probability	
Z_{n-1}	$P(Z_{n-1}, t) = p_{n-1}(t)$	System contains $n-1$ demands
Z_n	$P(Z_n, t) = p_n(t)$	System contains n demands
Z_{n+1}	$P(Z_{n+1}, t) = p_{n+1}(t)$	System contains $n+1$ demands

As for the simple loss system, the demand in the service station is also counted. Figure 2.37 shows a section of the state graph that can be arbitrarily extrapolated at the left and the right.

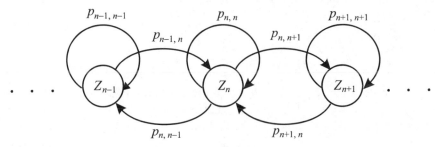

Figure 2.37. State graph of the M/M/1/∞ queue system (section)

Assuming that simultaneous events do not occur, a time interval Δt can always be found in which just one and only one state change takes place. Using the limit transition $\Delta t \to 0$ described later, the probability for the Z_n state can be calculated from the probabilities of the two neighbouring states Z_{n-1} and Z_{n+1} and the associated transition probabilities.

$$p_n(t + \Delta t) = p_{n-1}(t) \cdot p_{n-1,n}(\Delta t) + p_n(t) \cdot p_{n,n}(\Delta t) + p_{n+1}(t) \cdot p_{n+1,n}(\Delta t) \quad (2.89)$$

The transition probabilities in the time interval Δt yields:

$$\begin{aligned} p_{n-1,n}(\Delta t) &= P(T_A < \Delta t) & (2.90) \\ p_{n,n}(\Delta t) &= P((T_B \geq \Delta t) \wedge (T_A \geq \Delta t)) \\ p_{n+1,n}(\Delta t) &= P(T_B < \Delta t) \end{aligned}$$

Assuming that all random quantities are mutually independent and have an exponential distribution, this yields:

$$\begin{aligned} p_{n-1,n}(\Delta t) &= 1 - e^{-\lambda \cdot \Delta t} &= \lambda \cdot \Delta t + o(\Delta t) & (2.91) \\ p_{n,n}(\Delta t) &= e^{-\mu \cdot \Delta t} &= (1 - \mu \cdot \Delta t) \cdot (1 - \lambda \cdot \Delta t) + o(\Delta t) \\ p_{n+1,n}(\Delta t) &= 1 - e^{-\mu \cdot \Delta t} &= \mu \cdot \Delta t + o(\Delta t) \end{aligned}$$

When we now set the transition probabilities in Equation (2.89), after using the limit transition $\Delta t \rightarrow 0$ we obtain the differential equations of the system. The special case $n = 0$ (empty system) can be derived from the simple loss system.

$$\dot{p}_n(t) = \lambda \cdot p_{n-1}(t) - (\mu + \lambda) \cdot p_n(t) + \mu \cdot p_{n+1}(t) \qquad \text{for } n > 0 \qquad (2.92)$$
$$\dot{p}_0(t) = -\lambda \cdot p_0(t) + \mu \cdot p_1(t)$$

The steady state has a particularly simple solution. The system is stationary in this case, namely, the state probabilities P_n no longer change: $\dot{p}_n(t) = 0$. The system of differential equations (2.92) changes to a system of homogeneous linear equations that can be solved iteratively, which is easy to prove.

$$0 = \lambda \cdot P_{n-1} - (\mu + \lambda) \cdot P_n + \mu \cdot P_{n+1} \qquad \text{for } n > 0$$
$$0 = -\lambda \cdot P_0 + \mu \cdot P_1$$

First solve for P_1 by setting $n = 0$ in the equation, and then set the result in the equation for $n = 1$, *etc*. The result is a general equation for the state probabilities P_n depending on the probability P_0.

$$P_n = \rho^n \cdot P_0 \qquad (2.93)$$

Once again, the traffic value ρ plays a decisive role here, although note that Equation (2.93) is true only under the condition that the traffic value is less than 1. Otherwise the queue would grow continually and thus no stationary state could occur. The state probability P_0 can be easily eliminated using the total probability and the partial sum for the geometric series.

$$1 = \sum_{n=0}^{\infty} P_n = P_0 \cdot \sum_{n=0}^{\infty} \rho^n = P_0 \cdot \frac{1}{1-\rho} \qquad (2.94)$$

The result is finally a usable equation for the state probability P_n that depends only on the traffic value ($\rho < 1$).

$$P_n = \rho^n (1 - \rho) \qquad (2.95)$$

The practitioner is normally less interested in the actual state probability than the statistical characteristic values of the systems. If, for example, the average queue length is known $\bar{n}$, one has an important starting point for the dimensioning of the buffer storage.

$$\bar{n} = \sum_{n=0}^{\infty} n \cdot P_n = \frac{\rho}{1-\rho} \qquad (2.96)$$

Based on the differential equations that describe the system, obviously a number of further characteristic values, such as the average production time of a job depending on the time, the loading level of the service stations or the average wait time for manufacturing jobs can be calculated. The interested reader should consult the relevant and copious literature for queuing theory. [2–4]

2.5 Petri Nets

The idea of the Petri net is based on work by C. A. Petri, who submitted in 1962 at the Institute for Instrumental Mathematics at the Bonn University a dissertation with the title "Communication with Automatons". The concept developed by Petri is ideally suited for the modelling and analysis of discrete systems, thus also for manufacturing systems. In no other model are such different categories, such as resources and events, linked structurally with each other. Although the model modules are elementary, the mapping possibilities are almost unlimited.

As the name suggests, the theory of the Petri nets is based on graphical theory. Depending on the use, they can be assigned both to analytical models and to the simulation models. Indeed, simulations for manufacturing processes based on Petri nets exist, although these are far less common than those based on queuing theory. The reason may lie in the abstract importance of the system modules that the less experienced user cannot always easily grasp. The relationship between model and reality is less apparent than, for example, for the queue models.

2.5.1 Definition

Petri nets are directed graphs, which, in contrast to ordinary graphs, have two different classes of nodes:

- Places
- Transitions

It is assumed that both the set of places $S = \{s_1, s_2, s_3, \ldots\}$ and the set of the transitions $T = \{t_1, t_2, t_3, \ldots\}$ are nonempty, namely, the Petri net contains always both – transitions and places. The set of directed edges is defined by a *flow relation F* as follows:

$$F \subseteq S \times T \cup T \times S \tag{2.97}$$

Namely, an edge in the graph always connects a place with a transition or *vice versa*, never, however, places or transitions with each other. Using the mappings K and W, a natural number greater than 0 is assigned to each place s_i and each transition t_j of the Petri net:

$$K: \quad S \to N\backslash\{0\} \qquad\qquad (2.98)$$
$$W: \quad F \to N\backslash\{0\}$$

The value $K(s_i)$ is called the capacity of the place s_i and $W(s_i, t_j)$ is the weight of the edge from s_i to t_j. The importance of the capacity for a place is that it specifies the maximum number of markings that this place can accept. When one forms a relation between the places and the resources of the abstract manufacturing system from Section 1.2, the markings represent the number of references to the corresponding resource element. Markings form the dynamic element of the Petri net. Driven by the transitions, they move from place to place along and in the direction of the edges. Petri nets, however, differ from flow graphs through the absence of a conservation set. Markings are immaterial objects that may suddenly disappear or appear.

Preset and postset
A Petri net has $x, y \in S \cup T$ nodes. The set of all nodes from which edges lead to the nodes x is called the *preset* for x:

$$\bullet x = \{y \mid (y, x) \in F\} \qquad\qquad (2.99)$$

Correspondingly, the set of all nodes to which the edges leaving the nodes x lead is called the *postset* for x:

$$x\bullet = \{y \mid (x, y) \in F\} \qquad\qquad (2.100)$$

If a transition t_j fires, markings will be removed from every place of the preset and markings added to every place of the postset. The number of markings moved here corresponds exactly to the weight of associated edges. The transition itself does not decide when a transition fires, but the initiating event must arrive externally. Thus, the transitions indicate only the points in the system at which the events can act. The condition under which a transition activates is determined by the marking assignment of the preset and postset of the transition. Any missing markings in the preset or missing capacity in the postset hinder the activation of the transition. Places consequently perform the function of conditions under which the determined events can act and represent the state of the system.

State and marking
The initial state M_0 of a Petri net is determined by its marking – a mapping that each place s_i shows a number $M_0(s_i)$ of markings,

$$M_0: \quad S \to N \qquad \text{with} \quad 0 \le M_0(s_i) \le K(s_i) \qquad\qquad (2.101)$$

A Petri net can be represented as a graph where the places and the transitions are indicated as a circle and as a rectangle (or often simply as a bar), respectively. In addition to the edge evaluation and the capacity, the marking of the places is entered. Appropriate simulation systems also allow the animation of the marking flow with the significant increase in the clarity of the method. Figure 2.38 shows a simple example of a Petri net that consists of two places and transitions. Whereas the place s_1 has a marking, the place s_2 is empty.

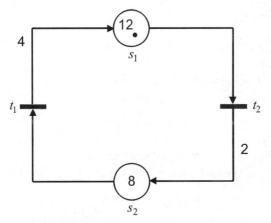

Figure 2.38. Example of a simple Petri net

Vector notation

Let us assume that the Petri net has $|S|$ places, then a *capacity vector* $\mathbf{k}$ and a *marking vector* $\mathbf{m}_0$ can be defined as follows:

$$\mathbf{k} = \begin{pmatrix} K(s_1) \\ K(s_2) \\ \vdots \\ K(s_{|S|}) \end{pmatrix} \quad \text{and} \quad \mathbf{m}_0 = \begin{pmatrix} M_0(s_1) \\ M_0(s_2) \\ \vdots \\ M_0(s_{|S|}) \end{pmatrix} \tag{2.102}$$

Similarly, for each transition $t_j \in T$ a vector $\mathbf{t}_j$ with $|S|$ components can also be defined in which the component $t_{j,i}$ specifies how many markings the transition t_j saves at the place s_i (positive value) or takes from the place s_i (negative value).

$$t_{j,i} = \begin{cases} -W(s_i,t_j) & \text{if } (s_i,t_j) \in F \\ +W(t_j,s_i) & \text{if } (t_j,s_i) \in F \\ 0 & \text{else} \end{cases} \tag{2.103}$$

$$\mathbf{t}_j = \begin{pmatrix} t_{j,1} \\ t_{j,2} \\ \vdots \\ t_{j,|S|} \end{pmatrix} \tag{2.104}$$

A *net matrix* (incidence matrix) N that describes the graph structure then can be created from the transition vectors.

$$\mathbf{N} = (\mathbf{t}_1, \mathbf{t}_2, \ldots, \mathbf{t}_{|T|}) \tag{2.105}$$

The net matrix, the capacity vector and the marking vector provide a complete description of a Petri net. For the example from Figure 2.38, then:

$$\mathbf{N} = \begin{pmatrix} 4 & -1 \\ -1 & 2 \end{pmatrix}, \ \mathbf{k} = \begin{pmatrix} 12 \\ 8 \end{pmatrix}, \ \mathbf{m}_0 = \begin{pmatrix} 1 \\ 0 \end{pmatrix}$$

Firing rules

A firing transition t_j takes from all places $s_i \in \bullet t_j$ markings each with $W(s_i,t_j)$ and takes in all places $s_k \in t_j \bullet$ markings each with $W(t_j,s_k)$. The so-called strong firing rule says that a transition is only activated when after firing the transition a negative marking assignment is not expected at any place and the capacity of the places is not exceeded anywhere. The firing of a transition in the net changes the marking *m* into the marking *m'*. This can be formulated elegantly using the vector notation:

$$\mathbf{m}' = \mathbf{m} + \mathbf{t}_j \tag{2.106}$$

A transition is activated only when the following equation applies to all places:

$$0 \le M'(s_i) \le K(s_i) \quad \text{with } i = 1 \ldots |S| \tag{2.107}$$

If several transitions are fired successively, this is called a *firing sequence*. The firing sequence $\sigma = t_1, t_2, \ldots, t_n$ is called usable for the marking M if all transitions of the sequence are activated.

Nets in which all edges and all places have the weight 1 and the capacity, respectively, are called *condition/event nets* or simply C/E nets.

Reachable set

A marking M of a Petri net is called reachable when a usable firing sequence σ exists that changes the initial marking M_0 into M. The set of all reachable markings is also called the reachable set $R_N(M_0)$.

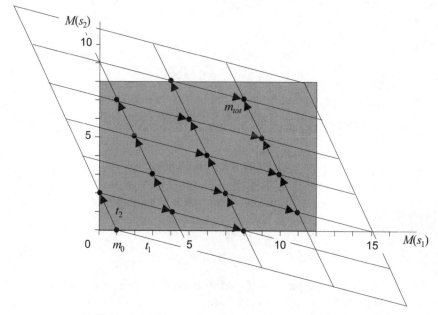

Figure 2.39. Vector representation of the reachable set

The vector representation (see Figure 2.39) can be used to derive the directed reachability graph, when the complete marking of each state, as shown in Figure 2.40, rather than the vector end points, are entered. The reachability graphs can be used to prove the important properties of the Petri net, such as reversibility or liveness. Major parts of the net analysis are based on graph-theory methods. However, in some cases, linear algebra methods can be used. Thus, Petri nets build a bridge between the purely simulative methods and the analytical methods.

Petri nets in the previously described form do not have any time properties, but restrict themselves only to events and the resulting system states. However, such nets can provide important conclusions about the behaviour of manufacturing systems, in particular, with regard to their control and the use of shared resources. The net analysis here concentrates on the reproducibility of system states and on the detection or avoidance of system deadlocks.

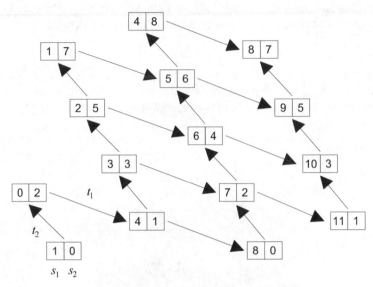

Figure 2.40. Reachability graph

Reversibility

A Petri net is reversible when:

$$\forall M_1, M_2 \in R_N(M_0): \quad M_1 \in R_N(M_2) \tag{2.108}$$

Any system state can be reproduced in reversible Petri nets. In contrast, a marking M_{tot}, from which no transition is activated, indicates a *total deadlock*. If a subset of the transitions still remains activated, this is called a *partial deadlock*.

The simple Petri net from Figure 2.38 shows a total deadlock for the marking (8,7). Generally, deadlocks in manufacturing systems have severe consequences. Their correction requires intervention in the execution control with the consequent significant potential for malfunction. The early detection and avoidance of such system states is thus very important.

Liveness

A transition $t_j \in T$ is called living when at any marking $M_1 \in R_N(M_0)$ always just one firing sequence σ exists that leads to a marking $M_2 \in R_N(M_0)$ for which t_j is activated. A living network thus describes systems in which none of the previously defined events can be excluded long term. With regard to manufacturing systems, reversibility and liveness of the net are fundamental requirements that the system design must meet.

2.5.2 Example

The modelling of a manufacturing system with a Petri net requires a very high degree of abstraction. This may well be one of the reasons why Petri nets are found comparatively seldom in industrial practice. If, however, the abstraction hurdle is tackled, this provides a very powerful instrument for the analysis, simulation and, to a limited extent, also for the calculation of discrete, event-driven systems. Furthermore, the Petri nets provide an insight into the structure of the interprocess communication, with a clarity that no other model concept permits. The following example from electronics production shows how a simple C/E net can be used to model a manufacturing process and visualise any conflict potential.

A job-lot manufacturer for electronic printed-circuit boards produces every day a large number of lots for different module types. Two SMT lines, each with several placement machines, are available for placing components on the circuit boards. The module types A and B are often produced in parallel, with type A on line 1 and type B on line 2. Both A and B require two setup groups, SG 1 and SG 2, which, however, are each present just once. A setup group is, for example, a feeder carriage on which a specific range of electronic components can be stored in belts and, when required, coupled to an automatic placement machine. A manufacturing batch is started autonomously by the two independent line controllers. The production can be started only when both setup groups, SG 1 and SG 2, are available. We also assume for the control algorithm that first the setup group SG 1 and then SG 2 is required. If either of the setup groups is busy, the lot will be suspended until the missing setup group becomes free. The following tables summarise the principle conditions and events of the manufacturing process.

Table 2.2. Conditions for the manufacturing process

Condition (place)	Description	Condition (place)	Description
s_1	Lot A reported at line 1	s_5	SG 1 made available for line 2
s_2	SG 1 made available for line 1	s_6	SG 2 made available for line 2
s_3	SG 2 made available for line 1	s_7	SG 1 free
s_4	Lot B reported at line 2	s_8	SG 2 free

Table 2.3. Events of the manufacturing process

Event (transition)	Description	Event (transition)	Description
t_1	Request SG 1 for line 1	t_4	Request SG 1 for line 2
t_2	Request SG 2 for line 1	t_5	Request SG 2 for line 2
t_3	Type A completed on line 1	t_6	Type B completed on line 2

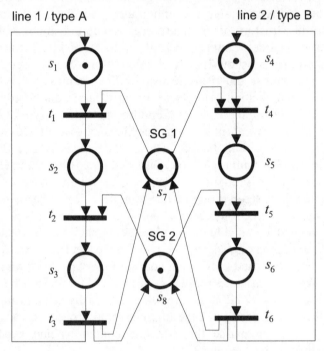

Figure 2.41. Petri net of the manufacturing system

Figure 2.41 shows the associated Petri net. The figure clearly shows the two independent processes on line 1 and line 2, and the setup groups SG 1 and SG 2 as shared resources.

The manufacturing process is simulated by the sequential activation of the individual transitions. This enables the marking flow in the Petri net. The reachability graphs from Figure 2.42 show that the two concurrent processes mutually exclude each other because of the shared use of the setup groups. Although the process that starts later must wait, the network is living.

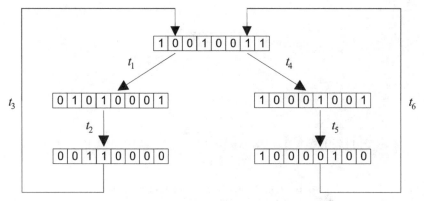

Figure 2.42. Reachability graph

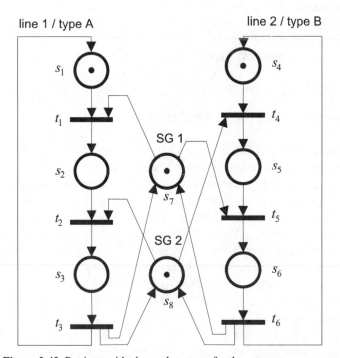

Figure 2.43. Petri net with changed strategy for the setup group request

Figure 2.43 shows a slightly changed net for the same manufacturing process. The only difference here is that line 2 first requests the setup group SG 2 and only then the setup group SG 1. This has the consequence that both processes can enter a start state and so no longer guarantee that they exclude each other.

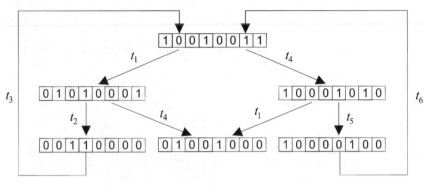

Figure 2.44. Reachability graph with deadlock

The associated reachability graph from Figure 2.44 shows that the system has a deadlock. There are two possible firing sequences that lead to a deadlock: t_1, t_4 and t_4, t_1. In both cases, each of the two processes has one setup group and waits for the other. Because neither of the two starting processes can be returned to its original state, the system cannot leave this state.

2.5.3 Elementary Links in Petri Nets

The example from Section 2.5.2 clearly shows the nature of the Petri nets. The manufacturing system is considered to be an environment in which several manufacturing processes can be embedded. In the example, these are the processes A and B that have a relationship with the circuit-board types A and B. In contrast, the SMT lines 1 and 2 can be considered as being only passive resources that are periodically required by the active processes. Both processes are controlled autonomously. Depending on the customer demand or the manufacturing manufacturer's schedule, they could be active successively or simultaneously. This form of independency is also known as concurrent processes. If we assume there was no conflict with the setup groups, both processes would be completely asynchronous and the Petri net would divide into two separate subnets. Only the use of shared resources provides a contact area between the two processes that partially cancels the asynchronous operation. The example net shown in Figure 2.41 is a classic alternative brought about by the mutual exclusion from critical process sections. Both processes compete for the same resources. The process granted the resource first suspends its competitor by the withholding of markings. The Petri net means *a priori* no organisational sequence is forced for the shared resources, rather, it is possible that either A or B may be activated successively several times or suspended for a longer period of time.

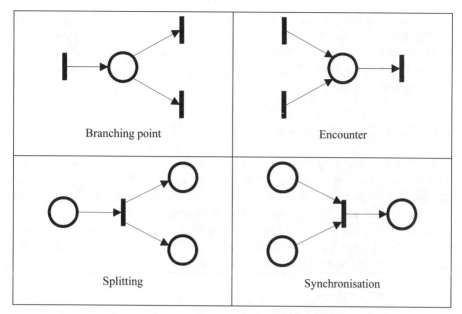

Figure 2.45. Elementary links in Petri nets

A characteristic of the example is that just one process can be active in the critical section. The manufacturing engineering obviously knows other forms of the process communication as the overview in Figure 2.45 shows. Branching point and encounter describe alternative processes. If we assume a simple C/E net, a branch produces a single process from n possible processes. An encounter links n processes to produce a single process so that the previously shared events are now performed sequentially. Branches are required, for example, to describe the distribution of products on the individual machines of a machine group. A company's central shipping department can be represented by an encounter.

As difference to the branching point, the splitting and the synchronisation are true concurrency. The splitting of a process is often found as lot splitting. In contrast to the branching point where the process instance is retained and only used for planning the use of the resources, the splitting creates new processes. Consequently, the parts of a split lot in a real manufacturing process are often managed with their own identification. The synchro-nisation is one of the most important elementary links of the manufacturing process. Both assembly tasks and all types of bill-of-materials relationships belong to this category. The nature of the synchronisation causes two original concurrent processes to wait for each other after which they then continue together or each separately.

The elementary links, especially, show that Petri nets are particularly suitable for representing the manufacturing process character. This becomes even more apparent when the state graphs of the manufacturing system are compared. Often, the individual process loops here fuse into a unit that is difficult to unravel and that hides the true structure. Conversely, the analysis of Petri nets is not possible without state graphs, both representation forms complement each other.

2.5.4 Extended Petri Nets

The individual markings in common Petri nets cannot be distinguished from each other. They are often all shown black in the graphical representation. Markings indicate only the state of the system, furthermore, they are not suitable for information transport. To make markings into true information carriers, they must be given specific properties, which, for example, could be achieved by colouring them. Consequently, nets with specific markings are also called coloured Petri nets or simply CP nets.

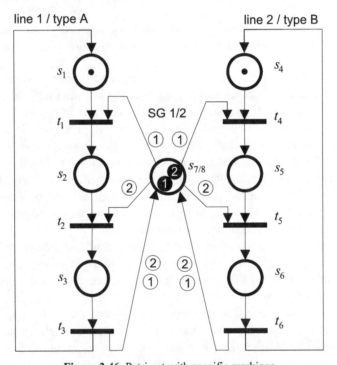

Figure 2.46. Petri net with specific markings

Figure 2.46 shows how a CP net through combining the two places s_7 and s_8 results from the example from Section 2.5.2. Both setup groups are now represented by the place $s_{7/8}$. Here the place has two specific markings "1" and "2" that indicate the assignment of the SG 1 or SG 2 setup group. It remains to specify which markings flow through the edges. In the example, the transition t_1 requests a marking of type "1" and the transition t_3 puts back a marking of type "1" and "2". In the same way as the two setup groups were combined, the processing steps of the two lines could be combined. In this case, the places s_1 and s_4 would become the shared place $s_{1/4}$, etc. The different products could be represented using markings of the type "A" and "B". Compared with normal Petri nets, the CP net requires significantly fewer places. The proximity to the real manufacturing system is also better because some of the specific markings correspond to the various manufacturing jobs [1]. Petri nets provide only information about which transitions are activated for a specified marking. In contrast, the firing of the transition itself depends on external factors.

The firing action and the associated marking transport are considered to be an event, namely it does not consume any time. However, for the operational planning it is desirable to also represent the time behaviour of the manufacturing system. References [4] and [5] introduce Petri nets with time-evaluated edges, transitions and places. However, the following general rule governs both the time-evaluated nets and the CP nets: Most extensions take some of the elegance from the theory of the Petri nets, which, in particular, impairs the analytical aspect of the theory.

References

[1]   ABEL, D.: *Petri-Netze für Ingenieure – Modellbildung und Analyse diskret gesteuerter Systeme.* Springer-Verlag, Berlin 1990

[2]   AMOSSOWA, N. N.; GILLERT, H.; KÜCHLER, U.; MAXIMOW, J. D.: *Bedienungstheorie – Eine Einführung.* BSB B.G. Teubner Verlagsgesellschaft, Leipzig 1986

[3]   CHURCHMAN, C.W.; ACKOFF, R.L.; ARNOFF, E.L.: *Operations Research – Eine Einführung in die Unternehmensforschung.* Verlag die Wirtschaft Berlin, Berlin 1968

[4]   KIENCKE, U.: *Ereignisdiskrete Systeme - Modellierung und Steuerung verteilter Systeme.* R. Oldenbourg Verlag, Munich, Vienna, 1997

[5]   KÖNIG, R.; QUÄCK, L.: *Petri-Netze in der Steuerungstechnik.* Verlag Technik, Berlin 1988

[6]   MÜLLER, P. H.: *Wahrscheinlichkeitsrechnung und Mathematische Statistik - Lexikon der Stochastik.* Akademie Verlag, Berlin 1991

[7]   WUNSCH; G.; SCHREIBER, H.: *Analoge Systeme (Grundlagen).* Springer-Verlag Berlin Heidelberg 1993

3 Simulation of Manufacturing Processes

3.1 Principles of the Simulation Methods

3.1.1 Definition of Terms

Many relationships use the term *simulation* in the analysis and synthesis of systems. This is certainly warranted when one considers the definition of the term. This has established itself for quite some time; [14] provides the most appropriate definition:

> "Simulation is the representation of a system with its dynamic processes in a model with experimental capability to obtain knowledge that can be transferred to the real system."

Other definitions can be found that are identical with the fundamental statement shown here. The main issue affects both analytical and experimental procedures. Without doubt, many well-known methods belong to the simulation category. Some examples include the method of finite elements (FEM) used for the simulation of heat or mechanical product properties, numerical methods for the simulation of electronic circuits, the interactive representation of 3D worlds as virtual reality (VR), such as for determining collisions for robot movements.

The use of a computer is not necessarily required, after all, investigations on a design model in a wind tunnel also belong to simulation, although the model here does not have any computer support. Other simulations can also be performed "manually" using special diagrams or graphs. The fundamental requirement, however, is always the ability to record the significant properties in the model that determine the behaviour of the investigated system. The action of the experiment with the model itself also belongs to the term of the simulation [14]:

> "Generally, simulation is considered to be the preparation, execution and evaluation of specific experiments with a simulation model."

Analytical concepts are severely limited in the obtainable system complexity. This fact, and not least the power of computing technology, has led to the development of experimental computer-supported procedures that permit the testing of defined system constellations. Typical for such procedures is the decomposition of the investigated system into understandable subprocesses that then can be regrouped to form a complete process for experimenting. Consequently, simulation is a tool for performing experiments. A simulator does not have any implicit intelligence and cannot itself make any improvement to the model [11]. Special optimisation methods, however, can be used in the interaction with simulators and so achieve partial improvements. Only the *targeted* search and variation for the experiments can, at best, lead to the desired results. It must be emphasised:

> Simulation is not optimisation.

3.1.2 Discrete Event Simulation

Manufacturing processes, more accurately, the manufacturing schedules, are normally discrete event systems (DES). The states of the manufacturing system change only at specific times; the duration of each state change can be assumed as being infinitely small. Only events cause state changes; such events occur at unpredictable times [4]. Such systems are not only *time-* and *action-discrete* but also *event-driven* in accordance with the classifications from Chapter 1. The representation can only be achieved by models of these classes. Where it has only limited importance whether the model states are value-discrete or value-continuous. The *discrete event simulation* is a suitable method for modelling manufacturing schedules that forms the kernel of the considerations in the following section. When this chapter mentions *simulation*, this always refers to the method of the discrete event simulation. Other modelling concepts are also listed in the Chapter 2. Process chains, data models, simulator modules, Petri nets, automatons, decision tables, simulation languages and programming languages represent such modelling methods. The degree of abstraction and suitability of the individual methods for the actual situation differ. In contrast to the level of detail, *abstraction* is considered to be the reduction of the complexity of a problem by removing those details not significant for the solution.

The model time t_m of the discrete event simulation can be organised using two different principles:

- event-driven or
- clock-driven

The states $Z = \{\mathbf{z}_1, \mathbf{z}_2, ..., \mathbf{z}_n\}$ of the real system in the simulation are caused by the events $E = \{E_1, ..., E_n\}$.

For the event-driven model time, the simulation events are assigned to the real times t_r. The time interval between the model events represents the time interval between the real events. The resolution of the time axis in the model can be made as detailed as required because the events are entered in the model events with exactly the times when they occur.

The use of a clock assumes fixed time intervals between which events can occur. In Figure 3.1b, the events E_1 and E_2 occur at time $t_{m,1}$. To achieve an adequate resolution, the time interval must be chosen as small as possible. This in turn causes a high computer loading because the event management must also poll at times when no entries are present.

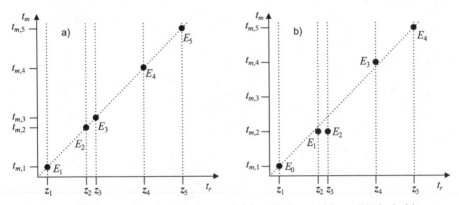

Figure 3.1. Organisation principles for the model time, (**a**) event-driven, (**b**) clock-driven

Models with event-driven model time can be given an additional clock cycle to achieve an artificial lengthening of the time interval. This is particularly necessary when the simulation run needs to be observed time-proportional, which is often the case for animation models. The process track can be illustrated in the individual forms in the static process model from Chapter 1. Figure 3.2 shows the process track for all three cases.

A simulation model always contains the complete system state $\mathbf{z}(t_i)$ at discrete time t_i. The subsequent state $\mathbf{z}(t_i)$ can be determined from state $\mathbf{z}(t_{i+1})$. State $\mathbf{z}(t_i)$ is transient and does not need to be saved. The saving of the states at all times, namely the recording of the chronological change of the states is a result of the simulation, the *result space*:

$$Z = \{z(t_0), z(t_1), \ldots, z(t_n)\} \tag{3.1}$$

The discrete event simulation always runs in the *state space* and so differs significantly from the scheduling method. The throughput scheduling is differentiated into forwards and reverse scheduling for which the time duration of the individual work steps are concatenated forwards beginning from a start date or backwards from an end date, even with execution-dependent lead times [18]. Such procedures based on the network planning are very common in production planning and, depending on the available computer power, capable of representing even complex relationships [2]. They always operate in the *result space Z* that is partially built. The larger the number of time-dependent system states to be used for the scheduling (capacities, setup states, job grouping, *etc.*), the larger the number of time-dependent state values that must be saved, incremented and evaluated in the result space. Thus, the mastering of the result space represents a significant limitation of this procedure.

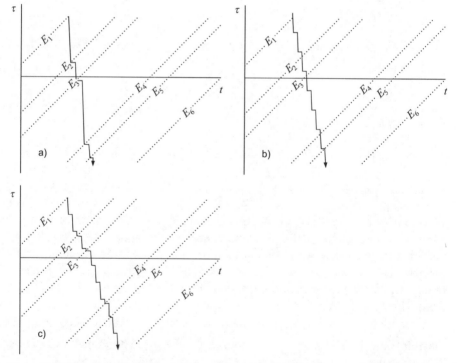

Figure 3.2. Process tracks in the simulation model, (**a**) event-driven, (**b**) clock-driven, (**c**) event-driven with clock

3.1.3 Uses for the Simulation

The original focus of simulation technology was aimed at planning of the processes. Today, the simulation is a common tool for planning, realisation and operation of systems [14]. Increasingly, the method is being used for control during operations. What just a few years ago required powerful workstations, can now be simulated on standard PCs. The trend in hardware and software development permits the more intensive use of simulators, which, conversely, has become ever more necessary because of the turbulent market conditions and increased demands in many areas. Simulative investigations for capital-intensive measures are the state-of-the-art. However, with the estimated 10% market penetration in Germany compared to CAD and PPC, each with more than 60%, is low [3].

As Figure 3.3 shows the *planning – realisation – operation* operational areas are also differentiated into *overall planning – detailed planning – design and realisation – operation* in accordance with [9]. For the individual planning levels, the use of simulation demands a different abstraction level. A higher level of abstraction always places high demands on the interpretation of the simulation results and may raise new problem situations that can possibly be answered with a detailed model. In addition to the planning levels, a classification is also made into the operational areas of the simulation. Depending on the primary use, a differentiation is made here into the *production planning and control, factory planning* or *technology planning*. Figure 3.3 shows examples of the individual areas and levels. It is advantageous here to have integrative simulators and simulation models that cover all areas and levels in order to quickly obtain objective criteria for all types of planning decisions.

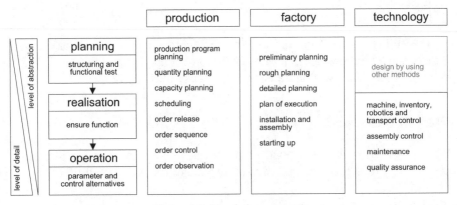

Figure 3.3. Simulation use areas

Increased demands in the real world lead unavoidably to more complex production systems and so require modelling methods that can record the complex relationships. In contrast to mathematical-analytical concepts that quickly reach their limits for representation of such systems, simulation can be used to exactly investigate the complex execution behaviour. However, analytical methods should always be preferred when they describe the problem sufficiently accurately. In particular, the investigation possibilities provided by simulation for the following situations should be emphasised

- real systems that do not (yet) exist,
- for planning variants without intervention in real systems,
- for alternatives with small effort,
- long intervals using time compression,
- for transition phases.

Obviously the high demands for the representation of complex production systems also place demands on the simulation tools themselves. For acceptance reasons, the object-oriented module concept, which supports a universal use, has established itself. However, a simulation system must be very carefully tested for its suitability for solving the associated problems, because real discrete event systems of various forms exist that were specially developed for their representation on various simulators.

3.2 Simulation Project

A simulation project can generally be divided into three phases that either can be performed in total or the individual phases repeated:

1. Preparation
2. Execution
3. Evaluation

The following sections discuss the three project phases in more detail. These considerations concentrate on the procedure for a simulation study. Section 3.6.2 discusses the transfer of the study into operational use.

3.2.1 Preparation

The preparation is the most comprehensive part of the project and, depending on its form, can involve up to 80% of the project effort [11].

Fundamental Decisions

It must always be decided at the beginning whether the pending problem can be solved using simulation. Reference [1] mentions 10 reasons why not to simulate that can help in making the decision. To summarise, simulation should not be used when

1. the problem can be solved using analytic means,
2. it is easier to perform an experiment on the real system,
3. the costs exceed the benefits,
4. the resources and time available for the execution do not suffice,
5. no data, not even estimates, exist,
6. the verification and validation of the model is not possible (see Section 3.2.1),
7. the expectations placed on the project cannot be met,
8. the system behaviour is too complex.

Goal

The specification of the project goal is very important in order to systematically develop the model and to perform the experiment. The global goal is divided into a suitable number of subgoals in order to better judge the attainability of the goal and a useful interpretation of the simulation results. Furthermore, in particular for time-critical projects, a time frame must be specified by defining deadlines by when the intermediate or complete results must be available for interpretation and decision.

Analysis and Development of the Data Base

An exact modelling requires the analysis of the real existing or of the planned system. This task can take a large part of the total project time. The data is then collected and used to build the data base for the simulation. The demands placed on the data for the problem definition are correspondingly higher than for other planning methods. The relevant production data includes job data, work-schedule data, bill-of-materials data, working-time data, resource data, topological data and defect data, and information about the workflow management.

To keep the data-input effort as small as possible, the data should be provided on data media in EDP-structured form. To avoid inconsistencies, and incompleteness and lack of accuracy in the data, the data should be tested and any necessary corrections made. The use of a relational data base can be very beneficial here. The data base is used to make an analytical estimate, often as a bottleneck analysis. This estimate permits an initial comparison with the objective and is used to validate the subsequent model.

Modelling

Before the modelling can be started, a suitable simulator must be selected for the use case. Criteria such as suitability, economics, ergonomics, interfaces, animation and result representation play important roles here [11].

The modelling is also performed in several steps. In the first phase, the knowledge gained from the analysis leads to the development of a conceptual model that is then used to build a logical or symbolic model, which by no means contains an operational simulation model. Rather, it serves to advance the abstraction of the real or planned production, namely the acquisition of the significant properties while ignoring unimportant details and the simplification of relevant details. The hierarchical modelling, in which the system is split into appropriate individual parts, is often useful here. A differentiation is made between the

- top-down concept and the
- bottom-up concept.

The top-down concept means that the overall structure of the model is determined first without considering the individual components in more detail and only later is the representation of the model segments completed. Conversely, for the bottom-up concept, the representation of the individual segments is pursued first and only later combined to produce an overall model [11].

The conversion to a software model takes place in a second modelling step. The actual model can be based on several concepts. A discrete event simulation model normally is an object-oriented representation, irrespective of whether a programming language or a simulator is used. The advantage of using the model world of a simulator is obvious because of the availability of the required modules in a library. The modules correspond to real and complex elements, many of which have been tested in applications. The verification of a programmed model can be formally omitted. The disadvantage lies in the specialisation of such modules for specific application areas, which, however, may be countered by the use of an embedded language.

Validation

The validation of the model is an important and difficult part of the project. Because only when the reality and the model agree suitably well can conclusions be drawn from the investigations that can be used in actual applications. The verification is the formal proof of the correctness of programs and program sections. Such a verification is normally supported with software [13] and that must be performed before the actual validation.

> The validation is the testing of the adequate agreement of the model and the system. With regard to the goals, care must be taken to ensure that the model reflects adequately and correctly the behaviour of the real system.

A complete coverage of model and reality can be achieved only to a limited extent. The guideline "so exact as necessary but so general as possible" also applies in this case. The validation is an iterative process that can return to any point of the preparation in order to improve the model quality leaving that point. The analytical general estimate from the preparation should agree with the first results. In the most unfavourable case, the project must be terminated at this place if the expected results cannot be achieved.

The validation must be tailored to the application, but systematically. The *deterministic control* by deactivating the random numbers during the experiment ensures deterministic event sequences, which, in particular with the help of animation (2D or 3D), can be followed on the computer monitor. The *average-value considerations* allows the agreement of the theoretical predetermined and simulated results to be checked. For *parameter tests*, only specific segments of the model are activated and so the model functionality is tested in detail. The *load-limit analyses* make it easy to provide reproducible general conditions for a simulation experiment. If a real system is available, it may be desirable to perform a *comparison* of the model with the real production. The model then can be modified in order to test new parameters or strategies [11].

3.2.2 Performing the Experiment

The simulation experiments are performed strictly goal oriented. Ensure for the experiment planning that as few simulation runs as possible are used to achieve the goal. A differentiation must be made whether a simulative system analysis or planning is involved. Whereas for the simulative system analysis that investigates the exact system behaviour, only various experiments with different random number flows need to be created in order to achieve statistically valid conclusions, for the planning of production systems, parameter, loading or rule variations are made in order to evaluate the effectiveness of various scenarios. Mathematical-heuristical optimisation methods that make variations based on a defined objective function are often used for the latter type. Important here is the filtering of relevant influencing factors in order to limit the search space in advance. The definition of the objective function is based on the *fitness* as a weighted combination of individual goals. Any contradiction of goals must be considered [17].

3.2.3 Evaluation

Statistics, monitoring and animation are used to evaluate the simulation. The results are obtained from the feedback from the simulation, which, when prepared appropriately, allow the interpretation of the results. Animation and monitoring are dynamic display methods, whereas mathematical statistics allow characteristic variables to be determined that describe the system behaviour as exactly as possible. Production systems are principally described by the *meeting of due dates*, *lead time*, *throughput* and *inventory* characteristic variables. Section 3.3 discusses the analysis principles and methods in more detail.

Because the quality of the evaluation results is a significant factor in determining the conclusions for the planned reality, the evaluation is very important. In particular, when the planners and the simulation experts are not the same persons, care should be taken to ensure exact documentation of the results, because the interpretation of the results is then left to the planners. The documentation can often use the export functions of the simulators that simplify the representation in popular evaluation programs. The interpretation of the results and the conclusions for the planned measures represent the last step of the evaluation.

3.3 Analysis of the Simulation Results

3.3.1 Event Trace

The events form the basis of the discrete event manufacturing, both in the real and in the modelled system. A date is assigned to each event; the events then form an order with regard to these dates. At the event time, certain states of the system change, which can also be designated as transitions. Thus events become the information carriers of the transitions. As for the acquisition of the real processes by the process data acquisition (PDA), the events with their transition information can also be saved for the simulation. Furthermore, certain system states assigned to the event can be saved.

> The recording of the events with their transition information during the simulation is designated as event trace. Recorded values of the state variables of the simulation model at the event times are called the trace data.

The event list term is used similarly. The event trace and the trace data can be used to generate various analyses, most of which are integrated in the simulators, but, however, can also be produced offline with other programs. Table 3.1 contains example data from an event trace; the trace data are indicated with an * (load – machine load, queue content). The trace can divide a more complex simulation event into several individual events, which then refer only to specific individual objects of the simulation model.

Table 3.1. Selection from an event trace

Date	Type	Station	Stype	Job	Load*	...
926185200	leave	queue3	queue	job1	8	...
926185200	enter	machine2	machine	job1	1	...
926185440	ready	machine2	machine	job1	1	...
926185440	leave	machine2	machine	job1	0	...
926185440	enter	queue2	queue	job1	1	...
926185440	leave	queue3	queue	job2	7	...
926185440	enter	machine2	machine	job2	1	...
926185440	leave	queue2	queue	job1	0	...
926185440	enter	machine1	machine	job1	1	...
926186160	ready	machine1	machine	job1	1	...
926186160	leave	machine1	machine	job1	0	...
926186160	enter	queue1	queue	job1	1	...
...	...	...	...	...	...	...

The event trace can be used directly in the animation in which the events are immediately transferred as state transitions to the model modules. The animation levels differentiate between state animation, 2D animation and 3D animation. The increased clarity, however, requires major modelling effort. The animation can be made either during the simulation run or reproduced later from the event trace. To obtain a time-proportional representation and to prevent discontinuities, the intervals between animated events must be extended "artificially". Although animations give high clarity, they provide only limited information about significant system characteristic values. They are particularly suitable for validation and presentation purposes.

The event list is difficult to analyse by itself. Depending on the scope of the model, such an event list can contain thousands or even millions of entries. Consequently, a filtering and sorting of the data can be done in an initial form. Useful here are considerations from the viewpoint of the products (jobs, lots, *etc.*) and the resources (machines, personnel, production facilities, queues, buffers, storage, *etc.*). The individual work steps are each characterised with a start and a stop event. The grouping of the activity events allows a transformation of date-pairs into time intervals (see Section 1.2.1) and the formation of the total duration for specific time intervals (total processing duration of a job, total wait time of a job, total processing duration of a machine, *etc.*) or their representation. The further compression of the time interval data using mathematical-statistical methods (average value, maximum, *etc.*) allows characteristic variables to be determined for specific equipment or classes of products from the event list. When stochastic models are used, it is important that the sampling space is sufficiently large in order to produce reliable conclusions. A simulation experiment can be repeated several times with different random values or simulated for a longer interval.

The logging of the values for the model states at the event times allows the chronological change of these state variables to be derived and displayed directly from the trace data. Particularly appropriate examples are the chronological changes of a queue content and a machine loading. The integration of state variable z_i over all time intervals Δt_j allows the calculation of average characteristic variables, which, together with the characteristic variables derived from the time intervals of the actions, provide important statements about the system behaviour:

$$\overline{z}_i = \frac{\sum_j z_{ij} \cdot \Delta t_j}{\sum_j \Delta t_j} \tag{3.2}$$

The evaluation of event traces is very flexible providing the appropriate trace data are available. The memory and the time required for calculating the results are, however, very high.

3.3.2 Online Evaluations

Even during the simulation, certain characteristic variables can be calculated online. This saves time compared with the evaluation of event traces. For example, the time durations for each individual object of the simulation model can be added during the run or state variables accumulated over specific intervals. This makes the calculated characteristic variables immediately available and so avoids the extensive filtering and sorting of the trace data. Typically, the calculation is performed in time series. The complete simulation run is divided into time intervals. A separate calculation is performed within each time interval. Statistical evaluations can be performed very quickly using the accumulated calculations to determine average values, totals or maximum values. This makes the lead time for each job, the loading of each machine, or the content of a queue immediately available.

An important use of the online evaluations is for monitoring. Selected state variables or accumulated state variables are displayed dynamically in diagrams during the simulation, normally with reference to the symbol of the object. Online evaluations are also fast and efficient for the representation of the results after the simulation. In the best case, it is not necessary to save event lists, which, in particular for very high number of events, should be a goal. In such cases, statistical characteristic variables are the better alternative for the detailed consideration of the process.

3.3.3 Characteristic Variables

Important characteristic variables for manufacturing are throughput, lead time, due-date observance and inventory (WIP). The analysis differentiates between products and resources, because each requires different information. This information can be considered for individual objects, object groups or all objects.

Product-oriented Viewpoint

Statements about lead time and due-date observance are important for products. A product here is considered to be a single, nondivisable element (job, lot, *etc.*) that must pass through a sequence of work steps, each of which has a time requirement.

Lead time
The flow begins with the release in the manufacturing at time t_S and ends with the completion at time t_E. The cycle time is thus determined as the date difference $d(t_S, t_E)$. Before each processing activity $V_{b,i}$ with the processing duration $d_{b,i}$, a wait time $d_{w,i}$ can occur during the simulation (see Figure 3.4).

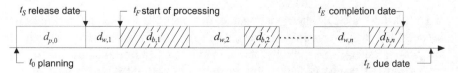

Figure 3.4. Times and durations of the product flow

The total wait time d_w and the total processing time d_b for a product can be determined as the sum of the individual wait times $d_{w,i}$ or the individual processing times $d_{b,i}$, respectively:

$$d_b = \sum_i d_{b,i}; \quad d_w = \sum_i d_{w,i} \tag{3.3}$$

The durations can be divided more finely. Explicitly, only active (processing) and passive (waiting) intervals are displayed here. Passive parts can be further divided into process-dependent wait time, pause-dependent wait time; and blocking times; active parts can be further divided into processing times and preparation times. However, this assumes that the associated system states are also represented appropriately in the simulation model and are recorded in the trace or the online evaluation.

At the individual product level, the lead time and the individual time components of the lead time are of interest in the evaluation. The summary places value on the individual cycle times of the products and the time components. Both the minimum and the maximum values that occurred can be recorded. The statistical evaluation shows the cycle times and the time components for specific groups of products (same product type, same completion interval, priority jobs, *etc.* or all), combined and as average values. The cycle time for a specific product set, calculated as the time difference between the release of the first product and the completion of the last product, is recorded.

Observance of due dates

The specification of a due-date t_L for the product allows the time difference d_L of the completion time t_F and the due date to be calculated. If the completion time lies before the due-date, this is an earliness d_{LF}, otherwise a tardiness d_{LS}:

$$d_L = d(t_F, t_L) \tag{3.4}$$

$$d_{LF} = \begin{cases} d_L; & \text{for } d_L \geq 0 \\ 0; & \text{for } d_L < 0 \end{cases}$$

$$d_{LS} = \begin{cases} 0; & \text{for } d_L \geq 0 \\ -d_L; & \text{for } d_L < 0 \end{cases}$$

To evaluate the due date observance, not only the individual earliness or lateness, but also the average values, totals, minimum and maximum values for all or specific products, are considered. Important are also the proportions of products with lateness $d_L < 0$ or due date observance $d_L \geq 0$ within a specific product set.

Resource-oriented Viewpoint

Statements about the loading, throughput and inventory are important for resources. *Resource* here is considered to be a single element (machine, machine operator, production facility, transport equipment, buffer, queue, storage, *etc.*) with an available capacity used to perform work steps or for storage.

The evaluation of the resource-related information is closely related to the capacity of the resource. The capacity is frequently specified as service capability and specified in service units for each time section or integrated over time sections, such as 1000 units/h [2]. In contrast, the capacity term here is considered as being the momentary acceptance capability of the resource (for example, 1 unit). A single resource available without interruption thus provides a capacity of

$$K = 24\,\mathrm{h}/1\,\mathrm{d} = 1 \tag{3.5}$$

in order to return to the service for each time section. The integral size of the capacity is designated here as work capacity in order to achieve a unique separation:

$$K_A = \int\limits_{t=t_S}^{t_E} K \cdot dt \tag{3.6}$$

The specific integral is solved at the begin t_S until the end t_E of the consideration interval. For a specific resource, this gives for a day $K_A = 1\ \mathrm{d}\cdot24\ \mathrm{h}/1\ \mathrm{d} = 24\ \mathrm{h}$.

Loading

The *available work capacity* K_{AR} of a resource over a specific time interval can be accumulated from the trace data or previously by the online evaluation, for constant capacity as the product of capacity and time interval duration. Similarly, the *active assigned work capacity* K_{APA} processed by the products can be accumulated in the time interval. The quotient of both work capacities represents the *loading* A_R of the resource:

$$A_R = K_{APA}/K_{AR} \tag{3.7}$$

The available work capacity for the resource can be reduced by the shift system, breaks, stoppages or maintenance. Correspondingly, unavailable parts can be determined by accumulation in which integration intervals are assigned to each of the availability states of the resource.

The resource can also be assigned by products in various states. Initially, a differentiation is made here into active (processing with time requirement) and passive (blocking without time requirement) states. The states can be further subdivided (setup, processing, stripping down, process-oriented blocking, pause, failure, operational pause, *etc.*) and the currently assigned work capacity can correspondingly be assigned to the assignment states.

At the individual resource level, the ratios of the available and the unavailable work capacities, and the ratios of the assigned capacity to the available work capacity are important. Here, the loading as active assigned work capacity can be related to the currently available or the total available work capacity of the resource.

Each of the ratios of all resources can also be compared and so determines the minimum and maximum loaded resources. For the division into time series, this can also be done for the ratios within the individual time series. In the statistical summary, the average loading can be determined for specific resources or all resources as the quotient of the total of the assigned resources and the total of the available work capacity of the resources.

Throughput

The throughput describes how many products leave a previously entered resource within a specific time interval. The calculation is performed by counting online during the simulation or by the evaluation of the event list. The chronological course of the throughput of a resource is realised by subdivision into time series. For the complete system, it suffices to consider just the last resource in the chain and to evaluate its throughput as the overall throughput. If necessary, the passed-through products can be differentiated into groups.

The inward movement of products to a resource also governs the throughput (outward movement). The inward movement, however, cannot be derived directly from the resource, because the incoming products are stored in front of the actual resource in another resource (queue, buffer, *etc.*). Consequently, the evaluation must assign the input buffer to the processing resources and consider the inward movement at products in the buffer as the ratio to the outward movement from the resource.

Inventory

The inventory (*active or passive assigned capacity*) of a resource can be represented as the chronological change of the trace data. For the statistical consideration, the average inventory B_R within a time interval is determined as the quotient of the *assigned work capacity* K_{AP} of the resource and the *interval duration d*:

$$B_R = K_{AP}/d \qquad (3.8)$$

Similar to the loading and the throughput, the time series of the average inventory can be developed and the average values over several resources determined as the quotient of the total of the average inventory levels of the individual resources and the number of resources. In contrast, the minimum, average and maximum inventory levels are of interest both for the comparison between the resources and also within time series.

Bottleneck

Bottleneck analysis does not represent a new form of the data preparation. Rather, it is the combination from the loadings, throughput and inventory analysis. To detect a bottleneck, the resources or the time intervals with the highest loading, a low ratio of inlet flows to outlet flows (throughput) and high inventory levels are investigated. Note that the bottleneck can move dynamically.

Material flow

To determine the material-flow analysis, the transactions of the products between the resources are observed. If the products are transported in the model itself using resources along prescribed paths, the throughput of the transport resources can be used for the material-flow analysis. Otherwise the transitions of the products between the resources in the time interval are counted.

To obtain a time-dependent picture, a subdivision into time series can also be made here. It is advantageous for the evaluation, to determine the maximum resulting material flow and calculate the ratio to all other material flows.

3.3.4 Representation Forms

Monitoring and animation, discussed previously briefly in Section 3.2.3, can be used for the simulation. For monitoring, the state quantities or the accumulated state quantities are normally displayed dynamically with regard to the model objects. The animation is a moving monitor diagram that can show the state changes or transactions for objects.

Figures 3.5, 3.6 and 3.7 show sample sections from the individual state, 2D- and 3D-animation animation categories. The increasing level of detail of the representations is bought with higher modelling effort. In the last stage, indeed, realistic movie-like sequences can be created.

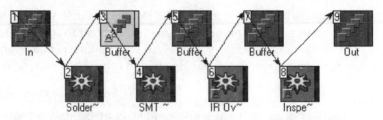

Figure 3.5. State animation of an SMT component placement

Tabular or graphical representations can be used for preparing the simulation results. The following section describes various graphical representation forms for several characteristic values. Certain visualisations are particularly suitable for acquiring and evaluating the values.

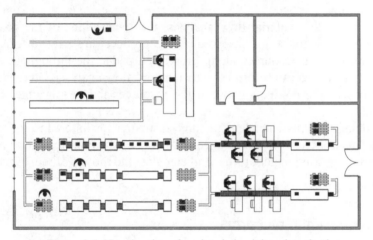

Figure 3.6. 2D animation of a printed-circuit board production

Figure 3.7. 3D animation of an SMT line

Circle Graph

The circle graph is particularly suitable for representing values with a fixed base value (100%). In contrast, the loading levels can best be prepared as relation to a fixed available capacity. The individual accumulated working capacities of the operating states are set as ratio to the base value and assigned an appropriately large circle segment as shown in Figure 3.8. However, the comparison of the diagrams of several objects has the disadvantage that the base value of the individual objects cannot be different.

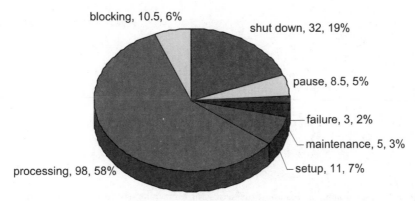

Figure 3.8. Machine loading shown as a circle graph

Another example of the use is the representation of the individual wait and processing time as ratio to the lead time of a product.

Bar Graph

Like the circle graph, the bar graph is also suitable for representing values with a fixed base value. However, in contrast to the circle graph, the bar height and the bar width can vary for the combination of several bars so that different base values can be illustrated. Thus, the bar graph is particularly suitable for comparing objects or time series.

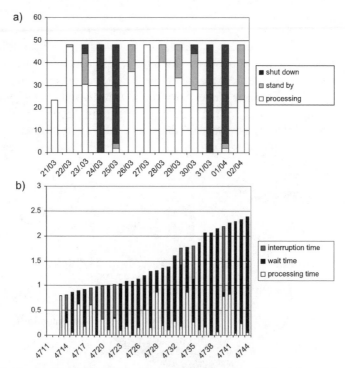

Figure 3.9. Bar graph, (**a**) loading-time series for a machine, (**b**) lead-time comparison for products

Figure 3.9 illustrates the use for a loading-time series of a machine and the cycle times of several products. Within a time series, the statistical average values of several objects can be formed and displayed as a bar graph, for example, to monitor average cycle times for specific products within time intervals.

Histogram

The histogram is primarily used for the statistical evaluation of samples. Chapter 5 provides a detailed discussion of this subject. For the evaluation of the simulation, a histogram is used for evaluating the stochastic value realisations, *e.g.*, dispersed processing times. The sample is divided into classes. Within the class limits, the number of realisations is counted and assigned as class frequencies of the class average (see Figure 3.10). In the result, a step function of the frequency density of the sample can be derived as an approximation for the probability density.

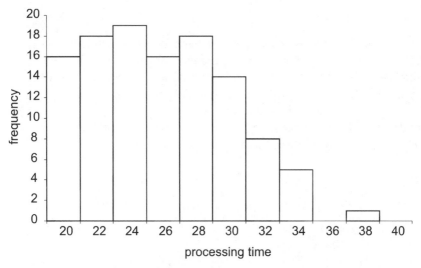

Figure 3.10. Histogram for 125 realised processing times in 10 classes

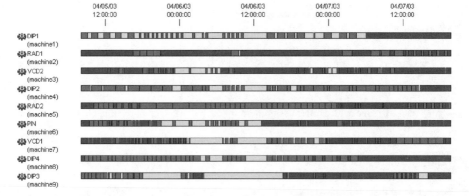

Figure 3.11. Gantt chart for states

Gantt Chart

A Gantt chart shows in a clear form the assignment of various resources by products as function of the time. The horizontal bar assigned to each resource shows the individual actions limited by their start and end times. The individual bar sections can be emphasised either according to states or separated into products (see Figure 3.11).

Line Chart

A line chart can be used to visualise dependencies of quantities. Several quantities can also be combined in a single diagram. In contrast to the bar graph, the ratio to a base value for a data point does not have priority here. The use is appropriate both for displaying dependencies within a simulation run and also for illustrating the dependencies of the factors for a repeated simulation. The analysis from Figure 3.12 can quickly determine which particular workplace causes the bottleneck.

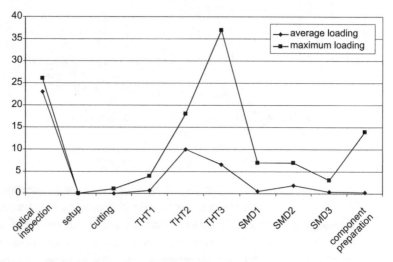

Figure 3.12. Buffer assignment for each workplace in the line chart

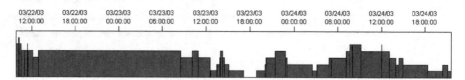

Figure 3.13. Assignment diagram

An assignment diagram is a special form of line chart that represents the inventory of a queue as a chronological graph (see Figure 3.13). The flow diagram can be used to show much more complex relationships [18].

Radar and Petal Diagram

Radar and petal diagrams can be used to consider several characteristic variables by using the area. In a radar diagram, the quantities of interest are assigned to arrows that have a constant angle to each other and that all start at the coordinate origin. The size of each quantity is characterised by its distance from the origin. The connection of the quantities with each other produces an area that is easy to understand visually.

In contrast, the quantities in the petal diagram are assigned according to their angular segments. The radius of the segment varies with the value. The resulting area appears optically proportional to the assigned value.

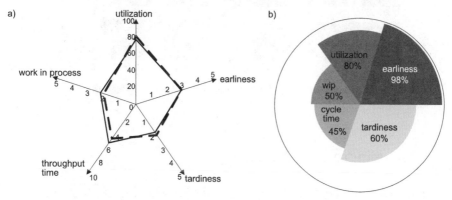

Figure 3.14. Radar and petal diagram, (**a**) Target system with several solutions in the radar diagram, (**b**) Normalised and weighted target system in the petal diagram

Radar diagrams are frequently used for considering the target. The values for the relevant target values – also several solutions – are entered in the diagram. The difficulty here lies in the representation of unnormalised quantities. If, however, the relationship to a standard value, the target value, for example, can be obtained, the resulting area can also be interpreted as a characteristic variable, which, however, depends on the order of the assignment of the characteristic variables to the arrows.

The problem for the petal diagram is similar. The angle of each segment can also be used here for the visualisation of the weighting.

Sankey Diagram

The Sankey diagram represents the flows between elements as directed arrows, which, at the element nodes can separate into several flows or receive the inflow from other flows. The width of the arrow is drawn proportional to the flow strength. This diagram form is particularly suitable for the representation of material, production and transport flows. The number of the moved elements between the fixed model elements is shown with an appropriately thick arrow. The thickness is best normalised to the strongest measured flow. Figure 3.15 shows an example for the flow of the manufactured products in which the model objects serve as nodes.

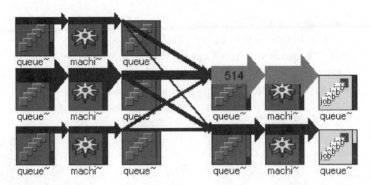

Figure 3.15. Sankey diagram used to represent a production flow

3.4 The simcron MODELLER

The simulator is the software program for representing the system. In the simplest case, it is a programming language. Various simulators with different concepts and scope have established themselves in the market. Refernce [9] provides a summary and a classification. Because not every instrument is suitable for the application, the simulator must be selected appropriately for the problem.

This section presents the *simcron MODELLER*, a simulation system primarily developed for use in the manufacturing planning [12]. Its basic philosophy and model world is based on research work at the Electronics Packaging Laboratory of Dresden University of Technology, Germany.

3.4.1 Overview

The simcron MODELLER is a simulator whose focus primarily lies in the *fast* calculation of manufacturing sequences. The modelling is performed object-oriented. As usual, the model structure is oriented on discrete event queuing systems. This concept guarantees a high degree of clarity and ensures the user acceptance. The simulator provides the user with predefined modules that can be combined to produce simulation models. The type and functional scope of the objects was designed to produce a good agreement to real objects. Provided an abstract way of thinking is adopted, it can be used for very general and theoretical situations.

The software system has a shell structure. The object world and the event dynamics are contained in the simulator kernel. For execution-speed reasons, this part is provided in compiled form. The next shell is an interface layer that provides the communication with the model world of the kernel. It is the first level with which the user can interact with the system. The communication is performed with an interpreter based on *Tcl* programming language [10]. The kernel and the interpreter together form an executable software component. This allows the simulation to be performed as a computing process, also without a graphical user interface. The graphical user interface that follows as an external layer permits user-friendly modelling. In addition, the simulator provides various interfaces to the external world. These interfaces, in particular, permit data exchange with other programs. Figure 3.16 illustrates the shell structure. Both the base software and the graphical user interface can run under all recent Windows platforms, Linux and other UNIX derivates.

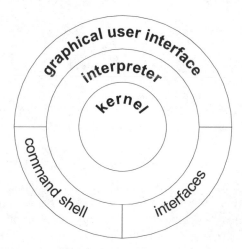

Figure 3.16. Shell structure of the simcron MODELLER

A large number of interfaces is available to permit both process communication and data exchange with other programs. The most important data-exchange standards supported by the simulator include:

- ASCII
- DDE or send
- COM
- ODBC
- XML
- Sockets and HTTP
- Program code in C/C++, Tcl.

3.4.2 Model Objects

The following discussions provide a short overview of the simcron MODELLER objects and their properties. The section explains the basic use of the objects for the modelling of manufacturing sequences without, however, discussing any detailed questions. Section 3.5 shows model examples, which, among other things, provide useful information for the method of representing manufacturing processes with the simcron MODELLER.

The objects can be divided into physical and logical objects. Whereas physical objects have a physical equivalent in the real world, logical objects represent control rules not primarily visible in reality. Physical objects include the resources (machine and queue) and the jobs. Control rules are recorded as technologies, branches, requirements objects, stochastics, schedules, shift schedules, methods, objective and control variables.

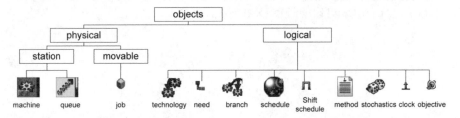

Figure 3.17. Model objects of the simcron MODELLER

Resources

Every manufacturing system consists of a set of resources used to different degrees during the manufacturing process. Such resources include personnel, machines, working hours, energy, consumption material, etc. Because some resources are available only to a limited extent while several manufacturing jobs currently in the system compete with each other, this produces the true operational planning problem.

Station: machine and queue

The simcron MODELLER does not represent all conceivable resources, but only the time- and space-related resources, in the form of machines and queues, the stations. All activities performed on the product, such as processing, transport, storage, *etc.*, are linked with the stations. Machines and storage areas have in common that they can accept one or more products (jobs) in accordance with the set capacity. A station is thus nothing other than a logistical object that provides space that can be queried and occupied by the jobs. However, what actually happens to a job in the station is not of interest. Only the length of stay of the job is considered. The critical difference between a machine and a queue is that whereas products require both time and space from a machine, they normally require only space from a queue. Namely, everything associated with storage is represented by queue modules and everything associated with the work time is represented by machines modules. There are no special modules for transport or personnel, because, in this consideration, they also provide only time- and space-related resources.

The capacity, namely the ability to accept, of the station is an important property of the object, is equivalent to the size of the job. A station is capable of accepting further jobs until the complete capacity of the station is occupied by jobs. The free capacity of the station and the size of a job decide whether or not the job can be accepted. Furthermore, the job requires a processing time (the minimum length of stay in the station) needed for the execution of the associated activity. Even after the processing, the processing sequence can require that the job remains in the station; in this case, a wait time occurs.

The factor that determines the processing duration for machines is the queuing discipline for the queues. The queue object is separated into a storage and a removal strategy. Whereas the storage strategy determines the position a job assumes when it enters the queue, the removal strategy specifies how many jobs, counted starting at wait position 1, can leave the station. If a job is stored in the sequence of its arrival and the first job is always removed, the station realises the FIFO (first in first out) discipline. The storage is oriented on well-known priority rules [2], which, when combined with the removal strategy, allow various disciplines to be implemented.

Table 3.2. Summary of the important properties of a station

Property	Explanation
Capacity	The acceptance capacity of the station specifies how many jobs of what size can be accepted.
Storage/removal	Queue discipline
Jobs	Jobs that are processed or stored in the station
Disturbances	Stochastical disturbances of processing times
Patterns/sinking	The station can contain patterns that produce new jobs using stochastical distributions (source function) A station can sink jobs
Availability	Stochastical or deterministic change of the availability of the station
Blockade	Blocking possibility as conveyor section when the first job cannot be passed on and the separation definition for jobs
Statistics	Online evaluations of the work capacity, inlet and outlet flows, durations, extreme values
Use	Machine, production facility, personnel, transport mechanism, conveyor section, queue, buffer, storage, *etc.*

In addition to the specific states that the station can assume from the processing of jobs, its availability state can also change. The availability is always binary, which, however, can be revoked time-controlled or randomly in order to create states such as maintenance, repair, pause or similar. If the station is not available, although it cannot accept, process or supply jobs, it can store jobs.

Job

The job has already been mentioned in conjunction with resources. It is the partner to the queue and machine resource modules. Whereas the latter provides the space and time logistical categories, the job acts as a space and time consumer and as a moveable object. Jobs primarily serve to represent products. Depending on the required representation accuracy, a job can represent a single product, a manufacturing lot or a complete year's contract. The module can, however, also be used to represent abstract objects, such as setup or maintenance orders. The job can also be used to define control states, such as Kanban with the size zero.

The job itself does not possess any activity information. The description of the movement of the job through the system and the logistical requirements become complete only in conjunction with a technology. The combination of the job parameters with those of the work step within the technology defines the space and time requirements of the job in the individual resources. For this reason, the module is relatively simple, because the link with the logical modules determines its operational capabilities.

The placement of a job in a station is always associated with a work step of a technology. A job can be located in several stations at any one time, in which case each connection of job and station forms a subordinate object. Strictly speaking, a station does not contain any jobs, but rather linkage objects with jobs, each of which references just one station and one job.

Table 3.3. Important properties of the job as overview

Property	Explanation
Logistical parameters	Determine on job level, space requirement, priority and quantity of the job
Specificator/identification	Differentiation of jobs in the technological respect
Activation	Analogue to the availability of the station in binary form
Dates	Release and due dates of the job
Stations	Linkages to stations, associated processing states, wait states and their parameters
Pattern	Linkages to sources in which the job serves as pattern for the creation of new jobs
Statistics	Online evaluations of durations and due dates
Use of the job	Job, lot, product, action, marking, *etc.*

The placement of jobs in stations can be created or removed both manually to establish an initial state and by the scheduler during the simulation run. The job assumes various processing and wait states within the station (the linkage of job and station). The synchronisation or the offset for the individual activities of the job can be used for various purposes.

Control

Whereas the physical objects of the simulator do not appear to be particularly complex, as often in reality, the execution rules constitute the difficulty of the system. A complete simulation model can be achieved only in conjunction with the logical modules that make the hidden logic of the real manufacturing visible to the user.

Technology

In contrast to other simulators, the simcron MODELLER does not define any direct relations between the resources. The technological processes are described by the module technology in combination with other objects. The technology defines a linear sequence of activities and can initially be interpreted as being a container for ordered work steps. Just one station is assigned to each workstep. This assignment determines the relations of the stations amongst themselves. Note that the buffers between the machine worksteps must also be represented as work steps with queues. Otherwise, this would produce an uncoupled system.

A single station can be included more than once in the technology. This allows finite loops to be represented. In addition to the station, the workstep contains other significant parameters that describe the activity. Some job and station parameters that can be specified at a detailed level in the workstep repeat, because, for example, the processing time of a job differs from that of another job on the same station. When a parameter is specified in the workstep, the parameter for the workstep always takes precedence over that of the job or the station. Primarily, the time and space requirements of the job in the workstep on the station are described, however, priority, release dates and states, due dates, stochastical disturbances for times, quantity acquisition and relations to other technologies and worksteps are other possible parameters. This allows the synchronisation with other worksteps, the alternative distribution to resources or the simultaneous assignment of several resources to be controlled. Other objects, the branch and the need-object, are used to establish these relations.

The specificators of a technology can be used for parameter variations for worksteps within a technological sequence. For each workstep, special parameters can be listed for each specificator. This produces a parameter matrix for each workstep and specificator. The assignment of the specificator to the job means that the corresponding technology parameter set governs this job. The placement of the job in a station is always associated with a workstep. The job, the technology workstep and the station have a triangular relationship with which the workstep parameters represent standard values for the singular realisation of an activity.

A machine or a queue in the workstep are used to request a specific resource. If several resources are to be used in parallel in the workstep, the technology acts as a workstep container. A second technology with the same number of worksteps as the number of required resources is added hierarchically to the workstep. Although the individual activities are all started together, their duration can vary and so the release of the resources will be staggered.

Table 3.4. Overview of the important properties of a technology workstep

Property	Explanation
Station	Resource requirement: machine, queue
	Alternatives are represented as branches
	Multiple resources are represented as additional technologies
Size, priority, identification	Job parameters in the workstep
Set, processing time, volume-dependent processing time, setup step	Determination of the processing time and state
Forwarding time	Early forwarding to the next workstep (multiple assignment of resources)
Gate time	Time separation between jobs in the station
Disturbances	Stochastical change of the workstep times
Need	Quantity request from other worksteps for the synchronisation of activities
Activation, release	State and time for the release of the workstep
Technology use	Linear workstep sequences, parallelisation of worksteps

Branch

To augment the linear technology, the branch allows the representation of alternative workstep sequences. Branches can be used to separate, sort or recombine job flows. In the latter case, the branch acts as a connector. Just like a queue or a machine, the branch can be embedded in a technology workstep. The cascading of branches is also permitted. The associated workstep links the job, depending on the branch type, with a new technology and a new workstep.

Thus, the job-technology assignment can be changed dynamically during the simulation run. Consequently, the branch can also be considered as being a technology operator. Although a branch can have any number of directions, it must contain at least one direction. Each individual direction has a reference to a workstep within a technology to which the job will be transferred when the corresponding direction becomes active. The particular direction activated by the branch depends on its type, the branch parameters and the job identification. Load-dependent, sorting or random branch strategies can be realised in various forms.

Table 3.5. Overview of the important properties of the direction of a branch

Property	Explanation
Technology/workstep	Workstep within the technology to which the branch should be made
Probability	The probability with which the branch is activated for a random branch type
Identification	Filter of the branch for specific jobs
Branch use	Alternative workstep sequence, combination, recursion

Branches can be used to nest technology workstep sequences without losing their linear structure. Because technologies can contain other technologies to describe parallel work steps, technologies and branches can be combined to describe any AND- and OR-linkages of resource requirements. When used together with the additional functions for the synchronisation and the early forwarding, very complex processes can be modelled. The recursive chaining of technologies using branches can be used to create endless loops.

Need

Technologies are used to define simple predecessor–successor relations between the worksteps within a single activity sequence of a product. Frequently, however, additional dependencies between arbitrary worksteps to other products are required, such as for an assembly described by a bill of materials. The simcron MODELLER represents such relations using a need-object. It is generally used to synchronise activities during the simulation run.

The need-object lists specific work steps for technologies, specifies a quantity relation for each relationship and is itself linked with a workstep. With the exception of the quantity relation, the created relations for $1:n$ worksteps have an equivalence in the pseudoactivity for network planning (see Section 2.2).

At the end of a processing activity during the simulation, the set of jobs is credited to the associated technology workstep. Worksteps that contain relations represented by need-objects can be started only when all required quantities have been booked (AND linkage). Grouping and debiting options of the need-object can be used to extend the relationship framework.

Schedule, shift schedule, clock

Schedule, shift schedule and clock are used for the specific creation of events during the simulation. Whereas the time and shift schedule are used to control the availability of stations, clocks initiate additional events, which themselves do not cause any state change of the model. In both cases, the switching sequences can be periodic or one-off.

A schedule consists of an overlaying of various shift schedules. Predefined shift schedules can be used by several schedules. For example, a model can consist of a works calendar (once), 2-shift and 3-shift systems (each periodic) saved in shift schedules, which are combined in schedules with 2-shift and works calendar, and 3-shift and works calendar; they can then be assigned to specific resources. Schedules can also be used to specify validity intervals for the individual shift schedules; this allows the availability of resources to be modified dynamically.

The shift schedule specifies a sequence of switching times with a switching direction (on or off). If the shift schedule is defined as being cyclical, the first and the last switch of the sequence specify the period duration. After the last switch, the switching sequence will be repeated infinitely. Otherwise, it will only be performed once. A shift schedule can only be used in combination with a schedule. The overlaying with other shift schedules (AND-linkage of the switching states of the shift schedules) means that not every defined switch of the shift schedule causes a switching of the schedule.

Similar to the shift schedule, the clock contains a sequence of times and can be defined as being cyclical. It acts independently and globally in the model. Each switching time defined by the clock causes a clock event.

Stochastics

Stochastical objects are required for the creation of random-number flows using specific distributions. A random number is requested for specific events and is produced by the conversion of an input value u into an output value v, which is created by multiplying the input value u by the value x from the random number stream X of the stochastical object. If no input value is passed to the object, $u=1$ will be assumed, and the value x output. The following distributions are available:

- Deterministic
- Uniform distribution
- Triangular distribution

- Exponential distribution
- Erlang distribution
- Normal distribution.

The random value for the processing times, failure and repair durations, intermediate arrival times, separation times or other parameter realisations can be affected by the use of the module.

Method

A language must be used if the complexity of the sequence to be represented exceeds the modelling capabilities of the predefined modules. As explained in Section 3.4.1, the simulator provides an interpreter that allows all model states and parameters to be varied. The interpreter uses the *Tcl* script language that has the complete functionality of a programming language.

Methods are containers for program code that can be attached to specific simulation events. They form the interface to the interpreter and do not place any restrictions on the program structure. Procedures and variables can be used at any time to influence specific states when certain events occur.

Objective variable

Objective variable objects perform the function of the statistical acquisition of characteristic variables of the simulated sequence. The simcron MODELLER provides with objective variables, specific characteristic variables from the categories: due-date observance, lead time, inventory and loading in accordance with Section 3.3.3; these objective variables relate either to all relevant or only selected model objects of a class as appropriate for the considered characteristic variable.

3.4.3 Simulation Run

A simulation can be started once all objects have been created, para-meterised and linked with each other. The simulator scheduler generates the events that cause the state changes in the model. To improve the tracing of the simulation run, the event processing can be performed in single steps. For the event-oriented execution of an experiment, the interruption-free event processing with a small time consumption can also be used. Events, which play a central role for the run, are represented in the simcron MODELLER as volatile objects with transition parameters. Events with the same model time t_m, so-called simultaneous events, are always processed sequentially using specific priorities.

The graphical user interface provides animation and monitoring functionality during the simulation. A state animation can initially be displayed on the display area in which the object states are assigned specific colours and the relations of the objects amongst themselves symbolised. The specification of layout data and transport paths (represented by stations, although animated differently) allows a 2D animation to be made. The animation speed and level can be varied. The stations are assigned monitoring graphics that dynamically represent assignments and loading characteristic values. A monitor window for the event and time tracking can also be displayed. The logging of the event list during the simulation run is optional; several logging levels are provided.

3.4.4 Report

The simcron MODELLER provides support for online reports and calculations based on the event trace, as described in Section 3.3. The characteristic variables of the simulated manufacturing sequence described there (meeting of due dates, lead time, inventory, loading and throughput) can be clearly displayed in tables and diagrams. The reports can be produced for individual objects, time series or groups of objects, and also exported for other programs. The graphical analytical tools include circle and bar graphs for loadings, inventory levels and throughput times, Gantt, assignment and material flow diagrams can be customised to display interesting relationships. The optimisation extension of the simulator also provides radar and petal diagrams.

3.5 Simulation Models

Subsequent model examples give an impression of the modelling capabilities of the simcron MODELLER and show operational possibilities of the simulation method in electronics production. The wide range of problems means only selected relationships can be shown, without, however, any particular details being discussed.

3.5.1 Flow Shop and Operating Curve

A manufacturing process in which all products pass through the same sequence of processing stations is called Flow Shop. Frequently, the material flow here is accompanied by buffer storage that sometimes even opens the possibility of changing the sequence of the products between the processing stations. The model example represents a simple Flow Shop manufacturing in which each lot spends 8 h at the SMT placement station and another 8 h in the manual populating/wave soldering. The processing times at the stations fluctuate with a uniform distribution between 80% and 120% of the standard value. In the manufacturing, new lots are dispatched with an exponential distribution in specific time intervals and should each be forwarded 2 days later.

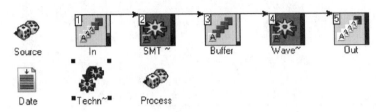

Figure 3.18. Flow Shop model

The *SMT* and *Wave* processing stations are represented by machines. A queue is present as buffer before each station. Finally, a queue also serves as output. A technology contains the activity sequence 1. *Input* – 2. *SMT* – 3. *Buffer* – 4. *Wave* – 5. *Output*. The processing time is specified as 8 h for the processing steps 2 and 4. The stochastic *processing*, with uniform distribution (0.8/1.2), is linked with *SMT* and *Wave* for the relative disturbance of processing times. A job is set as a pattern in the *Input*. The number of jobs to be created is specified as 1000 and the stochastic *creation* with exponential distribution used for the intermediate arrival times.

The characteristic variables acquired by objective variables are used to evaluate the simulation results. The average inventory level of the *input* and *buffer* queues, the loading of the *SMT* and *Wave* machines, the average delay, the average lead time and the average wait time of all jobs, and the throughput of the *Wave* machine are determined. The simulation experiment will be repeated for various intermediate arrival times, namely parameters of the stochastic *creation*, which can also be interpreted as a variation of the dispatching rate. The average total inventory is determined from the buffer inventory and loading, the average processing time is determined from the flow and the wait time. Table 3.6 shows the simulated values.

Table 3.6. Simulation results with changed dispatching rate

Dispatching rate [1/d]	Throughput [1/d]	Buffer inventory	Loading	Total inventory	Lead time [h]	Wait time [h]	Processing time [h]	Delay [h]
1.2	1.20	0.00	0.41	0.82	19.5	3.5	16	0.5
1.5	1.50	0.00	0.53	1.06	21.0	5.0	16	1.5
2.0	2.00	0.36	0.66	1.68	24.6	8.6	16	4.2
2.5	2.50	1.70	0.85	3.40	48.0	32.0	16	25.0
2.7	2.70	2.17	0.90	3.97	54.5	38.5	16	31.6
3.0	2.94	4.00	0.98	5.96	153.0	137.0	16	129.0
3.4	2.97	28.00	0.99	29.98	467.0	451.0	16	443.0
4.0	3.00	62.00	1.00	64.00	1009.0	993.0	16	984.0

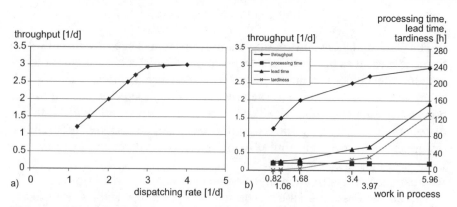

Figure 3.19. Flow Shop results, (**a**) throughput-dispatching dependency, (**b**) operating curve

If we initially consider the dependency of the throughput from the dispatching rate in Figure 3.19a, we can determine the load limit of the manufacturing system. If on average more than 3 lots are dispatched per day, the throughput does not further increase. The operating point of the system must be selected below this limit to prevent overloading. Figure 3.19b shows the operating curve of the manufacturing system determined from the characteristic variables in accordance with Wiendahl in possible operating points that display quantitatively the dependency of the characteristic variables for the inventory [18]. The simulated operating curve thus simplifies a decision regarding the required business conditions and can be used on all manufacturing systems.

3.5.2 Cluster in Semiconductor Production

Cluster tools are common in semiconductor production. In such a system, several machines are grouped in a closed cell and the wafer passes through a fixed sequence of processing steps. The wafers are docked in closed cassettes at the system entry. Automatic handlers transport the individual wafers to the machines. When the last processing step has completed, the wafers are returned to the cassette.

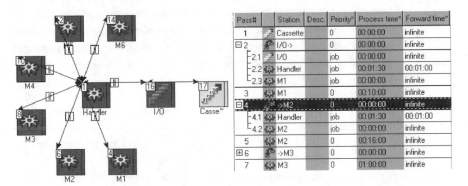

Pass#		Station	Desc.	Priority*	Process time*	Forward time*
1		Cassette		0	00:00:00	infinite
⊟ 2		I/O->		0	00:00:00	infinite
⊢2.1		I/O	job	00:00:00	infinite
⊢2.2		Handler	job	00:01:30	00:01:00
⊢2.3		M1	job	00:00:00	infinite
3		M1		0	00:10:00	infinite
⊟ 4		->M2		0	00:00:00	infinite
⊢4.1		Handler	job	00:01:30	00:01:00
⊢4.2		M2	job	00:00:00	infinite
5		M2		0	00:16:00	infinite
⊞ 6		->M3		0	00:00:00	infinite
7		M3		0	01:00:00	infinite

Figure 3.20. Cluster model, layout and technology section

The model consists of the *M1* to *M6* machines and the *handler*, which is also a machine. The entry and exit, the *cassette*, and the *I/O* transfer channel are represented by a queue; not only the transfer channel, but also the machines and the *handler*, can only accept one wafer. The attempt to represent the process with a linear technology *cassette – I/O – handler – M1 – handler – M2 etc.* leads to a deadlock because the simulator would transfer the jobs without considering the assignment of the other stations. This, in turn, would lead to a situation in which the handler must pass a job to a machine, which itself requires the handler for forwarding a job. The solution involves modelling with parallel technologies: *cassette – (I/O->: I/O AND handler AND M1) – M1 – (->M2: handler AND M2) – M2 etc.* In this case, before a job can be accepted by the handler, it must also be ensured that the subsequent machine is free. A processing time occurs for the transport only for the *handler*. Each transport activity takes 1 min, of which the *handler* requires a further 30 s to return to its start position. This behaviour is represented as an early forwarding. Although the "processing time" of the *handler* is always 1.5 min, the forwarding time is only 1 min because the job then can be passed to the next station, while the assignment of the *handler* is held for another 30 s. The processing time in the machine work steps varies from process step to process step.

The section from the Gantt chart for the machines of the model in Figure 3.21 illustrates the operation. Because of the short processing times, wait times arise for *M1* and *M2* in which the job cannot be passed to the *handler*. The transfer of a job to the downstream process takes 1 min, whereas the currently emptied machine will be reassigned after 2.5 min at the earliest.

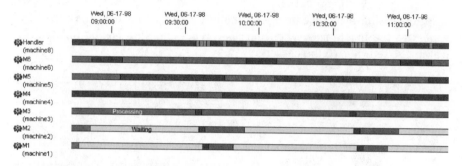

Figure 3.21. Section from the Gantt chart

This deterministic model allows the execution in the system to be investigated very exactly. Very good planning forecasts or improved cycles can be achieved by considering the stochastic deviations of the individual times and the execution for parameter variations for such manufacturing systems.

3.5.3 Quality Processes of Printed-Circuit Board Production

The example shown here shows the general possibilities for using the simulation technology in quality assurance. It is a simplified model of the printed-circuit board production that consists of the *THT – SMT – AOI – reflow – manual populating – wave soldering – ICT* technological sequence, where primary attention is placed on the *AOI* and *ICT* inspection processes. Each technological process can have associated inspection and rejection processes, as discussed in detail in Chapter 6. After the final inspection, the *ICT* inspection process, any defects found from all processes will be corrected. A differentiation is made according to the defect classes that result from the individual processes. In this case, a differentiation is made according to the repair processes for defects from *THT* and *SMT*, *reflow*, and *manual populating* and *wave soldering*, respectively. The defect rates for the individual processes are known. The total defect rate for the *THT* and *SMT* process steps is $p_1 = 8\%$, for the *reflow* process step $p_2 = 1\%$ and for *manual populating* and *wave soldering* in total $p_3 = 5\%$. This raises the question whether the *AOI* inspection process is cost effective [6].

The model uses branches that create random defect rates. The model consists of the machines for the individual process, inspection and repair steps; each machine has a queue. The basic technology contains the technological sequence with the queues and the machines. A technology is also created for the repair after *AOI* with queue, machine and a branch for the *reflow* queue. The repair technologies after *ICT* contain the queue and the machine for the repair, and a repetition of the last 3 work steps (*ICT*, *ICT*, *exit* queue) of the basic technology. In the basic technology, two branches are inserted, after *AOI* with 2 directions to the *AOI* repair technology and to the *reflow* queue of the basic technology, and after *ICT* with 4 directions, of which three exit to the two repair technologies and one exits to the basic technology. The technologies contain the required times for the process steps. Figure 3.22 shows the model with the possible branches. There are no recursive passes here; a printed-circuit board can be tested and repaired just once. The probability that defects from *THT* and *SMT* are found in the *ICT* is zero, because such defects have already been detected and corrected.

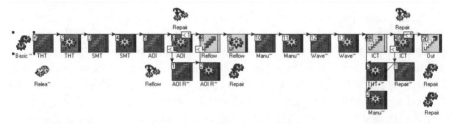

Figure 3.22. Model of the quality processes for a printed-circuit board production

The experiment simulates the flow of 10,000 printed-circuit boards and the machine assignment times in the inspection and repair steps are measured. If a machine hourly cost rate is assumed, the individual inspection and repair costs can be determined as described in Chapter 6. The experiment is then repeated without *AOI*. The associated defects must now be filtered out and corrected at the end in the *ICT*. This means the defect probability at the *ICT* increases for these defects.

As the results from Table 3.7 show, the unit costs for the performed quality strategies differ. From the cost viewpoint, it saves cost to omit the *AOI* inspection step and find and correct the defect only at the end.

The previously discussed variants can certainly be analytically recalculated. It becomes more complicated, however, when individual defect classes of the process steps differ and must be handled differently for the inspection and the rejection or time-dependent quality behaviours need to be investigated. The simulation provides the possibility of the acquisition of all characteristics and, in conjunction with the *time* factor, forms a reliable basis for the investigation of quality strategies, in particular for operational planning.

Table 3.7. Inspection and repair costs for 10,000 printed-circuit boards

	Cost rate	With AOI		Without AOI	
		Machine assignment	Unit costs	Machine assignment	Unit costs
AOI inspection	94.92 €/h	55.5 h	52.7 ct	0.0 h	0.0 ct
AOI repair	66.29 €/h	6.4 h	4.2 ct	0.0 h	0.0 ct
ICT repair	94.92 €/h	88.5 h	84.0 ct	95.3 h	90.5 ct
Repair THT, SMT, reflow	81.32 €/h	3.8 h	3.1 ct	32.6 h	26.5 ct
Manual soldering, wave soldering repair	81.32 €/h	12.6 h	10.2 ct	11.6 h	9.4 ct
Total			€ 1.54		€ 1.26

3.5.4 Job Shop in Electronics Manufacturing

A large product range and the customer orientation in electronics manufacturing complicate the planning conditions. The technological process sequences that vary from product type to product type, generally designated as *Job Shop*, provide additional difficulties. The model shows typical conditions in electronics manufacturing, in which almost all modelling objects of the simcron MODELLER are used.

All processing stations, both manual and mechanical, are represented by machines with the appropriate input buffers. They provide workplaces not only for the component preparation, axial, radial and SMT placement, but also visual workplaces and a manual workplace. Several alternative workplaces are available in some cases (*preparation, axial placement, SMT placement, screening*). Each product type receives its own technology that contains the appropriate branches for the alternative workplaces. Additional production facilities (acceptance units and setup equipment), of limited availability, used for the placement activities (*radial, axial, SMT*) are represented as machines and included in the relevant work steps using parallel technologies. For simplicity, technologies, branches and production facilities are omitted from Figure 3.23.

The individual workplaces are served using various shift systems (one, two and three shift). Each shift system has a schedule that is assigned to the associated workplaces. The system is preassigned with approximately 100 work orders using jobs that provide a work reserve for more than 1 week. Dependencies between the jobs, provided they are present as an order bill-of-materials, will be recorded as need-objects. Material dependencies are controlled in the model using order and work step release. Early forwarding is used to represent the overlapping of sequential work steps.

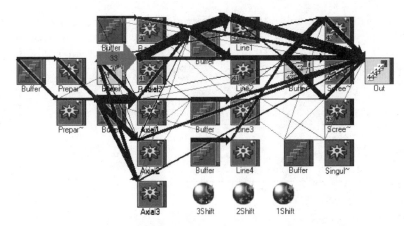

Figure 3.23. Model of an electronics manufacturing process

The network of possible passages of work orders through the system is very difficult to visualise. If, however, one considers the material flow in Figure 3.23 at the end of the simulation to illustrate the performed transactions, some main flows become apparent.

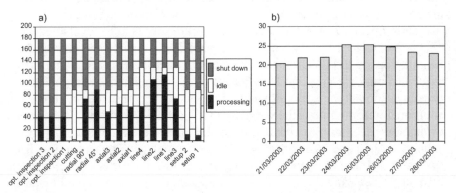

Figure 3.24. Analysis of the Job Shop, **(a)** loading of the workplaces, **(b)** time series of the average number of orders on hand for screening

The simulation exactly predetermines a manufacturing week and this can then be used as a planning basis. Some work orders show a delay already at the beginning. After the planning, the delay further increases because of the required throughput times.

If the loading and the orders on hand for the workplaces are analysed in Figure 3.24, it can be determined that although the screening workplaces form the bottleneck, they have reserve work capacity because they are only available in the one-shift system. We now investigate whether a capacity increase is sensible by changing the shift system with regard to the due-dates observance. Successively, one each of the additional three screening workplaces is converted to two-shift operation. Figure 3.25 shows the effect on the average and the maximum delay; the number of workplaces converted to two-shift operation is shown on the x-axis. The delay reduction can be considered as being linear.

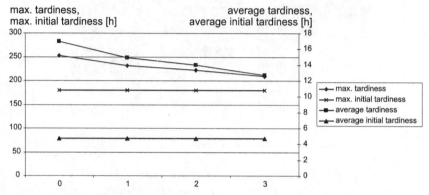

Figure 3.25. Dependency of the delay on the shift system for the screening

The improvement provided by the possible measures is tested as a ratio to the base delay. The conversion of an individual workplace from one-shift to two-shift operation for the screening workplaces can be evaluated as a 33.3% increase of the working time. Table 3.8 compares the cost and the benefits for the individual increment steps. It must be appreciated that from this viewpoint, the cost and benefits are poorly correlated and the measures possibly should be discontinued. For an exact inspection, however, additional information should be included, which permits, for example, evaluation based on the costs, such as the labour costs compared with the saving through the avoidance of the payment of penalties resulting from delays in meeting the contract. However, because the latter cost element is difficult to determine in practice, the decision, despite the exact simulation results, is often subjective.

Table 3.8. Costs and benefits for the screening

Workplaces changed to two shifts:	0	1	2	3
Labour costs	0%	33.3%	66.6%	100.0%
Benefits for maximum delay	0%	29.0%	42.3%	59.3%
Benefits for average delay	0%	16.6%	24.4%	34.8%

3.6 Process-accompanying Simulation

There exist large gaps between the modelling performed by a simulation expert and the daily operational planning. On the one hand, the use of the simulation method requires comprehensive specialised knowledge, abstraction competence and experience with model creation, which the planner normally does not possess. Conversely, the simulation expert certainly does not know the operational processes as well as the planner, who, after all, also makes the decisions. However, the simulation method provides promising possibilities that should not remain unused during the planning. A planning as shown in Figure 3.26 is desirable, in which the conclusions for making decisions are based not just on processes realised in the past, but that can be simulated during the business games, where the simulated results serve as the basis for the actual decision-making process.

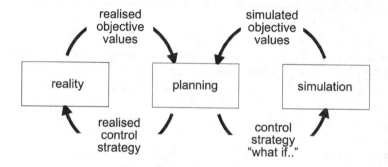

Figure 3.26. Planning with inclusion of simulation

For such a concept, the integration of the simulation in the operational processes is an essential prerequisite, not only in organisational respects, but, in particular, also in the technical respect.

> *Process-accompanying simulation* is defined as being the simulation that runs in parallel to the real process and that serves to control the real process.

Consequently, the associated questions of integration, acquiring the model, comparison of the real and the simulated process and the model adaptation will be discussed in this section. The principles for the control of processes are discussed in Section 1.2.3.

3.6.1 Template Model

Every simulation requires a model that represents the reality with adequate accuracy. Time considerations in the process-accompanying use mean that it is not possible to manually build a complete model in a simulator. Rather, a model that reflects the current situation must be created automatically. Because the basic structures and behaviour patterns of the model describe the execution principles of the real process they must be predefined. If each pattern is described accurately just once, this produces a template model, which, however, is not an executable model by itself. Only the creation of instances of the described pattern and the assignment with values make a simulation model that can run. To speak of a parameter model is only partially correct. Each model instance is an object-oriented model, the presence of the model objects and the relations of the objects amongst themselves, ignoring the number, is difficult to comprehend as a parameter.

The simplest template model is the module library of a simulator and the linkage possibilities of the modules amongst themselves. Model instances can be created and parameterised automatically by the creation and linking of modules, provided the simulator permits program-controlled modelling. Normally, however, in reality, processes occur that cannot be modelled by the modules alone. Thus, the template model is augmented with additional strategies or object extensions in a unique form. As previously, a template model must be created by a simulation expert, the reusability, however, is large.

The template model is an empty wrapper that becomes an executable model only when it is filled with data. Thus, an important part of the templates is also an interface for the acceptance of data. The data are provided from external sources and can be repeatedly loaded into the model in order to create an instance with current data.

Because patterns not represented in the template model also cannot occur in the model instance, it should always be possible to extend the template model, even when initially all occurring patterns are included. Template models tailored to the simcron MODELLER have been developed for electronics manufacturing and have a high level of reusability. However, it is still possible that a template may need to be extended for a special application.

3.6.2 Integration

Data Origin

The inclusion in the operational processes is a significant prerequisite for process-accompanying simulation. The template model and the simulator provide a framework, which, however, is not capable by itself of performing simulation runs under changed conditions. The automated data acquisition precedes the instantiation of the model. A differentiation is made between master data and transaction data. Master data contain all information about system elements (resources, product types, *etc.*) that have mid- to long-term validity. In contrast, transaction data are subject to frequent changes (jobs, inventory, due dates, *etc.*). The data management in companies is closely related with the terms *PPC* (*production planning and control*) and *ERP* (*enterprise resource planning*).

> The *production planning and control* covers the organisational planning, control and monitoring of all processes that result during the order processing starting with the tender preparation and ending with the shipping. This planning and control considers quantity, due date and capacity aspects.

As a supplement to the definition from [7], production program planning should also be mentioned. The extension of the PPC system in the form of the ERP system provides an overall planning and control of the complete supply chain of a company. ERP systems consist of several applications used for purchasing, materials management, production planning and control, warehouse management, human relations, quality assurance and financial management. ERP systems, or at least PPC systems, are used as standard software in almost every electronics company. As the primary planning system, they handle the management of the master and transaction data for the production using central databases [5].

Consequently, it is sensible to establish a connection between the PPC/ERP and the simulation. The master and transaction data serve as the data source for the simulation and fill the template model. The simulation tasks associated with the production planning and control are also combined under the term PPC simulation and can be covered using the process-accompanying simulation [15]. The overview in Table 3.9 summarises which data are requested from the primary planning system.

Table 3.9. Data for the process-accompanying simulation

Master data	Transaction data
Resources (machines/machine groups, workplaces, production facility, personnel)	Jobs (quantities, due dates)
Time model (shift system, works calendar)	
Work schedules (work step sequences, resource assignment, alternatives, *etc.*)	Order work schedules
Bill of materials	Order bills of materials
Execution organisation	Inventory (warehouse, production)
Maintenance data, defect data	Order progress

Technological process sequences and their execution-relevant parameters are stored for each product type in work schedules of the master data. Each job can then be assigned its own copy of the work schedule as an order work schedule. This order work schedule can also have deviations from the master data for a special job. The bills of materials, which are organised at the master data level and describe the structure of each product from subassemblies, behave similarly. For quantity planning (PPC), it is determined for each individual job how many subassemblies are taken from the warehouse or covered by other jobs in the production. The resulting order bill-of-materials represents the relations to other jobs.

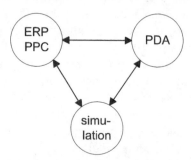

Figure 3.27. Connection of the ERP/PPC, PDA, simulation systems

The PDA systems (*process data acquisition*) that record the order progress are frequently integrated in the PPC or ERP systems or provided as additional solution. From the real production, specific events are reported to the system, either manually, semi- or fully automatically. These are the start and end events of the actions (workstep, maintenance, repair, pause, *etc.*). Similar to the event trace for the simulation, the transformation can be made for time-pairs that permit the determination of the real activity durations. Important for the process-accompanying simulation is the developed state of the real system recorded by the PDA. ERP/PPC, PDA and the simulation are connected with each other in the integrated solution (Figure 3.27).

The data availability must differentiate between the PPC and ERP systems. It must be clarified for the specific case whether all data relevant for the modelling can be obtained from the primary system. If this is not the case, the base data must be extended. This affects primarily the master data, which, for example, can be stored in supplementary databases.

Implementation

The planner initially only has contact with the primary system. The planner's view of the manufacturing is recorded in the ERP or PPC system. In any case, it is considered to be useful when the planner can continue to work with this view, which normally significantly differs from that of the simulation world. The extent to which this is possible depends principally on the quality of the primary system. Depending on the available data, three implementation forms are differentiated:

1. The ERP/PPC system provides all required data. Data interfaces allow the direct exchange with the template model and the simulation system. The primary system controls the simulation run.
2. The ERP/PPC system provides some of the required data. An extension of the primary system (database and user interfaces) can add the missing data. Interfaces to the simulation are created analogously to the first case.
3. The ERP/PPC system provides some of the required data. A second system mirrors the primary data augmented with the missing part. The secondary system has its own data bases and interfaces not only to the primary system, but also to the template model and to the simulation. The secondary system controls the simulation run.

Figure 3.28 illustrates the 3 variants of the implementation. In the first two cases, as before, only the planner has contact with the ERP or PPC system. The simulation is integrated in the higher-level system so that the user perceives the simulated business games as a function extension. In the last case, the planner changes the system and performs his planning task in the secondary system. The simulation is a functional component of the secondary system. In this case, the usual consideration of the planner takes place in the secondary system. This means no fundamental rethinking is necessary. The term *simulation-supported control station* is also used for this approach in the detailed planning [8].

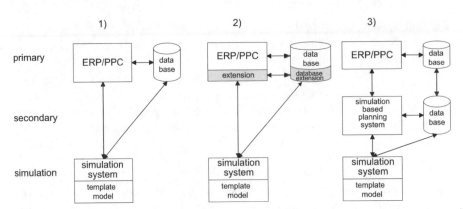

Figure 3.28. Implementation forms

ERP and PPC systems have client–server architectures. The user intervenes with the user interfaces of the distributed clients on the application server that provides the system functions and manages the data bases. Because several users can participate in the planning and the planning-relevant tasks, the secondary systems require a similar structure. In all variants, the simulation should be sufficiently integrated in the planning systems so that it not longer appears as a system but rather is provided as an embedded function. There are two ways of controlling the simulation process:

1. The simulation is controlled by the client and can run on several client computers. This requires that the simulation application is installed the appropriate number of times or provided over the network. The simulation itself then communicates with the server.
2. The simulation is controlled by the server. This requires that the simulation application is installed just once. The client initiates a server function.

Standard interfaces are used for the data exchange. Although ODBC is widely supported for the interaction with data bases, other exchange formats, such as XML, are possible. ERP and PPC systems have their own interfaces, which certainly must be supported for integrated solutions.

The sequence for a simulation experiment generally consists of the steps:

1. Invocation of the simulation function
2. Data transmission from the planning system
3. Create an instance of the template model
4. Run the simulation
5. Transmission of the results data to the planning system.

The same interfaces are used for the reverse transmission of the results data to the planning system. The results of a run can be assigned to a single experiment and initially compared with each other in the planning system before an alternative is selected and used as a realisation template. This assumes that the planning system can manage alternative plans and contains not just a single active realisation schedule.

The simulation results can be evaluated in the planning system. Both the events from the trace transformed into activity pairs and the data from the online reports (see Section 3.3) from the results are passed to the planning system and displayed there appropriately so that the planner can evaluate the planning variants.

A selected planning variant can be set active for controlling the manufacturing process. The assignment planning is made visible at certain points in the manufacturing process. Suitable media are terminals that only permit specific client functions of the planning system and display planned order sequences with start and end dates on the individual workplaces, and that then serve as handling instructions for the machine operator. This makes any planning changes visible online and also helps to reduce the volume of manufacturing papers [16].

3.6.3 Comparison of Reality and Simulation

The quality of the planning results is evaluated using the process-accompanying simulation, it is important here to determine a distance between the simulated and the real manufacturing process. Two procedures are suggested.

Procedure I
Procedure I evaluates the deviations between the simulation and the reality using the interested objective variables. The current valid planning run P represents an objective realization $\mathbf{T}_P$ that was previously determined using simulation. During the course of the real process, a new simulation run C will be performed in intervals, ideally for each arriving acknowledge; this simulation run is based on the previously realised sequence R and the current state $\mathbf{z}(t_C)$ of the manufacturing system. The simulation end is specified as being the simulation end of the previous planning run P. This simulation also produces an objective vector $\mathbf{T}_C$.

Figure 3.29a shows in the static process model (see Section 1.2.3) the process tracks of the planning run P at time t_P and of the comparison simulation C that is started at time t_C and accepts the information of the real sequence R. $\mathbf{T}_C$ is the projection of R and C in a vector of scalar quantities. Figure 3.29b shows the graph of the distance $D_i(t)$.

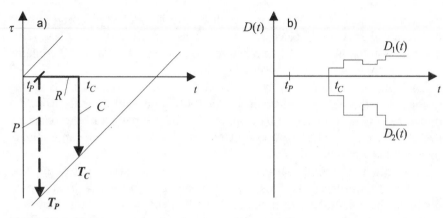

Figure 3.29. Distance – procedure I, **(a)** process tracks of the planning and comparison simulation, **(b)** distance development

The first deviation is measured at time t_C. Further comparison simulations determine additional distances that are applied to the time.

The distance can be easily normalised because each of the vectors with n scalar values can be compared with each other:

$$D_i(t) = \frac{T_{P,i}(t) - T_{C,i}}{T_{P,i}} \quad i = 1, 2, \ldots, n \tag{3.9}$$

This procedure has the advantage that deviations in the sequence that do not affect the targets are not recorded in the distance. Because it is easy to normalise the distance, a relation can be established to other systems. The disadvantage lies in the use of the simulation model itself for evaluating the distance. The model is, however, critical for producing the deviations between the simulation and the reality.

The accuracy of the procedure is estimated as being adequate, but, however, not as being very high. Furthermore, the computing effort required to determine a separation is high.

Procedure II
Procedure II is based on the evaluation of the deviations for simulated and realised activity dates. During the course of the real process, each simulated planned action has an equivalent in the reality. Two deviation types are differentiated:

1. Deviations of the start dates D_s of the individual actions (simulated start date $t_{s,j}$, realised start date $t_{r,j}$ of the activity j)
2. Deviations of the complete duration D_D of the individual actions (simulated duration $d_{s,j}$, realised duration $d_{r,j}$ of the activity j).

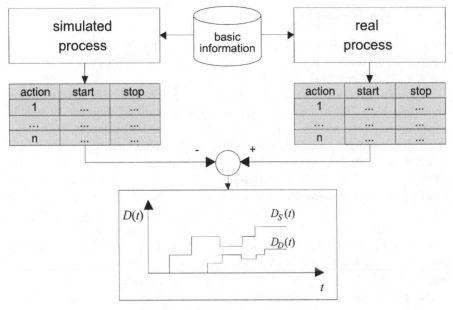

Figure 3.30. Distance – procedure II

Each distance $D(t)$ is calculated from the total of the individual deviations measured at time t as shown in Figure 3.30. Depending on whether the values are added as signed values or as absolute values, the fluctuations may or may not be compensated:

$$D_S(t_j) = D_S(t_{j-1}) + d(t_{s,j}, t_{r,j}) \qquad (3.10)$$
$$D_D(t_j) = D_D(t_{j-1}) + (d_{r,j} - d_{s,j})$$

or

$$D_S(t_j) = D_S(t_{j-1}) + \left| d(t_{s,j}, t_{r,j}) \right|$$
$$D_D(t_j) = D_D(t_{j-1}) + \left| (d_{r,j} - d_{s,j}) \right|$$

The advantage of the procedure lies in its accuracy. Each deviation of the real sequence from the simulated sequence is projected in the distance, irrespective of its effect. Although the required computing effort is low, the normalisation to allow comparison with other processes is not unique. Each sum of all planned activities durations can be used as the reference quantity on which the distance $D(t)$ can be based.

3.6.4 Model Adaptation

Procedures I and II provide means for monitoring the deviation between planning and reality. The model must be changed if the deviations between the simulation and reality become too large.

A *synchronisation* is performed in a first step. The model is adapted using the same base information so that the current state of the real system is represented in the model. This is then used for a new simulation. This is the procedure used for procedure I to determine the distance for the comparison simulation C. Thus, the simulation run C becomes the new plan. The difference of C to itself is thus zero at the point of synchronisation. In addition to the synchronisation, a new planning can also be performed if the planning conditions, namely the base information, have changed.

Synchronisations can be performed in fixed time intervals, manually or automatically depending on the distance. If the model contains modelling errors, the separation in the further course will increase again. Although repeated synchronisations can cleat the separation, it will not be permanently corrected. Consequently, an *adaptation* is performed to make model adjustments. In this case, specific behaviour patterns from the execution information (*PDA*) of the real process are filtered and processed.

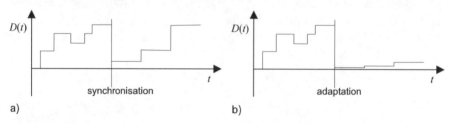

Figure 3.31. Change of the distance over time, (a) with synchronisation, (b) with adaptation

In the simplest case, real activity durations and the associated model parameters are determined. Mathematical-statistical methods, such as a moving-average value, moving median or exponential smoothing, can be used to change the corresponding model parameters. The parameter adaptation produces an improved model, which, in the most favourable case, can remove the modelling error and correct the deviation. The more complicated structure or rule deviations in the model can, at best, be adapted with difficulty.

The difference between the synchronisation and the adaptation becomes apparent when one considers the graph of the distance after a synchronisation and after an adaptation in Figure 3.31.

The execution of a theoretical experiment allows a manufacturing process to be reconstructed exactly as often as required. If this condition applies, various changes of the distance can be made when a synchronisation or an adaptation is performed at a different time during the process. The measured changes can be transformed into the static process model and, as in Figure 3.32, can be entered as a three-dimensional function over the complex time domain.

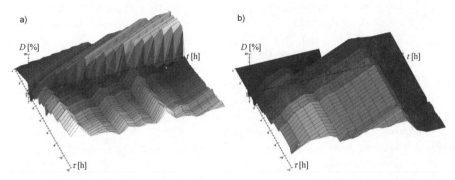

Figure 3.32. 3-dimensional representation in the static process model, (**a**) synchronisation, (**b**) adaptation

The simulation model for the shown investigations represents the flow of a flexible manufacturing system for placement on circuit boards. We do not discuss any further details here. The synchronised model (Figure 3.32b) clearly shows that the deviation D for all t increases with increasing prediction horizon ($\tau < 0$), even when the deviation can be eliminated along the real time axis t ($\tau = 0$). For a fixed $\tau < 0$, more or less constant prediction errors $D(t, \tau = const) = const$ can be determined. The prediction errors generally change proportional to $\tau : D \sim \tau$.

In contrast, the adapted model (Figure 3.32b) shows that the deviation D for increasing t rises less steeply for $t = const$ along $\tau < 0$. Thus, the prediction error decreases during the course of the real process. The graph $D(t, \tau = const)$ can be interpreted as being a learning curve that provides information about the success of the adaptation.

References

[1]   Banks, J.; Gibson R.: *Don't Simulate When…10 Rules for Determining when Simulation is Not Appropriate.* In: IIE Solutions, September, 1997

[2]   DANGELMAIER, W.; WARNECKE H.-J.: *Fertigungslenkung. Planung und Steuerung des Ablaufs der diskreten Fertigung.* Springer-Verlag, Berlin, Heidelberg, 1997

[3]   http://www.emplant.de, Zugriff 3.6.2003

[4]   KIENCKE, U.: Ereignisdiskrete *Systeme, Modellierung und Steuerung verteilter Systeme.* R. Oldenbourg Verlag, Munich, Vienna, 1997

[5]   KURBEL, K.: *Produktionsplanung und –steuerung. Methodische Grundlagen von PPS-Systemen und Erweiterungen.* R. Oldenbourg Verlag, Munich, Vienna, 1995

[6]   LINß, G.: *Trainingsbuch Qualitätsmanagement. Trainingsfragen. Praxisbeispiele. Multimediale Visualisierung.* Fachbuchverlag im Carl Hanser Verlag, Munich, Vienna, 2003

[7]   MEINBERG, U.; TOPOLEWSKI, F.: *Lexikon der Fertigungsleittechnik. Begriffe. Erläuterungen. Beispiele.* Springer-Verlag, Berlin, Heidelberg, 1995

[8]   MERTINS, K.; RABE M. (HRSG.): *The New Simulation in Production and Logistics. Prospects, Views and Attitudes.* IPK Berlin Eigenverlag, Berlin, 2000

[9]   Noche, B.; WENZEL S.: *Marktspiegel Simulationstechnik in Produktion und Logistik.* Verlag TÜV Rheinland, Cologne, 1991

[10]   OUSTERHOUT, J.: *Tcl und Tk. Entwicklung grafischer Benutzerschnittstellen für das X-Window-System.* Addison-Wesley, Bonn, 1997

[11]   SCHMIDT, U.; KUHN, A. (HRSG.): *Angewandte Simulationstechnik für Produktion und Logistik.* Verlag Praxiswissen, Dortmund, 1997

[12]   SIMCRON: *Der simcron MODELLER 3.1. Benutzerhandbuch.* Simcron GmbH, Dresden, 2003.

[13]   VDI-RICHTLINIE 3633: *Simulation von Logistik-, Materialfluss- und Produktionssystemen. Begriffsdefinitionen.* VDI-Verlag, Dusseldorf, 2000

[14]   VDI-RICHTLINIE 3633, BLATT 1: *Simulation von Logistik-, Materialfluss- und Produktionssystemen. Grundlagen.* VDI-Verlag, Dusseldorf, 2000

[15]   VDI-RICHTLINIE 3633, BLATT 5: *Simulation von Logistik-, Materialfluss- und Produktionssystemen. Integration der Simulation in die betrieblichen Abläufe.* VDI-Verlag, Dusseldorf, 2000

[16]   WEIGERT, G; WERNER, S.; KELLNER, M.: Fertigungsplanung *durch prozessbegleitende Simulation.* In: Frontiers in Simulation. Anwendungen der Simulationstechnik in Produktion und Logistik. Conference Proceedings for the 10th ASIM Conference. SCS, Gent, 2002, p. 42–51

[17]   WEIGERT, G.; WERNER, S.; HAMPEL, D.; HEINRICH, H.; SAUER, W.: *Multiobjective Decision Making – Solutions for the Optimization of Manufacturing Processes.* In: 10th International Conference FAIM2000. Flexible Automation & Intelligent Manufacturing. Maryland, USA, Juni 2000, Conference Proceedings p. 487–496

[18]   WIENDAHL, H.-P.: *Fertigungsregelung. Logistische Beherrschung von Fertigungsabläufen auf Basis des Trichtermodells.* Carl Hanser Verlag, Munich, Vienna, 1997

4 Optimisation of Technological Processes

4.1 Calculation of Extreme Values

Many problem definitions concerned with electronics process technology lead to mathematical relationships that can be solved using the maxima or minima of objective functions. We use the example of the "classic" sample-size problem to demonstrate the procedure.

A technological process with intensity I produces demand B for which it requires duration D.

$$D = \frac{B}{I} \tag{4.1}$$

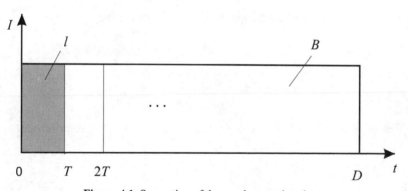

Figure 4.1. Separation of the requirement into lots

The requirement is divided into n lots each of size l. For the manufacturing of a lot, the lot production duration T is required, *i.e.*

$$B = n \cdot l, \qquad D = n \cdot T, \qquad T = \frac{l}{I} \tag{4.2}$$

Three cost elements arise:

1. Lot-size-independent costs C_I, also the setup costs per lot. They are independent of l. (dimension: € per lot).
2. Process costs for a product c_U, also unit costs. (dimension: € per unit).
3. Storage costs c'_S: the completed products are stored until the lot size is reached and so cause costs (dimension: € per unit and hour).

The number of stored products $b(t)$, i.e. the inventory, grows linearly (actually a staircase graph):

$$b(t) = I \cdot t = \frac{l}{T} \cdot t \tag{4.3}$$

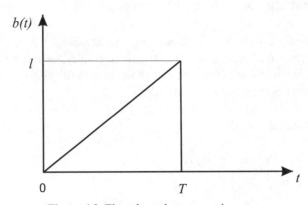

Figure 4.2. Time-dependent storage inventory

The total storage costs that arise during the manufacturing of a lot are

$$C_S = \int_0^T c'_S \cdot b(t)\, dt = c'_S \cdot I \cdot \frac{T^2}{2} = c'_S \cdot \frac{l^2}{2I} \tag{4.4}$$

This produces the complete costs for manufacturing the requirement

$$C_{sum} = n \cdot C_I + n \cdot l \cdot c_U + n \cdot c'_S \cdot \frac{l^2}{2I} \tag{4.5}$$

$$= B\left(C_I \cdot \frac{1}{l} + c_U + \frac{c'_S}{2I} \cdot l \right)$$

These costs have a minimum at l^*.

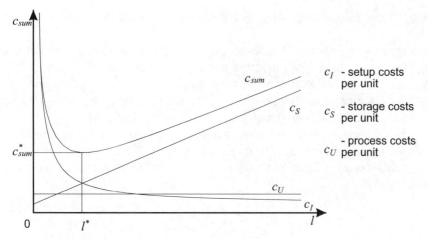

Figure 4.3. Complete costs c_{sum} for 1 unit depending on the lot size

Calculation of l^*:

$$\frac{dC_{sum}}{dl} = B\left(-C_I \cdot \frac{1}{l^2} + \frac{c'_S}{2I}\right)$$

This gives

$$l^* = \sqrt{2I \cdot \frac{C_I}{c'_S}} \tag{4.6}$$

The optimum (economic) lot size l^* thus depends neither on the requirement nor on the unit costs. This is a noteworthy conclusion.

4.2 Linear Programming

In the planning and configuring phases of manufacturing systems, linear optimisation is an important method for the dimensioning of manufacturing quantities taking account of the cost, quality and time aspects.

The mathematical formulation of the problem definition requires two things. Firstly, a linear objective function that should achieve a minimum or a maximum. Conversely, restrictions exist that generally form a system of linear inequalities. The example of the optimisation of manufacturing quantities is used to illustrate the procedure.

4.2.1 Optimisation of Product Volumes – Problem

An operation produces M different product types E_m with different volumes x_m ($m = 1, 2, \ldots M$). The operation involves N technological processes T_n ($n = 1, 2, \ldots, N$) that each product must pass through. The process durations d_{mn} are known. A maximum daily usage duration a_n is available for each process. Finally, each product yields a profit of g_m (per unit).

The matrix D of the process durations, the vector a of the maximum chronological process use durations, the vector x of the product volume and the profit vector g describe the task. We require the vector x^* for which the total profit

$$G^* = g^T \cdot x^*$$ (4.7)

has a maximum.

A simple numerical example is used to demonstrate the problem definition:

Numerical example: manufacturing of electronic components
Number of products: $M = 2$; number of processes: $N = 3$

Matrix of the processing durations: $D = \begin{pmatrix} 4 & 6 & 9 \\ 8 & 3 & 3 \end{pmatrix}$ in seconds (per unit)

Vector of the process use duration: $a = \begin{pmatrix} 32,000 \\ 30,000 \\ 27,000 \end{pmatrix}$ in seconds (per day)

Vector of the production volumes ("production vector"):

$$x = \begin{pmatrix} x_1 \\ x_2 \end{pmatrix} \text{ in unit (per day)}$$

Profit vector:

$$g = \begin{pmatrix} 0.4 \\ 0.2 \end{pmatrix} \text{ in euros (per unit)}$$

Required: x^*, G^*

4.2.2 Optimisation of Product Volumes – Solution Method

First step: formulation of the restrictions
Initially, the restrictions are stated as a system of inequalities. This uses the process balances:

Numerical example:

$$4x_1 + 8x_2 \le 32,000 \quad (1)$$
$$6x_1 + 3x_2 \le 30,000 \quad (2)$$
$$9x_1 + 3x_2 \le 27,000 \quad (3)$$

where the following inequalities apply:

$$x_1 \ge 0$$
$$x_2 \ge 0$$

These 5 inequalities represent an area in the x_1-x_2 diagram (Figure 4.4). Each point of the marked surface provides a possible solution of the production vector.

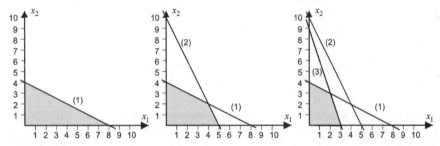

Figure 4.4. Permitted area of the production vector, (left: just straight line 1 – centre: straight lines 1 and 2 – right: all 3 straight lines)

Generalisation of the restrictions:
By generalising the numerical example, it is easily to show:

$$\mathbf{D}^T \cdot \mathbf{x} \le \mathbf{a} \qquad \mathbf{x} \ge \mathbf{0} \qquad\qquad (4.8)$$

This system of inequalities represents a closed area in which the permitted production vector x can lie. Mathematically, this area is a convex linear area, also known as a *simplex*.

Second step: formulation of the objective function
Initially, the numerical example is used to show the principle.

Numerical example:
The profit G results from the relationship
$$G = 0.4x_1 + 0.2x_2 \quad \text{in euros (per day).}$$

It is desirable to present this representation as a family of straight lines, where each straight line specifies a constant profit (see Figure 4.5).

Generalisation:

With regard to the following solution method, the objective function must always be formulated so that the minimum is found. Because the maximum profit is required in the shown case, the objective function is formulated as negative profit:

$$z = -\mathbf{g}^T \cdot \mathbf{x} \tag{4.9}$$

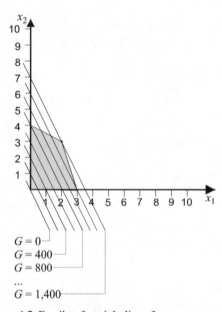

Figure 4.5. Family of straight lines for a constant profit

**Third step: determination of the minimum value
of the objective function**

Again, initially a numerical example is used to illustrate the solution method.

Numerical example:

It is easy to see the solution method from Figure 4.5. The parallel "downward" displacement of the straight lines for constant profit from the coordinate origin increases the profit. The profit has its largest value at the intersection point of the straight lines (2) and (3). Because this point belongs to the permitted range, it is the optimum production point and leads to the maximum profit, *i.e.*

$x_1{}^* = 2,000$ units (per day)
$x_2{}^* = 3,000$ units (per day)
$G^* = 1,400$ euros (per day)

Generalisation:
Initially, the system of inequalities for the restrictions is transformed into a system of equations by introducing an N-dimensional vector y that specifies the nonassigned usage duration for each technological process. Again, an example is used initially to show the method.

Numerical example:

$4x_1 + 8x_2 + y_1 = 32{,}000$
$6x_1 + 3x_2 + y_2 = 30{,}000$
$9x_1 + 3x_2 + y_3 = 27{,}000$

Generalisation:
This equation can be easily generalised:

$$\mathbf{D}^T \cdot \mathbf{x} + \mathbf{y} = \mathbf{a} \quad \text{with} \quad \mathbf{x} > \mathbf{0}; \quad \mathbf{y} > \mathbf{0} \tag{4.10}$$

Numerical example:
When one solves the equations for y_1, y_2 and y_3, this produces the system of equations

$y_1 = -4x_1 - 8x_2 + 32{,}000$
$y_2 = -6x_1 - 3x_2 + 30{,}000$
$y_3 = -9x_1 - 3x_2 + 27{,}000$

Generalisation:

$$\mathbf{y} = \mathbf{A} \cdot \mathbf{x} + \mathbf{a} \quad \text{with} \quad \mathbf{A} = -\mathbf{D}^T \text{ and } \mathbf{x} \geq \mathbf{0},\ \mathbf{y} \geq \mathbf{0} \tag{4.11}$$

This system of equations, together with the objective function (Equation (4.9)), is now brought into a schema, the so-called first simplex tableau (ST1).

ST1	x^T	1
y	A	a
z	$-g^T$	0

Figure 4.6. Simplex tableau for Equations (4.11) and (4.9)

The simplex tableau should also be specified for the associated numerical example.

ST1	x_1	x_2	1
y_1	-4	-8	32,000
y_2	-6	-3	30,000
y_3	-9	-3	27,000
z	-0.4	-0.2	0

Figure 4.7. Simplex tableau 1 for the numerical example

The simplex tableau is thus an appropriate representation of the 4 equations
$$y_1 = -4x_1 - 8x_2 + 32,000$$
$$y_2 = -6x_1 - 3x_2 + 30,000$$
$$y_3 = -9x_1 - 3x_2 + 27,000$$
$$z = -0.4x_1 - 0.2x_2 + 0$$
and is used to calculate the optimum using the exchange method. For the further procedure, it is desirable to number the fields of the simplex tableau (Figure 4.8).

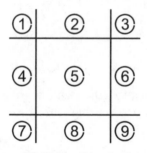

Figure 4.8. Fields of the simplex tableau

The exchange method is now used successively to change the complete simplex tableau until the coefficients in field 8 do not have any negative values. In this case, z is then a minimum value when the vector in field 2 is set to zero. Thus, the optimum solution then can be read from field 4 with 6 or 7 with 9.

4.2.3 Exchange Method and Simplex Algorithm

The system of equations

$$y = A \cdot x + a$$

permits other equivalent representations. When, for example, we solve the equation for y_i for the variable x_k, this produces

$$x_k = \frac{1}{a_{ik}} \cdot y_i - \sum_{\substack{n=1 \\ n \neq k}}^{N} \frac{a_{in}}{a_{ik}} x_n - \frac{a_i}{a_{ik}} \tag{4.12}$$

$$
\begin{aligned}
y_1 &= a_{11}x_1 + a_{12}x_2 + \ldots + a_{1k}x_k + \ldots + A_{1n}x_n + \ldots + a_{1N}x_N + a_1 \\
y_2 &= a_{21}x_1 + a_{22}x_2 + \ldots + a_{2k}x_k + \ldots + A_{2n}x_n + \ldots + a_{2N}x_N + a_2 \\
&\;\;\vdots \\
y_i &= a_{i1}x_1 + a_{i2}x_2 + \ldots + a_{ik}x_k + \ldots + a_{in}x_n + \ldots + a_{iN}x_N + a_i \\
&\;\;\vdots \\
y_m &= a_{m1}x_1 + a_{m2}x_2 + \ldots + a_{mk}x_k + \ldots + a_{mn}x_n + \ldots + a_{mN}x_N + a_m \\
&\;\;\vdots \\
y_M &= a_{M1}x_1 + a_{M2}x_2 + \ldots + a_{Mk}x_k + \ldots + a_{Mn}x_n + \ldots + a_{MN}x_N + a_M
\end{aligned}
$$

	x_1	x_2	$\ldots$	x_k	$\ldots$	x_n	$\ldots$	x_N	1
y_1	a_{11}	a_{12}		a_{1k}		a_{1n}		a_{1N}	a_1
y_2	a_{21}	a_{22}		a_{2k}		a_{2n}		a_{2N}	a_2
$\vdots$									
y_i	a_{i1}	a_{i2}		a_{ik}		a_{in}		a_{iN}	a_i
$\vdots$									
y_m	a_{m1}	a_{m2}		a_{mk}		a_{mn}		a_{mN}	a_m
$\vdots$									
y_M	a_{M1}	a_{M2}		a_{Mk}		a_{Mn}		a_{MN}	a_M
*	b_{i1}	b_{i2}		-		b_{in}		b_{iN}	b_i

	x_1	x_2	$\ldots$	y_i	$\ldots$	x_n	$\ldots$	x_N	1
y_1	b_{11}	b_{12}		b_{1k}		b_{1n}		b_{1N}	b_1
y_2	b_{21}	b_{22}		b_{2k}		b_{2n}		b_{2N}	b_2
$\vdots$									
x_k	b_{i1}	b_{i2}		b_{ik}		b_{in}		b_{iN}	b_i
$\vdots$									
y_m	b_{m1}	b_{m2}		b_{mk}		b_{mn}		b_{mN}	b_m
$\vdots$									
y_M	b_{M1}	b_{M2}		b_{Mk}		b_{Mn}		b_{MN}	b_M

Figure 4.9. Exchange method

When one inserts this new equation instead of the previous equation in the system of equations, where the variable x_k is also replaced in all other equations, this produces a system of equations, which, although it has a different notation, has the same relationships between the variables.

For the special case $M = N$ and $y_m = 0$ for all m, the exchange method can be used to solve a linear system of equations.

The exchange method is shown in Figure 4.9.

1. Specify pivot element	$p = a_{ik}$
2. Calculate the element in the pivot column and row intersection point	$b_{ik} = \dfrac{1}{p}$
3. Calculate the elements of the pivot column (outer intersection element)	$b_{mk} = \dfrac{a_{mk}}{p}$
4. Calculate the elements of the pivot row (outer intersection element)	$b_{in} = \dfrac{a_{in}}{-p}$
5. Calculate all the other elements	$b_{mn} = a_{mn} + a_{mk} \cdot b_{in}$

Figure 4.10. Exchange rules

The coefficient a_{ik} for the exchange of y_i and x_k is called the pivot element. With regard to the described exchange of the equations, five computation rules (exchange rules) can be specified for determining the new coefficients (see Figure 4.10).

An important problem is the specification of the *pivot element*. This is not done arbitrarily, after all, the optimum solution should be achieved with the smallest possible number of exchange steps. The descent method is used to select the pivot element:

The *pivot column* is obtained as follows:

Select the smallest value in the field 8 (*i.e.* the "most-negative" value); the associated column is selected as pivot column k.

The *pivot row* is obtained as follows:

Divide all values a_{mk} of this column (in field 5), form the associated quotients with the coefficients a_n in the field 6 and find the minimum.

$$\left| \frac{a_m}{a_{mk}} \right| \Rightarrow \text{Min} \qquad (4.13)$$

The minimum is contained in row i. This is then selected as pivot row i. The pivot element lies at the intersection point of the pivot row and the pivot column. This descent method is performed each time.

Numerical example

The optimum solution can be obtained from ST 3:

$x_1 = 2{,}000$ units per day for E_1

$x_2 = 3{,}000$ units per day for E_2

$y_1 = 0$ (process T_1 is fully utilised)
$y_2 = 9,000$ (seconds per day, process T_2 not fully utilised)
$y_3 = 0$ (process T_3 is fully utilised)
$G^* = 1,400$ euros per day

ST1	x_1	x_2	1	
y_1	-4	-8	32,000	
y_2	-6	-3	30,000	$p = -9$
y_3	-9	-3	27,000	
z	-0.4	-0.2	0	
*	$-$	$-\dfrac{1}{3}$	3,000	

ST2	y_3	x_2	1	
y_1	$\dfrac{4}{9}$	$-\dfrac{20}{3}$	20,000	
y_2	$\dfrac{2}{3}$	-1	12,000	$p = -\dfrac{20}{3}$
x_1	$-\dfrac{1}{9}$	$-\dfrac{1}{3}$	3,000	
z	$\dfrac{2}{45}$	$-\dfrac{1}{15}$	$-1,200$	
*	$\dfrac{1}{15}$	$-$	3,000	

ST3	y_3	y_1	1
x_2	$\dfrac{1}{15}$	$-\dfrac{3}{20}$	3,000
y_2	$\dfrac{3}{5}$	$\dfrac{3}{20}$	9,000
x_1	$-\dfrac{2}{15}$	$\dfrac{1}{20}$	2,000
z	$\dfrac{1}{25}$	$\dfrac{1}{100}$	$-1,400$

Figure 4.11. Simplex tableaux ST 1 to ST 3

4.3 Dynamic Programming

4.3.1 Introduction

A technological complete process that consists of several steps and for which a linear objective function can be specified can be analysed using dynamic optimisation. The process can assume different states in each step. Transfer equations, generally known from system theory, serve here as restrictions between the steps and their states. We show a simple example, which, however, does not originate from electronics process technology, to illustrate the problem specification.

4.3.2 Example

Problem definition

A missile has height h_0 and speed v_0 at time t_0. It should climb to a height h_1 and reach a speed v_1 there. Because this does not involve an incremental process, the process will be divided into steps. In the example there are 7 steps, *i.e.* 3 height steps and 4 speed steps. The missile can attain different speeds and heights in each step. The possibilities for the transition from step to step are specified in Figure 4.12.

The control determines which state can be reached in the following step, although either only the height or only the speed can be increased. The objective function is the energy consumption. The associated energy for each step transition is known (indicated by the number on the arrows). The "path" from the beginning to the end that has the minimum overall consumption is to be found.

The incremental process corresponds to a phase diagram. The optimisation leads to the determination of the optimum path of the associated required optimum control and the optimum value of the objective function.

Solution method

Start with the last step, determine the energy consumption still required until the end and write this value in the end node – obviously the value zero results for this end node.

Then analyse the penultimate step (step number 6). Two states exist. Write the associated required energy consumption until the end in the corresponding nodes (4 and 3). Three states (nodes) exist in the 5th step. There are 2 possibilities for the middle nodes, *i.e.* there are two paths to the end. Obviously select the most favourable path, mark it with an arrow and write the associated energy consumption in the node (7). Continue this process until the start node is reached. Then mark the optimum path.

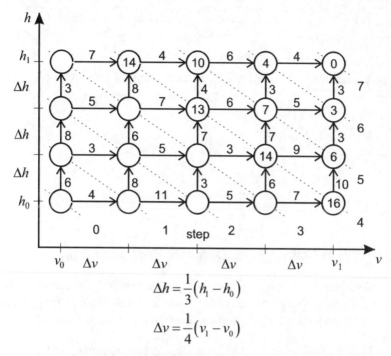

$$\Delta h = \frac{1}{3}\left(h_1 - h_0\right)$$

$$\Delta v = \frac{1}{4}\left(v_1 - v_0\right)$$

Figure 4.12. State diagram with values for ψ and values for S from step 7 to step 4

The example shows that the optimum control from step to step does not necessary follow the path that initially has the lowest energy consumption (*e.g.*, from the start node to step 1). Conversely, the trajectory path to the end also does not necessary lead in each step through the state with the lowest energy consumption.

The dynamic programming achieves an overall optimum because only the permitted transitions from step to step are considered, *etc.* If, however, one added the sum of all individual minima from step to step to produce an overall minimum, this can possibly violate the restrictions and so produce an incorrect solution.

Indeed, the nature of dynamic programming involves determining for each step and in each step for each state the associated optimum path for the objective function until the end. It can also determine the energy consumption. This is specified in Figure 4.13.

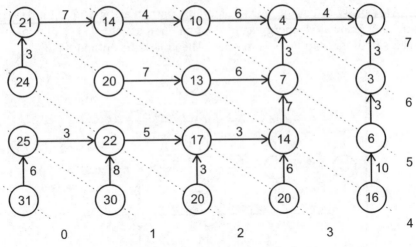

Figure 4.13. Conditional optimum paths

4.3.3 Problem Definition for Dynamic Programming

A complete process consists of steps or can be divided into steps. Each step n is characterised by a vector of state quantities (state vector) $x[n]$. A control process with the vector $u[n]$ permits the transition from step n to the next step $n + 1$.

The law, *i.e.* the "rule" with which the state quantities and the control quantities for step n create the state quantities for the next step $n + 1$, is called the state-transition equation (in vector notation).

$$\mathbf{x}[n+1]=\varphi\left(\mathbf{x}[n],\mathbf{u}[n]\right) \tag{4.14}$$

These equations represent the restrictions of dynamic programming. The objective function is a linear expression

$$z = \sum_{n=0}^{N-1} \psi\left(\mathbf{x}[n],\mathbf{u}[n]\right) \tag{4.15}$$

i.e. it is the sum of the costs required for the transition from one step to the next step, summed over all steps.

In this example

$$\mathbf{x}[n] = \begin{pmatrix} h[n] \\ v[n] \end{pmatrix}$$

The sequence of states $\{x^*[0], x^*[1], x^*[2], \cdots, x^*[7]\}$ for which the target quantity z assumes the optimum is to be found.

4.3.4 Bellman Recursion Equation

A Bellman function $S_n(x[n])$ is introduced for each state $x[n]$ of the step n. This function specifies as optimum value the size of the still-required energy consumption on the path from the associated state until the end of the process. S_n is consequently a *conditional* optimum energy consumption from the associated state $x[n]$ of the step n until the end state $x[N]$ of the step N and will be calculated for *each* state of the step n, independent of the state through which the final optimum process passes.

Derivation of the recursion equation (using the example in Figure 4.12).

First step $N = 7$
We are located at the process end (step N, state $x[N]$). The energy consumption still required is zero, *i.e.*

$$S_N(x[N]) = S_7(x[7]) = 0$$

Second step $N - 1 = 6$
Two states exist (numbering of the states, see Figure 4.14)

$$\mathbf{x}_1[N-1] = \mathbf{x}_1[6] = \begin{pmatrix} h_0 + 3\Delta h \\ v_0 + 3\Delta v \end{pmatrix}$$

and

$$\mathbf{x}_2[N-1] = \mathbf{x}_2[6] = \begin{pmatrix} h_0 + 2\Delta h \\ v_0 + 4\Delta v \end{pmatrix}$$

From these two states, one reaches the step N using the associated control and with the energy specified on the arrows. The graphs (Figure 4.12)

$$\psi\left(\mathbf{x}_1[N-1], \mathbf{u}_1[N-1]\right) = 4$$

and

$$\psi\left(\mathbf{x}_2[N-1], \mathbf{u}_2[N-1]\right) = 3$$

provide the associated energy consumption that is then written for both states in the corresponding nodes, *i.e.* these numeric values are the same as the values for the Bellman function:

$$S_{N-1}\left(\mathbf{x}[N-1]\right) = \psi\left(\mathbf{x}[N-1], \mathbf{u}[N-1]\right)$$

or with numbers:

$$S_6\left(x_1[6]\right) = 4$$

$$S_6\left(x_2[6]\right) = 3$$

Now proceed to the next step.

Third step N – 2 = 5
Three states exist here (numbering of the states, see Figure 4.14)

$$\mathbf{x}_1\left[N-2\right] = \mathbf{x}_1[5] = \begin{pmatrix} h_0 + 3\Delta h \\ v_0 + 2\Delta v \end{pmatrix}$$

$$\mathbf{x}_2\left[N-2\right] = \mathbf{x}_2[5] = \begin{pmatrix} h_0 + 2\Delta h \\ v_0 + 3\Delta v \end{pmatrix}$$

$$\mathbf{x}_3\left[N-2\right] = \mathbf{x}_3[5] = \begin{pmatrix} h_0 + \Delta h \\ v_0 + 4\Delta v \end{pmatrix}$$

To derive the value for the Bellman function, the state $x_2[N - 2]$ is initially selected.

The energy required to go from state $x_2[N - 2]$ to state $x[N]$ depends on the selected path

1st path: $\mathbf{x}_2\left[N-2\right] \rightarrow \mathbf{x}_1\left[N-1\right] \rightarrow \mathbf{x}[N]$

2nd path: $\mathbf{x}_2\left[N-2\right] \rightarrow \mathbf{x}_2\left[N-1\right] \rightarrow \mathbf{x}[N]$

Now calculate the required energy cost for both paths using values for the Bellman function already calculated for step $N - 1$ and by adding the associated values

$$\psi\left(\mathbf{x}_2[N-2], \mathbf{u}_2[N-2]\right)$$

To produce a numeric value for the Bellman function in step $N - 2$ for the state $x_2[N - 2]$ (in the figure, the average node of step 5), form the minimum using two control possibilities and then add each of the associated consumptions until the next step $\psi(x_2[N - 2], u_2[N - 2])$ to the associated value of the Bellman function. In numbers:

$$S_5\left(\mathbf{x}_2[5]\right) = \text{Min}\left\{3+4\,;5+3\right\} = 7$$

Thus, path 1 forms the conditional optimum.

For all states of step $N-2$, this produces the following recursion equation

$$S_{N-2}\left(\mathbf{x}[N-2]\right) = \underset{\mathbf{u}[N-2]}{\text{Min}}\left\{\begin{array}{l} \psi\left(\mathbf{x}[N-2],\mathbf{u}[N-2]\right)+... \\ ...+S_{N-1}\left(\mathbf{x}[N-1]\right) \end{array}\right\} \qquad (4.16)$$

If one now generally sets the step n for all states in Equation (4.16) for the step $N-2$, this produces the Bellman recursion equation

$$\underset{\mathbf{u}[n]}{\text{Min}}\left\{\psi\left(\mathbf{x}[n],\mathbf{u}[n]\right)+S_{n+1}\left(\mathbf{x}[n+1]\right)\right\} = S_n\left(\mathbf{x}[n]\right) \qquad (4.17)$$

Because not all states are represented in this equation, but rather only those states reached from a specific $x[n]$ using the associated control $u[n]$, write the state transition equation (Equation (4.14)) in the recursion equation

$$S_n\left(\mathbf{x}[n]\right) = \underset{\mathbf{u}[n]}{\text{Min}}\left\{\psi\left(\mathbf{x}[n],\mathbf{u}[n]\right)+S_{n+1}\left(\varphi\left(\mathbf{x}[n],\mathbf{u}[n]\right)\right)\right\} \qquad (4.18)$$

For each state $x[n]$, the calculation of the Bellman function also produces *the* control $u^b[n]$, the conditional optimum, *i.e.* the Bellman function, permits,

$$\mathbf{u}^b\left[n\right] = \mathbf{u}^b\left(\mathbf{x}[n]\right) \qquad (4.19)$$

This equation takes into account that only *one* control is selected for each state, namely *that* which specifies the path that produces the value of the Bellman function.

Setting $n=0$ in the Bellman recursion equation produces the optimum overall consumption

$$S_O\left(\mathbf{x}[0]\right) = S_{sum} \qquad (4.20)$$

This also produces

$$\mathbf{u}^b\left[0\right] = \mathbf{u}^b\left(\mathbf{x}[0]\right) = \mathbf{u}^*\left[0\right]$$

Thus, the control for step 1 that produces the optimum path is known for the process start. Because the state at the process begin $x[0]$ is also always an optimum state, *i.e.*

$$\mathbf{x}[0] = \mathbf{x}^*\left[0\right],$$

the optimum state for the first step can also be calculated:

$$x^*[1] = \varphi\left(x^*[0], u^*[0]\right)$$

From the relationship

$$u^b[1] = u^b\left(x^*[1]\right)$$

by setting the optimum state $x^*[1]$ produces

$$u^b\left(x^*[1]\right) = u^*[1]$$

This produces the general equation

$$x^*[n+1] = \varphi\left(x^*[n], u^*[n]\right) \tag{4.21}$$
$$= \varphi\left(x^*[n], u^b\left(x^*[n]\right)\right)$$

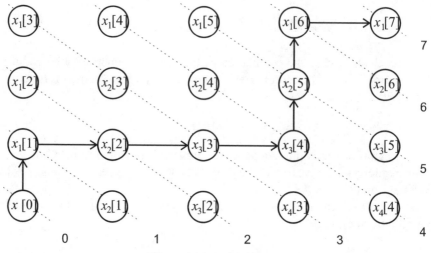

Figure 4.14. Optimum path

4.3.5 Dynamic Programming Strategy

For an incremental process with a start point and an end point, states and controls are defined for each step and rules for the transitions determined from step to step. Each possible transition is characterised with a numeric value that corresponds to a cost (or a benefit). The objective function for the optimisation is the total cost. The optimisation consists of 2 parts.

The first optimisation begins at the end point. Use Equation (4.14) to calculate backwards for each state in all steps the minimum value for the cost until the end point. In addition, the control that specifies the path is "marked" for each state. The first part is called the reverse optimisation.

The direct optimisation is then performed. The initial state and the conditional optimum control determined in the first part are used to determine the optimum state of the next step, *etc*. The second part is called the forwards optimisation. Namely, if an incremental process is to be optimised, the first part is conditionally optimised *against* the sequence and then optimised in the flow direction. This requires, obviously, that a rule and an evaluation of all step transitions exists.

4.4 Simulation-based Optimisation

4.4.1 Principles

For manufacturing optimisation, one must always differentiate between target quantities and the influencing variable. The *influencing variables* act as input quantities for the manufacturing system, which responds at the output with the corresponding target quantities. Each element s from the set S, the so-called control space, corresponds to an assignment of the factors that influence the requirements with concrete values. The manufacturing system itself can be described with the transformation $f: S \rightarrow Z$ with Z as target space. The control space can be restricted using secondary conditions, namely, not every value assignment for the influencing variables is permitted.

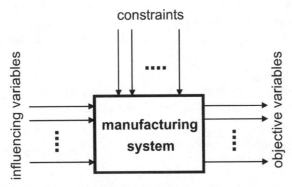

Figure 4.15. Control of the manufacturing system using factors that influence the requirements

Because the control space is represented in the target space, the elements of the control space can be compared with each other, provided at least one order relation is declared on the target space. A control $s^{(i)}$ is optimum when $z^{(i)} \geq z^{(j)}$ applies to all $s^{(j)} \in S$. Because the relation operator can be reversed, it does not matter for the optimisation algorithms whether a maximum or a minimum is required. In the following discussion, the maximum is always assumed as optimum.

Both the control space and the target space can be envisaged as being a space with several dimensions, when, for example, a dimension of the control space is applied to each influencing factor. In a multidimensional target space, an order relation must be provided, such as by defining a distance. The structure of the control space depends primarily on the type of the associated factors that influence the requirements. In general, three types of factors that influence the requirements can be differentiated:

1. Manipulated variables
2. Selection (1 of n or m from n)
3. Permutation (from n elements).

The *manipulated variables* can be used, for example, to control the capacity of buffer stores or the number of the provided machines. Because manipulated variables can be represented as integers or real numbers, one can define a distance in a control space that consists just of manipulated variables. The selection can be used, for example, to find those priority rules from the n possible priority rules that provide the optimum target value. In most cases, the selection can be formally reduced to integers, for example, by simply numbering the individual priority rules. The distance and the order relation can be used only to a limited extent here, although they formally could be used. Permutations are always appropriate when jobs or any other objects of the manufacturing process are to be exchanged with each other. As with selection variables, they cannot be ordered. Although the definition of a distance is formally possible, it remains unusable until a relationship to the separations can be established in the target space. This is further complicated by the fact that the number of the possible permutations, namely the number of possible values that a permutation variable can assume, grows exponentially with n. Stirling's formula provides a good approximation value for the number of permutations:

$$n! \approx \left(\frac{n}{e}\right)^n \sqrt{2\pi n} \left(1 + \frac{1}{12n} + \frac{1}{288n^2} + \ldots\right) \tag{4.22}$$

In general, it can be shown that control spaces can be well structured provided no selection or permutation variables need to be considered. The permutation variables normally greatly complicate the optimisation problem, not least because of the rapidly increasing number of variable values.

Structure of the solution

Simple optimisation problems exist in which the solution involves answering a question with Yes or No. This type of problem is also called a *decision problem*. In contrast, a *sequence problem* and a *travelling-salesman problem* require that the solution can be specified as a permutation or as an itinerary, respectively. Because a simple Yes-No answer no longer suffices here, this type of problem no longer belongs to decision problems.

In many cases, however, the complicating optimisation problems can be transformed to the simpler decision problems. This allows the travelling-salesman problem to be also formulated as follows: does a journey exist that lies below a specified target value $z^{(min)} \in Z$? The answer is once again either "Yes" or "No".

The size of a problem

Each optimisation problem has a size N generally determined by the number of variable parameters. For the sequence problem, it is the length of the sequence in which exchanges are to take place. For the travelling-salesman problems, it is the number of towns or stations that must be visited. The number of variables may not be increased unnecessarily, but must remain limited to the essential number to avoid distorting the proportions.

Cost

The cost involved in solving the problem can be measured both in time and in memory capacity. Because lack of memory can also be replaced by time and, conversely, time can also be replaced with memory, the following discussion considers only the time cost. To measure the time expenditure independent of the associated computing system, the number c of computing steps needed to reliably solve the problem is measured. The cost estimate must not consider any special cases or any solutions found by chance. A special case for the sequence problem is the 2-machine problem with n jobs that can be solved under specific conditions[1] in n steps using the Johnson algorithm. However, because the algorithm fails for three or more machines, it cannot be used for the cost estimate for solving the general sequence problem.

---

[1] All jobs have the same technological sequence. The jobs may not overtake each other within the system and the two machines are fully uncoupled from each other using a sufficiently large buffer storage.

Complexity
The cost required to provide the reliable solution of a decision problem can be generally differentiated into two problem classes. The class P (P polynomial time) includes those problems for which the number c of computational steps on a deterministic computer is not larger than an arbitrary polynomial in N. Although the time required to solve such a problem in specific cases can be very long, this type of problem is called efficiently solvable because the required time increases comparatively slowly with the problem size.

$$c \leq N^k$$

Problems that can be solved on a nondeterministic computer with polynomial-dependent effort belong to the class NP (N nondeterministic, P polynomial time). A problem that belongs to P always also belongs to NP:

$$P \subset NP \qquad (4.23)$$

Stephen Cook discovered in 1970 a general transformation that maps every decision problem from NP to a single problem, the so-called satisfiability problem[2]. This transformation has the property that the satisfiability problem can only be answered with Yes when the associated problem from NP was also answered with Yes. Furthermore, the transformation is itself efficient, namely can be performed in a polynomial time. For the solution of the satisfiability problem itself, no efficient decision algorithm has yet been found. Rather, for all currently found algorithms, the cost for the decision increases exponentially with the size of the problem:

$$c \leq a^N$$

This is also the case for the sequence problem, for the partitioning problem, for the travelling-salesman problem and for very many other problems. This class of problem is known as NP-complete and so designates a class of decision problems that are particularly difficult to solve. As [1] shows, there are currently already several thousand known NP-complete problems and their number continues to increase. Their complexity appears to agree somehow with that of the satisfiability problem. Even the efficient solution of just one of the NP-complete problems would have the consequence that one could use Cook's general transformation to also efficiently solve at once all other NP problems. This means mathematically, that instead of Equation (4.23), the following relation would apply:

$$P = NP \qquad (4.24)$$

---

[2] If a given Boolean expression with n variables for a specific assignment of the variables with 0 or 1 gives the value 1, the expression is satisfied.

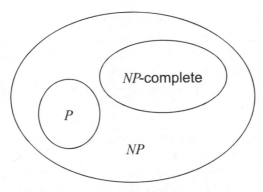

Figure 4.16. *NP*-completeness

Figure 4.16 shows as a Venn diagram the relations between the *P*, *NP* and *NP*-complete classes. If one considers not only decision problems, but also general optimisation problems, this produces the class of *NP*-hard problems that also cannot be solved efficiently. For obvious reasons, the equality of *P* and *NP* or the discovery of an efficient decision algorithm for one of the *NP*-complete problems would be a near sensation. Because current experience indicates that this is not to be expected, the solutions for *NP*-complete problems can only be approximated [2].

4.4.2 Optimisation Cycle

The actual optimisation task can be divided roughly into two steps:

1. *Find* a (permitted) control
2. *Evaluate* the control in the target space.

With regard to the optimisation, the control space in the following discussion is also called the *search space*. Figure 4.17 shows the general optimisation cycle, which contains the two steps: find and evaluate. Each loop passage returns a suggested solution that consists of the complete control $s^{(i)}$ and the associated evaluation $z^{(i)}$. Except for the classic optimisation method, several cycles are normally required to find the optimum, where each cycle generates a complete suggested solution[3]. The action ends when the termination criterion is reached. The best-available value then must not necessarily agree with the theoretical optimum value, in particular, when the cycle had to be terminated for time reasons. Irrespective of this, the process associated with the optimisation cycle in the following discussion will be designated as optimisation and the result as the optimum.

---

[3]   The optimisation cycle does not describe any iteration procedure.

Classic optimisation methods (extreme-value calculation, linear or dynamic optimisation) require only a single cycle and also normally determine the theoretical optimum. The iteration steps, such as for linear optimisation, do not have any effect on the cycle motion. The find and evaluate steps can then fuse to form a single step and then no longer need to be considered separately. All classic procedures, however, place high demands on the search space, which, in particular, cannot be satisfied when selection and permutation variables also need to be considered.

A simulation-supported optimisation is involved when the evaluation step shown in Figure 4.17 is performed using a simulation model. The primary advantage of the simulation method compared with the analytical method is that it can represent reality much more accurately. This must be bought at the cost of an often higher computing time, which, however, can only be partially compensated through the use of a fast simulator.

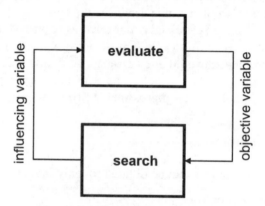

Figure 4.17. General optimisation cycle

The simulation system evaluates neutrally a solution suggestion without any reference to the previous search step. Conversely, the search algorithm should not require any additional information about the system to be optimised and relies only on the evaluation performed by the simulator. The only information available to the search algorithm are data about the previous solution suggestions.

As Figure 4.18 shows, the simulation and the optimisation system are so shielded from each other that no information, other than the goal and factors that influence the requirements, can pass over this threshold. Perhaps, we can best compare this strategy with a blind flight in which the pilot (the search algorithm), because of adverse weather conditions, must place his complete trust on his instruments (simulation).

Under certain conditions, a single optimisation cycle suffices for the simulation-supported optimisation. This allows, for example, different priority rules to be modelled in a simulation model. Let us assume that comprehensive experiments in the past have shown that in a specific buffer storage, the use of the "shortest operation time" (SOT) priority rule almost always lead to success; then this rule would be "hard wired" in the model. During the running production, this model can be used at any time to calculate an execution sequence that under the pertaining conditions with some reliability is the optimum solution or at least nearly the optimum solution. Although this is not optimisation in the true sense of the word, this procedure in practice is certainly viewed as such. Not infrequently, this misunderstanding leads to an incorrect use of the terms simulation and optimisation as synonyms.

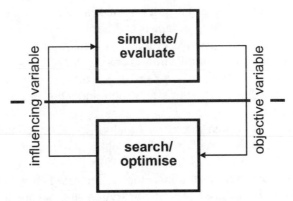

Figure 4.18. Simulation and optimisation systems are shielded from each other

A simulation-supported optimisation in the true sense involves the optimisation cycle being performed several times until the termination condition is satisfied. If the example of the priority rules is used again, this means for a given order supply, different priority rules are set for the critical buffer storage and then the complete execution sequence calculated using simulation. The SOT rule or some other priority rule can eventuate as optimum here [6].

Although there is a large number of priority rules, their number is manageable. If we assume that the simulator is sufficiently fast, we could certainly systematically test all known rules and so obtain the true optimum execution sequence with the given factors. However, for the case that the influencing factors contain a sequence with n elements, the situation changes fundamentally. The number of possible permutations is then $n!$. If we wanted to consider all values, the cost would indeed increase exponentially with the size of the problem, as Equation (4.22) shows.

To get an impression of the growth speed, the reader should successively substitute for n values such as 5, 10, 15, 20, *etc.* The result will strengthen the conviction that a solution in the form of the complete enumeration is practicable only for very small problems. Unfortunately, because the sequence problem belongs to the *NP-hard* problems, no other significantly faster optimisation algorithms are available. This means that an approximation solution must be used for the optimisation problem.

4.4.3 Search Algorithms

Heuristic and, in particular, stochastic procedures, have gained particular importance for the search algorithms used for the simulation-supported optimisation. As the designation *heuristic* implies, all these procedures have only a limited exact mathematical base. In particular for stochastic procedures, one relies on very quickly finding the optimum by chance, irrespective of whether the problem is *NP*-hard. Obviously the probability that the optimum of an *NP-hard* problem is actually already found after a very small number of optimisation steps is infinitesimal. Indeed, the stochastic searching has no advantages under these conditions, unless the problem does not have just one single optimum, but rather a very large number of optima.

If the number of optima increases equally fast as the complexity of the problem, the probability with which an optimum is found can certainly assume an acceptable order of magnitude. This is especially the case, as in practice, not the optimum itself, but rather an approximation value that is as good as possible is required. As an illustration, this means a target space with a single peak is not suitable for stochastic search algorithms. If, however, the target space more resembles a churned-up water surface that has then been frozen, stochastic algorithms can be used to advantage.

A simple consideration shows that most operational planning problems must be of the "water-surface" type. For problems of this type, although the control space increases exponentially, the target space, in contrast, grows slower by orders of magnitude. This has the consequence that cardinal number equivalence classes form in the control space. We can use as an example a 2-machine system for which an input sequence is to be determined so that the cycle time for n jobs assumes a minimum. We assume that the processing times are integer values[4] and the smallest and largest values for the processing time are 0 and 10, respectively. It is easy to show[5] that there is a lower barrier for cycle time 0 and an upper barrier for $n \cdot 10$. Longer cycle times assume delayed execution sequences and are certainly not optimum. The number of sequences in the control space is then

---

[4]   The computer simulation always uses discrete values, even when they are not integers.

[5]   All processing times are 0 or all processing times are 10 and executed sequentially.

$n!$; the number of sequences for the equivalence classes is not larger than $n \cdot 10$. Although we do not know the cardinality of each equivalence class, we can determine the average cardinality:

$$\bar{m} = \frac{n!}{n \cdot 10} = \frac{(n-1)!}{10}$$

In other words, if one actually determined the cycle time for all $n!$ possible sequences, this must produce the same value for a very large number of sequences. If one further assumes that all equivalence classes have the same cardinality, this would produce a probability that an arbitrarily selected sequence lies in a specific equivalence class:

$$P = \frac{1}{n \cdot 10}$$

It can be assumed that the optimum itself, or at least values very near the optimum, can be produced by very many different sequences. Consequently, under these circumstances, the optimisation with stochastic search algorithms can produce the required result relatively quickly.

Although the *blind search* is one of the simplest stochastic search methods, notable successes have been achieved with its use, such as for the optimisation of execution sequences. The designation "blind" indicates that no knowledge of the search space is necessary. The starting point is an initial solution $(s^{(0)}, z^{(0)})$. The algorithm then can be described as follows:

1. $z^{(max)} := z^{(0)}$ and $s^{(max)} := s^{(0)}$
2. $i := i+1$
3. *Search*: select an arbitrary point $s^{(i)}$ from S.
4. *Evaluate*: determine the associated target value $z^{(i)}$.
5. If $z^{(i)} > z^{(max)}$, then $z^{(max)} := z^{(i)}$ and $s^{(max)} := s^{(i)}$.
6. Go to 2) or exit.

The successively found solutions $(s^{(i)}, z^{(i)})$ do not have any relationship with each other. The same solutions can also be produced several times. Statements about the structure of the control space are not necessary. If a distance can be defined in the control space, the *blind search* method can be modified. If we assume a neighbourhood $S^{(i)} \subset S$ that contains all elements whose separation compared with $s^{(i)}$ is not larger, then ε can be specified for each point $s^{(i)}$ in the control space S. This allows the modified algorithm to be formulated as follows:

1. $z^{(max)} := z^{(0)}$ and $s^{(max)} := s^{(0)}$
2. $i := i+1$
3. *Search*: select an arbitrary point $s^{(i)}$ from $S^{(max)}$ in the neighbourhood of $s^{(max)}$.

4. *Evaluate*: determine the associated target value $z^{(i)}$.
5. If $z^{(i)} > z^{(max)}$, then $z^{(max)} := z^{(i)}$ and $s^{(max)} := s^{(i)}$.
6. Go to 2) or exit.

The difference between the two algorithms is that whereas in the first case the complete search space is always available, in the second case, only part of the search space, namely the neighbourhood of the current control point, is considered in each cycle. The first is called a *global search* and the second is called a *local search*. The definition of a suitable neighbourhood or a vicinity in the search space is, however, no simple task, in particular, when permutation variables belong to the influencing factors.

In the past years, a number of heuristic optimisation algorithms have been developed and tested. They can generally always be reduced to the local search axiom. Most of these algorithms are surprisingly simple and consequently easy to program. The more intelligent the algorithm, the more information that is exchanged between the individual optimisation cycles.

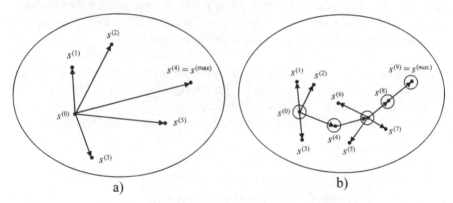

Figure 4.19. Example of (**a**) global and (**b**) local search

Care should be taken before making any statements about the capability of the individual algorithms. The achieved results are principally determined by the structure of the specified problem and also depend on the experience of the user. Some of the most important procedures are only briefly listed here. If necessary, the reader should consult the many publications for more detailed information about this topic [4]. :

- Genetic algorithms
- Evolution strategies
- Tabu search
- Simulated anealing
- Great Deluge algorithm
- Threshold acceptance.

4.5 Optimisation with Several Goals

Many optimisation problems in practice do not have just a single goal, but rather involve a goal complex that consists of several individual goals. For example, a requirement can be to minimise the average lead time for all manufacturing jobs of a specific lot while ensuring the maximum machine loading. This case has two target quantities:

- z_1: average machine loading
- z_2: average lead time.

The extent to which the two goals conflict with each other cannot be determined without exact knowledge of the manufacturing system. The two target quantities can be combined to form a target vector:

$$\mathbf{z} = (z_1, z_2)^T$$

Because no order relation in the classic sense is declared in the vector space, the individual solutions cannot be compared with each other and the use of most heuristic optimisation methods would be questionable. A solution involves using some suitable method to map the vector to a numeric value Ψ, which, following the genetic algorithms, is frequently also designated as *fitness*. The scalar product from Equation (4.25), also called the *method of weighted sums*, can be used as a mapping function.

$$\Psi = \sum_{k=1}^{K} \omega_k \cdot z_k = \boldsymbol{\omega}^T \cdot \mathbf{z} = |\boldsymbol{\omega}| \cdot |\mathbf{z}| \cdot \cos \alpha \qquad (4.25)$$

The components of the weighting vector $\boldsymbol{\omega}$ can be used as factors with which the importance of the associated goals will be subjectively judged. If, for example, the importance of the machine loading is estimated as 80% and that of the lead time as 20%, this would give $\boldsymbol{\omega} = (0.8; 0.2)^T$. The intention to emphasise the target value for the average machine loading is apparent. If the lead time is to be minimised, it suffices to give the associated weighting factor a negative sign.

If one substitutes a constant value for Ψ in Equation (4.25), the equation describes a level in the K-dimensional vector space. For the case $K = 2$, simple straight lines in the 2-dimensional space will be generated from the levels, as Figure 4.20 shows. The gradient of the straight lines or the position of the K-dimensional levels in the target space is determined only by the weight vector. This means that the weight vector has a decisive effect on the optimum solution.

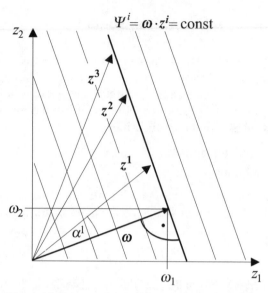

Figure 4.20. Lines with the same fitness value depending on the weighting vector

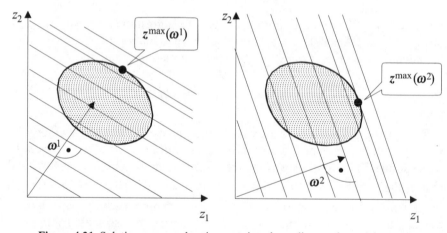

Figure 4.21. Solution range and optimum points depending on the weight vector

Figure 4.21 shows an example in which the largest possible value should be found for both z_1 and z_2. Changing the weight vectors from $\boldsymbol{\omega}^1$ to $\boldsymbol{\omega}^2$ moves the optimum point from $\mathbf{z}^{\max}(\boldsymbol{\omega}^1)$ to $\mathbf{z}^{\max}(\boldsymbol{\omega}^2)$. In contrast to the single goals, the optimisation for several goals always leads to a compromise solution. Neither the goal vector $\mathbf{z}^{\max}(\boldsymbol{\omega}^1)$ nor the goal vector $\mathbf{z}^{\max}(\boldsymbol{\omega}^2)$ contains the reachable maximum for the component z_1 or z_2. In contrast, the evaluation for z_1 with 0 and z_2 with 1 leads again to an optimisation task with a single goal whose optimum solution is provided by z_2^{max} .

This leads to the derivation of the set of all solutions that cannot be improved, also called the *compromise set* or *Pareto set* (Figure 4.22) [7].

The critical disadvantage of the method of the weighted sum lies in the fact that the optimum solution is influenced not only by the weights but also by the absolute values of the individual target components [8].

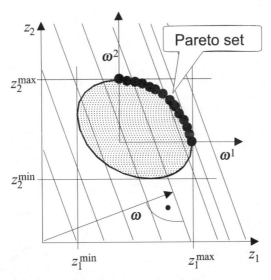

Figure 4.22. Pareto set of a solution range

Let us assume in the initially mentioned example that the lead time is measured in minutes and lies in the general order of 24 h, then we get values for z_2 around 1000. The machine loading is specified as a percentage and cannot be larger than 100. After evaluating with the weight vector, we get for z_1 values that cannot be larger than 80, whereas the values for z_2 lie in the general order of 200. This strongly forces the optimum solution in the direction of the minimum lead time. This counters the original intention to give the machine loading a higher weight. This also follows immediately from Equation (4.25), because the scalar product is commutative and thus the goal and weight vector can be exchanged. We can solve the problem by normalising the individual goal quantities z_k to r_k so that the largest and the smallest value of the normalised goal quantity has the value 1 and 0, respectively (Equation (4.26)).

$$r_k = \frac{z_k - z_k^{min}}{z_k^{max} - z_k^{min}} = \frac{z_k - z_k^{min}}{\Delta z_k} = \frac{z_k}{\Delta z_k} - \frac{z_k^{min}}{\Delta z_k} \tag{4.26}$$

The normalised goal vector yields:

$$\mathbf{r} = \mathbf{A} \cdot \mathbf{z} - \mathbf{a} \tag{4.27}$$

The following relation applies to the matrix A and the column vector $\mathbf{a}$:

$$\mathbf{A} = \begin{pmatrix} \left(\Delta z_1\right)^{-1} & 0 & \cdots & 0 \\ 0 & \left(\Delta z_2\right)^{-1} & \cdots & 0 \\ \vdots & \vdots & \ddots & \vdots \\ 0 & 0 & \cdots & \left(\Delta z_K\right)^{-1} \end{pmatrix} \text{ and} \tag{4.28}$$

$$\mathbf{a} = \begin{pmatrix} z_1^{\min} \cdot \left(\Delta z_1\right)^{-1} \\ z_2^{\min} \cdot \left(\Delta z_2\right)^{-1} \\ \vdots \\ z_k^{\min} \cdot \left(\Delta z_k\right)^{-1} \end{pmatrix} = \text{const}$$

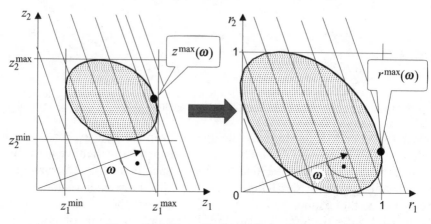

Figure 4.23. Normalisation of the goal vectors

The optimum normalised goal vector $\mathbf{r}^{\max}(\boldsymbol{\omega})$ is in general not identical to the original optimum, namely without normalisation, determined goal vector $\mathbf{z}'^{\max}(\boldsymbol{\omega})$. The new fitness value can be calculated using Equation (4.29) as follows:

$$\begin{aligned} \Psi'(\mathbf{z}, \boldsymbol{\omega}) &= \boldsymbol{\omega}^T \cdot (\mathbf{A} \cdot \mathbf{z} - \mathbf{a}) \\ &= (\boldsymbol{\omega}^T \cdot \mathbf{A}) \cdot \mathbf{z} - (\boldsymbol{\omega}^T \cdot \mathbf{a}) \end{aligned} \tag{4.29}$$

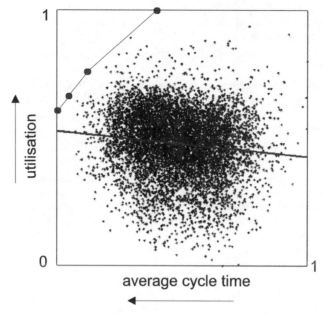

Figure 4.24. Solution set with regression line and Pareto set

Because the scalar value $(\omega^T \cdot a)$ only moves the solution set in the target space, it does not have any affect on the relative position of the optimum within the target space. Thus, the transformed weight vector alone determines the new optimum. The normalisation is thus equivalent to the linear transformation of the weight vector.

Figure 4.24 shows a solution set for the two goal quantities, average lead time and loading, for example, as the result of using a stochastic search algorithm after several hundred cycles. The Pareto set is highlighted at the left-hand upper diagram edge. The possible optimum points are marked. Which of these solutions is actually classified as being the optimum depends from the weights. Such weights can be assigned subjectively [3, 5, 9].

References

[1]   DEWDNEY, A.K.: *Der Turing Omnibus – eine Reise durch die Informatik mit 66 Stationen.* Springer-Verlag, Berlin, Heidelberg 1995

[2]   DOMSCHKE, W.; SCHOLL, A.; VOSS, S.: *Produktionsplanung – Ablauforganisatorische Aspekte.* Springer-Verlag, Berlin et. al. 1997

[3]   DRESZER, J.: *Mathematik Handbuch – für Technik und Naturwissenschaft.* Fachbuchverlag Leipzig, Leipzig 1975

[4]   DUECK, G.; SCHEUER, T.; WALLMEIER, H.-M.: *Toleranzschwelle und Sintflut: neue Ideen zur Optimierung.* Spektrum der Wissenschaften, Digest: Wissenschaftliches Rechnen, Heidelberg 1999, p. 22–31

[5]   ESTER, J.: *Systemanalyse und mehrkriterielle Entscheidung.* Verlag Technik, Berlin 1987

[6]   HAMPEL, D.: *Simulationsgestützte Optimierung von Fertigungsabläufen in der Elektronikproduktion.* Verlag Dr. Markus A. Detert, Templin 2002

[7]   PESCHEL, M.; RIEDEL, C.: *Polyoptimierung – eine Entscheidungshilfe für ingenieurtechnische Kompromisslösungen.* Verlag Technik, Berlin 1976

[8]   WEBER, M.; KRAHNEN; WEBER, A.: *Scoring-Verfahren, häufige Anwendungsfehler und ihre Vermeidung.* Der Betrieb, 48(1996) H. 33, p. 1620–1627

[9]   WEIGERT, G.; WERNER, S.; HAMPEL, D.; HEINRICH, H.; SAUER, W.: *Multi Objective Decision Making - Solutions for the Optimization of Manufacturing Processes.* In: 10th International Conference FAIM'2000, Flexible Automation & Intelligent Manufacturing, Maryland, USA, June 2000, Proceedings p. 487–496

5 Quality Assurance

5.1 General Goals and Terms

The quality of a product is the "nature" with regard to its "suitability to satisfy specified and assumed requirements (quality requirement)" (DIN 55350, quality assurance and statistics terms, part 11, basic terms for quality assurance [2]). Other important terms that affect the quality are the material unit (parts, product) or immaterial units (service, software), the defect (the nonfulfilment of a requirement) and the quality characteristic (the characteristic that determines the quality).

A quality element contributes to the quality. These elements are combined to form a quality circle. The most important *quality elements* include:

- Market research
- Concept
- Design
- Testing
- Manufacturing planning
- Characteristic

- Manufacturing
- Final inspection
- Storage
- Shipping
- Maintenance
- Disposal

The *quality assurance* covers all tasks of quality management, quality planning, quality control and quality tests. Quality control is understood to be the preventative, monitoring and corrective tasks that are realised for the manufacturing of products with the goal of satisfying the quality requirements. Quality tests serve the goal of determining the extent to which a product satisfies the quality requirements.

Quality assurance is performed together with the required resources in the form of a specified operational and organisational structure. A differentiation is made between company-specific and contract-specific quality-assurance systems.

Ever more companies allow their quality-assurance system to be evaluated by a certification authority and so confirmed with a *certificate*. This normally involves the production of a quality manual. Approximately 5000 companies have such a certificate in Germany. In general, the certification is made in accordance of the international standard DIN ISO 9000-9004.2000 (in general, also DIN ISO 9000.2000 ff. [3]). Companies with such a certificate have improved sales chances compared with other companies.

The quality data are used to make conclusions for the control of the technological process. If the quality requirements are not met, it must be decided where interventions need to be made in the technological process and which measures need to be performed. This realises a quality-control loop that forms the basis for a stable robust manufacturing process.

Samples are frequently used to obtain statements about the quality. The values of such quality characteristic values must often be interpreted as probabilities. Statistical process control (SPC) is a very powerful element for process control.

The classic quality control tests the product in order to make a statement about its quality. In contrast, *process control* is used for monitoring the individual manufacturing processes in order to ensure the required product quality. In mechanical engineering, many process parameters are also monitored using process metrology (*e.g.*, pressure, oil temperature, cutting tool wear). In electronics manufacturing, this process control normally cannot be performed by measuring process quantities, but rather by evaluating the product quality of one sample taken from a lot. This means in order to make a statement about the process situation, individual products are investigated and their quality documented. Thus, this sample test is not only a resource for quality control but, in particular, process control.

To attain appropriate improvements for an investigated process, it is necessary to make maximum use of the information that a taken sample can provide. Such information (measured/tested process or product characteristics) is in practice more or less characterised by chance. This means there is the possibility that the conclusions drawn from the taken and evaluated sample may be incorrect. In the extreme case, a well-running process can be evaluated as being a bad process. The reverse situation is also possible. There is the danger that measures are adopted that can cause a further deterioration of the process. This will certainly be noticed sometime and so the question of the usefulness of such methods then immediately arises. The task is now to minimise such incorrect decisions, because a total avoidance of such decisions is impossible. To achieve this goal, the type and way chances can affect electronics manufacturing must be investigated in order to avoid at least incorrect decisions that result from incomplete knowledge.

5.2 Description of Quality Characteristic Variables

5.2.1 Quality Characteristic Variables as Random Value

Quality assurance includes many characteristic variables subject to fluctuations. Such variables are said to *disperse*. Such characteristic variables can be, for example, the value of an ohmic resistor, the diameter of a spindle, the number of reject parts in a lot. Each actual measured or counted value of such a characteristic variable is the result of the effects of systematic and random factors and will be designated as a *random event* in the following discussion. Terms from probability theory are used to describe such quality characteristic variables. Correspondingly, we will first consider random events and their handling.

A random event is an event that has occurred but under the given conditions did not need to have occurred or an event that has not occurred but under the given conditions could have occurred. Random events are written as lowercase letters (normally with indices). The counterpart to the random event is the mandatory event. An event was necessary when, under the given conditions, it must have occurred or will occur.

Another term is the *random variable*. It is always written with *uppercase* letters. The random variable is the generalisation of all associated actual and possible random events. The random events are the actual and possible realisations of the associated random variable. Thus, the number of faulty products determined while taking a sample from a lot is a random variable. The actual number is then a realisation.

Random variables can be combined. The addition, subtraction, multiplication, *etc.*, of random variable produce new random variables. Similarly, random events can also be combined. A mathematical theorem says that events can be combined as sets. This means that the mathematical apparatus from the set theory can also be used on events.

Union of events
The union of the two events x_i and x_j to form the event y is designated as

$$y = x_i \cup x_j \quad (x_i \text{ or } x_j)$$

This means that the event y occurs when at least one of the events x_i or x_j occurs. The union of events is also called the sum of events.

Intersection of events

$$y = x_i \cap x_j \qquad (x_i \text{ and } x_j)$$

y occurs when both x_i and x_j also occur. The intersection of events is also called the product of events.

Certain event Ω

Ω is used to designate an event that always occurs, namely is certain to occur.

Impossible event Φ

Φ is an event that never occurs.

Complementary event

The complementary event for x_i occurs only when x_i does not occur. In general, the complementary event for x is designated with $\overline{x}$.

Disjoint events

The two events x_i and x_j are said to be disjoint (incongruous) when

$$y = x_i \cap x_j = \Phi$$

The set theory rules apply to random events, for example, commutative law, the associative law, the distributive law and de Morgan's rules.

Elementary event

An elementary event cannot be divided. For example, the "the lot contains 5 reject parts" event is an elementary event. All other events are thus composite events. Elementary events are disjoint to each other.

Population

The population is the entirety of identical elements (units, products, *etc.*) for which a probability or statistical conclusion is to be drawn. Examples are lots, batches, *etc.* A subset formed from the population is called a *sample*. A sample characterises to some extent the properties of the population. The number of elements in the population is generally designated as N. The number of elements in the sample is called the *sample size n*.

5.2.2 Calculation of Probabilities

The occurrence of random events is always uncertain. One only knows with certainty that some event will occur. Some random events occur more frequently, whereas others occur less frequently. Random events have a degree of possibility of their occurrence. This degree is assigned a measured value designated as a *probability* in the following discussion. $P(X = x_i)$ is the probability of the random event x_i, namely that the random value assumes the realisation x_i.

Other possible notations are $P(x_i)$ or just p_i. Such a probability is also designated as the absolute probability. The following base definitions must be observed:

- $P(x_i)$ can only assume values between 0 and 1.
- $P(\Omega) = 1$
- For arbitrary disjoint events x_i and x_j:

  $P\left(x_i \cup x_j\right) = P\left(x_i\right) + P\left(x_j\right)$ (addition theorem for probability theory).

The following laws also exist:

$$P(\Phi) = 0$$

$$P\left(\overline{x}_i\right) = 1 - P\left(x_i\right)$$

$$P\left(x_i \cup x_j\right) = P\left(x_i\right) + P\left(x_j\right) - P\left(x_i \cap x_j\right)$$

$\sum_{i=1}^{n} x_i = 1$ (The events $x_1 \ldots x_n$ are a decomposition of the certain event.)

Determining probabilities

There are two basic methods of determining the probabilities. The relative frequency of observations or empirical methods can be used to deduce the probability. The relative frequency $y(x_i)$ of the random event x_i is defined by

$$y\left(x_i\right) = \frac{k\left(x_i\right)}{n}$$

where $k(x_i)$ is the absolute frequency of the occurrence of the random event x_i when the action that initiated the random event has run n times. Experience shows that with increasing n, $y(x_i)$ ever more "approaches" a specific numeric value. Thus, $y(x_i)$ can be used as an approximation for $P(x_i)$, where, in particular, larger inaccuracies are possible for small n.

If the internal structure of the object in which the random action runs is known, the unknown probabilities can also be calculated. The classic probability equation uses the fact that the possible results of the random activity n are equally probable elementary events x_i (e.g., the occurrence of a specific number when a dice is cast). In this case:

$$P(x_i) = \frac{1}{n}$$

Other random events x can be described using linkages from various elementary events (number m). If both prerequisites apply, the classic probability equation then can be written as:

$$P(x) = \frac{m}{n} = \frac{\text{number of "favourable" elementary events for } x}{\text{Total number of "possible" elementary events}} \qquad (5.1)$$

This equation, can, for example, be used to calculate the probabilities for a lottery. Another simple example is the dice ($n = 6$).

Another way of determining probabilities is the geometric probability interpretation. It can be used when the number of elementary events is infinite, such as for geometric dimensions. The classic interpretation then can be modified so that the total number of elementary events forms a total surface (line segments and volumes are also possible) and a finitely large subarea represents the random event x. The sought probability is then the ratio of the areas to each other.

5.3 Discrete Quality Characteristics and their Distributions

5.3.1 Discrete Quality Characteristic Variables

A quality characteristic variable X (quality characteristic variables are always written as uppercase letters) is called discrete only when it can assume uniquely distinguishable values (realisations; designated as x_1, x_2, ... x_i, ...x_n (the indices can also start with zero) in the following discussion, are written as lowercase letters). In other words, the quality characteristic variable X can only assume countable distinct values. This number is normally designated with n in the following discussion. In electronics manufacturing, the number of defect modules X in a lot (of size N) is a typical discrete quality characteristic variable. This characteristic variable X can assume $n = (N + 1)$ different realisations:

$x_0 = 0$; $x_1 = 1$ ..... $x_n = N$

namely x_0 means all modules in the lot are OK

x_1 just one module is defect

x_n all modules are defect

It is important that the quality characteristic variable has a unique definition. In the described case, this means that a module is defect when it has at least one defect (*e.g.*, soldering defect). Thus, *for this definition!*, it does matter how many defects are on *one* subassembly. In general linguistic usage, one speaks of the number of defect units (for the assembly production, a module is thus a unit).

Another definition would be, for example, defects (*e.g.*, soldering defect) per module. The general formulation here is then defects per unit. A third definition would be defects (*e.g.*, soldering defects) in the lot. The number of possible realisations would now be $N \cdot k$ (where N is lot size and k is the number of solder joints).

In practice, the *discrete* term is used less often. Instead, the *attributive* quality characteristic variable term is used, where normally the characterisations "good" or "bad" occur here.

5.3.2 Discrete Probability Distribution

Each realisation x_i of the quality characteristic variable X has a specific probability p_i.

$$P(x_i) = P(X = x_i) = p_i$$

Because the individual realisations x_i can be distinguished (disjoint), then:

$$\sum_{i=1}^{n} p_i = 1 \qquad \text{respectively} \qquad \sum_{i=0}^{n} p_i = 1$$

All the following representations assume that the indexing starts with 1. The corresponding equations can be changed appropriately to start the indexing with zero. The possible realisations x_i and the associated probabilities p_i are called a discrete probability distribution.

$$\begin{pmatrix} x_1 & x_2 \cdots x_i \cdots x_n \\ p_1 & p_2 \cdots p_i \cdots p_n \end{pmatrix}$$

A typical representation form is the column representation. Figure 5.1 shows an example.

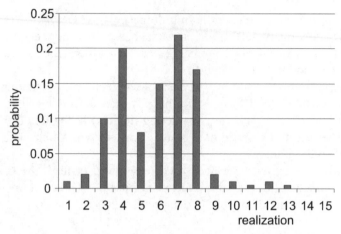

Figure 5.1. Graphical representation of a probability distribution

5.3.3 Distribution Function

A distribution function $F(x)$ of a discrete quality characteristic variable X designates the probability that the quality characteristic variable X is smaller than (but not equal to) a specific numeric value x.

$$F(x) = P(X < x)$$

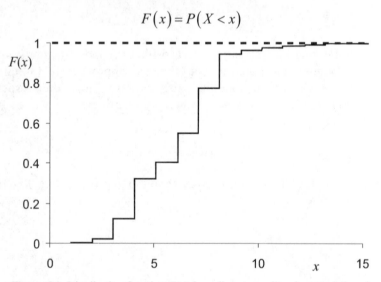

Figure 5.2. Distribution function $F(x)$ for a discrete quality characteristic value

The distribution function has the following general properties:

- $F(-\infty) = 0$ $\qquad$ $F(\infty) = 1$
- $F(x)$ is monotone not decreasing.
- At the places x_n, $F(x)$ is continuous from the left (namely $F(x)$ assumes the smaller value at these places!).

The distribution function can be represented graphically as a staircase curve (Figure 5.2 with the same number of numeric values as in Figure 5.1). For a discrete probability distribution, $F(x)$ can be calculated as follows:

$$F(x) = \sum_{i=1}^{k} p_i \qquad \text{for } x_k < x \tag{5.2}$$

5.3.4 Parameters for Discrete Probability Distributions

The knowledge of a probability distribution or the distribution function of a quality characteristic variable suffices for its complete description. Often, however, this completeness is not necessary and indeed sometimes cannot even be comprehended or indeed not known. In such a case, parameters have been defined that represent the most important properties of such a distribution and so permit a good, but incomplete, evaluation of the quality characteristic value, even without knowledge of the complete distribution function.

The most important parameter is the *expected value E{X}* (expected value of the quality characteristic variable X, often designated as μ in the following discussion). It has the following definition for a discrete quality characteristic value:

$$E\{X\} = \mu = \sum_{i=1}^{n} x_i \, p_i \tag{5.3}$$

The expected value μ is a weighted average value of all possible realisations of the discrete quality characteristic value. This value is somewhat comparable to the centre of gravity of a system. For quality assurance, the expected value, for example, can be the expected average number of defect products in a goods item or the operating point of a technological process. This average number or this operating point ensues *only* over a large number of realisations.

For many discrete quality characteristic values, the expected value can assume a value that can never occur for a specific realisation, for example, on the average, 1.3 faulty products in a goods item. This actual numerical value allows, for example, to plan for the required repair capacities, which, however, affect the consideration only after quite some time. Underloading or overloading will occur in a specific case. Another simple example is the average number on the face of a dice (= 3.5) when it is thrown. In no circumstances may the expected value be interpreted as being the most frequent value.

Another important parameter is the *variance* $D^2\{X\}$. In the following discussion, σ^2 is normally used for the variance. It is defined by:

$$D^2\{X\} = \sigma^2 = E\left\{(X - \mu)^2\right\} = \sum_{i=1}^{n}(x_i - \mu)^2\, p_i = \sum_{i=1}^{n} x_i^2\, p_i - \mu^2 \qquad (5.4)$$

The variance is the average squared deviation from the average value and is a measure of how the individual realisations as squared average differ from the average value. For practical purposes, however, the variance is not very useful. If the quality characteristic variable has a dimension (*e.g.*, Ω), the value could be $\sigma^2 = 400\ \Omega^2$. However, at first sight, one can scarcely use it. A practicable derivation from the variance is the *standard deviation* σ:

$$\sigma = \sqrt{\sigma^2} = \sqrt{D^2\{X\}} \qquad (5.5)$$

For quality assurance, the standard deviation, together with the expected value, is the most important parameter. The standard deviation expresses the extent to which individual values of a random variable can deviate from the expected value. It is easier to comprehend than the variance (for the above example $\sigma = 20\ \Omega$!). However, we must emphasis that for determining the standard deviation, the variance must always be calculated first.

A large standard deviation is generally associated with a poor quality evaluation. If the expected value for the proportion of defect modules can be used to make a plan for the repair capacity, the associated standard deviation is a measure of how strongly in the specific case the actual loading deviates from the planned loading (both overloading and underloading). This then depends only on chance. Thus, the standard deviation is, in particular, a measure for random deviations.

Another characteristic value is the *variation coefficient v*, defined as

$$v = \frac{\sigma}{\mu} \qquad (5.6)$$

It is a relative standard deviation and thus does not have any unit. It is particularly useful for comparing different processes, procedures, *etc.*

Other useful parameters are the skewness γ_1 (statements about the symmetry of the distribution) and the excess γ_2 (statements about the width of the distribution).

5.3.5 Important Discrete Probability Distributions

Binomial Distribution

The binomial distribution is the discrete distribution most frequently used for quality assurance. The following model shows the derivation.

We consider a machine that continuously produces a specific product (*e.g.*, resistors). When some arbitrary product is picked, the inspection can produce one of two results. If the resistance value does not lie within specified limits (tolerance limits), event A occurs (the resistor is a reject). If the resistance value lies within these limits, the $\overline{A}$ event (complementary event for A) occurs.

The probability of the occurrence of a reject is p. Consequently, the probability for the production of a useable resistor is $q\ (= 1 - p)$.

After now taking a sample of n units (sample size) from the running production, a search is made for the $P(X = k) = p_k$ with which exactly $k\ (= 0 .. n)$ reject resistors are contained in this sample. An arbitrary sample size n has the probability distribution:

$$P(X = k) = p_k = \binom{n}{k} \cdot (1-p)^{n-k} \cdot p^k = g(k; n; p) \tag{5.7}$$

where $(1-p)^{n-k} \cdot p^k$ is the probability for a specific sequence of reject resistors and the binomial coefficient $\binom{n}{k}$ is the number of the possible "ordering" of k reject resistors in n picked resistors (combination without repetition). Another possible short notation is $g(k; n; p)$.

The distribution function calculates to:

$$P(X < k) = \sum_{i=0}^{k-1} \binom{n}{k} \cdot (1-p)^{n-k} \cdot p^k = G(k-1; n; p) \tag{5.8}$$

Such a distribution is designated as a binomial distribution. The expected value and the variance can be calculated using the following equations:

$$\mu = E\{X\} = \sum_{k=0}^{n} k \cdot P(X = k) = n \cdot p \tag{5.9}$$

$$\sigma^2 = D^2\{X\} = \sum_{k=0}^{n} k^2 \cdot P(X = k) - \mu^2 = n \cdot p \cdot (1-p) \tag{5.10}$$

Hypergeometric Distribution

For the binomial distribution, in the ideal case, the sample should be taken from an infinitely large population. A running production comes very near to this condition. An independency of the events (namely the probability p is constant for each removal), which normally is the case, was assumed here.

If, however, we consider the removal of a sample from a closed goods item that contains exactly N products (N item size) of which exactly M products are defect, then this independency no longer applies, because each removed product changes the defect rate of the remaining items.

The calculation of the probability $P(X = k)$ finding exactly k faulty products in n removed products is a purely combinatorial problem. The classic probability Equation (5.1) is used.

The number of possibilities of removing a sample n from a goods item of size N is: $\begin{pmatrix} N \\ n \end{pmatrix}$ (= total number of possibilities). The number of favourable cases consists of drawing exactly k defect products from M defect products *together with* the drawing of exactly $(n - k)$ defect-free products from the $(N - M)$ defect-free products. This gives the following equation for calculating the sought probability $P(X = k)$:

$$P(X = k) = p_k = \frac{\begin{pmatrix} M \\ k \end{pmatrix} \cdot \begin{pmatrix} N - M \\ n - k \end{pmatrix}}{\begin{pmatrix} N \\ n \end{pmatrix}} \tag{5.11}$$

Instead of the number of defect products M, the defect rate of the item p ($=M/N$) is also used. The probability distribution (5.11) is designated as a *hypergeometric distribution*. The expected value and the variance can be calculated from the following two equations:

$$\mu = E\{X\} = n \cdot p = n \cdot \frac{M}{N} \tag{5.12}$$

$$\sigma^2 = D^2\{X\} = n \cdot \frac{N - n}{N - 1} \cdot p \cdot (1 - p) \tag{5.13}$$

The hypergeometric distribution is used in quality assurance for testing goods items by taking samples. This is especially popular for goods arrival and goods shipping control. This gives the problem specification, find information about the unknown defect rate of the goods item by taking a sample.

Poisson Distribution

The Poisson distribution represents a special case of the binomial distribution. If the sample size is very large (theoretically tending to infinity) and the probability p is very small (tending to zero), the limit theorem for Poisson yields

$$P(X = k) = \lim_{\substack{n \to \infty \\ p \to 0 \\ np=\text{const}=\lambda}} \binom{n}{k} \cdot (1-p)^{n-k} \cdot p^k = \frac{\lambda^k}{k!} \cdot e^{-\lambda} \qquad (5.14)$$

the Poisson distribution with λ as the major parameter. The expected value and the variance can be obtained from the following two equations:

$$\mu = E\{X\} = \lambda \qquad \sigma^2 = D^2\{X\} = \lambda \qquad (5.15)$$

Hypergeometric distribution $$p_k = \frac{\binom{M}{k} \cdot \binom{N-M}{n-k}}{\binom{N}{n}}$$	General mathematical model for the removal of sample (of size n) taken from a population (of size N) with a defect rate $M=p \cdot N$
$\Downarrow \quad N \to \infty$	
Binomial distribution $$p_k = \binom{n}{k} \cdot (1-p)^{n-k} \cdot p^k$$	General mathematical model for the removal of a sample taken from an infinitely large population or for independent sample results; it provides a good approximation for the hypergeometric distribution for $n < 0.1\,N$
$\Downarrow \quad \left.\begin{array}{l} n \to \infty \\ p \to 0 \end{array}\right\} np = \lambda$	
Poisson distribution $$p_k = \frac{\lambda^k}{k!} \cdot e^{-\lambda}$$	General mathematical model for stochastic arrival and serving tasks; it provides a good approximation for the binomial distribution for $np < 0.05$ or $p < 0.1$.

The Poisson distribution is used particularly in queuing and server theory. The λ parameter is then the arrival or serving rate. The derivation of the Poisson distribution using the limit theorem (5.14) shows that this distribution applies to small probabilities of p. Consequently, it is also known as the "distribution of the seldom events". The hypergeometric, the binomial and the Poisson distribution have relationships that are particularly suitable for numeric calculations.

Other important discrete distributions are the two-point distribution, the uniform discrete distribution and the geometric distribution.

5.3.6 Uses in Module Production

The discrete quality characteristic variables are used, for example, in module production to make estimates for the expected quality of new modules. An important characteristic variable is the first-pass yield (*FPY*), namely the yield of modules after manufacturing without any repair steps being necessary. Initially, a *unique* definition when a module is defect free must exist. In the following discussion, we define that a module is defect free when it does not contain any defect solder joints *and* no defect inserted components *and* no electrically faulty components. Although a clear definition for "defect" is also necessary, it is not considered here.

To make estimates for the *FPY*, the following data must be known or determined:

Solder joints

n_l number of solder joints

p_l defect rate (or probability) for a solder joint

Placement

n_b number of components

p_b defect rate for placing a component

Components

n_k number of components (normally $=n_b$)

p_k probability of an electrically defective component

For the consideration of the solder joints, we can now make the following statements:

The module is free from soldering defects when from n_l "successive" (in analogy to a successive removal) produced solder joints are all defect free. The previously mentioned model of the binomial distribution (5.7) can be used here. The probability y_l of a module without soldering defects and thus the yield is specified by the following equation:

$$y_l = \left(1 - p_l\right)^{n_l} \approx e^{-n_l p_l}$$

For the approximation, the binomial distribution has been replaced by the Poisson distribution. The conditions required for this approximation normally ($p_l \ll 10\%$) apply.

The probabilities for a module without placement faults (p_{ob}) and without component faults (p_{ok}) can be determined similarly. The first-pass yield can be obtained by overlaying the individual faults with the multiplication of the corresponding individual yields:

$$FPY = (1 - y_l) \cdot (1 - y_b) \cdot (1 - y_c)$$

(5.16)

This multiplication is permitted when no dependencies exist between the individual fault types. Such a dependency would be present when the occurrence of a fault type changes the probability of another fault type. At least the occurrence of a placement fault (an excessive offset) can cause an increased number of soldering faults, whereas relationships between the electrical function of a component to soldering faults certainly do not exist. The following discussion ignores such dependencies, because the determination of the required additional data is extremely difficult.

Table 5.1 shows some simple numerical examples. The fault probabilities have been assumed to be equal for all three components (in the table header). The first column contains the number of components. To simplify the calculation for the number of the solder joints, it has been assumed that each component has 20 connections. It can generally be shown that higher defect probabilities and larger number of components lead to a rapid reduction of the first-pass yields.

Table 5.1. Sample calculation for the FPY (in %)

N =	p (in DPM)					
	1	5	10	20	50	100
10	99.98	99.90	99.79	99.58	98.96	97.92
20	99.96	99.79	99.58	99.16	97.92	95.89
50	99.90	99.48	98.96	97.92	94.89	90.03
100	99.79	98.96	97.92	95.89	90.03	81.07
200	99.58	97.92	95.89	91.94	81.07	65.73
500	98.96	94.89	90.03	81.07	59.19	35.08

This model can be refined as required. The next necessary step is the differentiation of the component types, because a chip, a QFP and a BGA do not have the same probability for the occurrence of a soldering defect. However, the determination of the required data is always difficult. If these data are incorrect, the results are also incorrect, where, in particular, the effects of such incorrection can magnify for the estimated individual defect rates.

One should proceed very carefully when drawing possible conclusions from an estimated first-pass yield. We must also emphasise that comparisons of such numbers are permitted only for identical (or at least very similar) modules, because the effect of the number of the components is relatively easy to see.

5.4 Distribution Functions of Continuous Quality Characteristics

5.4.1 Continuous Quality Characteristic Variables

A quality characteristic variable is called continuous if it can assume infinitely many arbitrary realisations from an interval. Typical continuous characteristic variable for the quality assurance are the actual dimensions of products (resistance value, but also placement displacement, print displacement for solder paste deposits, *etc.*). One assumes that no two products exist that are fully identical with regard to a specific characteristic. This statement is not affected by the fact that a limited measuring accuracy can make two such characteristics appear to be identical.

Instead of the mathematical term *continuous*, it is also usual to speak of a variable or measurable quality characteristic variable.

When tolerance or specification limits are used, it is possible to convert a continuous quality characteristic variable into a discrete quality characteristic variable (within the tolerance limits = "good", outside = "bad"). However, this should never be done unnecessarily, because this conversion normally also means a loss of information. If a measured value lies near the tolerance limit, even a small measurement error can mean a "tipping" between good and bad.

5.4.2 Density Function

For a continuous quality characteristic variable, because of their infinite number, no probabilities can be assigned to the individual realisations. The density function $f(x)$ with the following properties is used to characterize the distribution

$$f(x) \geq 0 \qquad \text{for any } x$$

$$\int_{-\infty}^{\infty} f(x)\,dx = 1$$

$$f(x) = \frac{P(x \le X < x + dx)}{dx}$$

$f(x)$ is consequently the probability that the quality characteristic variable X assumes a realisation in the differentially small interval $[x, x+dx]$. If $f(x)$ is known, the probability can be calculated with which X assumes a realisation in a specified interval $[a, b]$. Figure 5.3 illustrates this behaviour.

$$P(x_{ab}) = P(a \le X < b) = \int_a^b f(x)\ dx \tag{5.17}$$

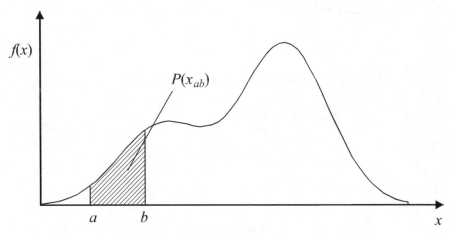

Figure 5.3. Calculation of the probability $P(x_{ab})$

5.4.3 Distribution Function

The $F(x)$ distribution function for a continuous quality characteristic variable is defined by:

$$F(x) = P(X < x) = \int_{-\infty}^x f(\xi)d\xi \tag{5.18}$$

Thus, the following is also true: $P(X > x) = 1 - F(x)$

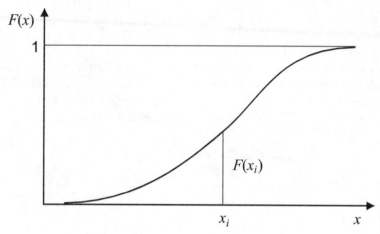

Figure 5.4. General form of the distribution function

The distribution function can be interpreted as the area under the distribution density to the left of x. The following equation specifies the relationship to the distribution density:

$$f(x) = \frac{d\,F(x)}{dx}$$

Instead of Equation (5.17), the following equation can be used for calculating the probabilities:

$$P(x_{ab}) = P(a \le X < b) = F(b) - F(a) \qquad (5.19)$$

5.4.4 Parameters

Similar to the discrete quality characteristic variables, the continuous quality characteristic variables can also be described using parameters. The knowledge of these parameters often suffices for the evaluation of the corresponding quality characteristic value. The following two equations describe the expected value μ and the variance σ^2:

$$\mu = E\{X\} = \int_{-\infty}^{\infty} x\,f(x)\;dx \qquad (5.20)$$

$$\sigma^2 = D^2\{X\} = \int_{-\infty}^{\infty} (x-\mu)^2\,f(x)\;dx = \int_{-\infty}^{\infty} x^2\,f(x)\;dx - \mu^2 \qquad (5.21)$$

The meaning of the expected variable and the deviation is similar to that for the discrete quality characteristic variables.

5.4.5 Important Continuous Distributions

Uniform Distribution

The uniform distribution, or also rectangular distribution, has the following distribution density:

$$f(x) = \begin{cases} \dfrac{1}{b-a} & \text{for } a \leq x \leq b \\ 0 & \text{otherwise} \end{cases} \tag{5.22}$$

The quality characteristic value X is distributed evenly over the interval $[a,b]$. The following equations yield the expected value and the variance:

$$\mu = \frac{a+b}{2} \qquad\qquad \sigma^2 = \frac{(b-a)^2}{12} \tag{5.23}$$

The uniform distribution is used in the quality assurance, such as for sorting processes, when from a widely dispersed population, a screening of the products is performed by extraction. After the sorting, one normally has a uniform distribution of the considered characteristic in the associated class. Furthermore, in practice, a uniform distribution is often assumed when absolutely no knowledge is available about the quality characteristic variable. However, care must be exercised here when other conclusions are drawn from this assumption. Such conclusions can very quickly become incorrect.

Normal Distribution

For the manufacturing of products, a large number of factors act on the considered characteristic. This normally means that each such characteristic can be described by a quality characteristic variable. The realisations will deviate more or less from each other. If one analyses a large number of processes, it can be shown that a large proportion (approx. 70–80%) of the quality characteristics can be described using the normal distribution.

The cause can be found in the central limit theorem of probability theory:

The sum of n independent quality characteristic values has a normal distribution for $n \rightarrow \infty$ under relatively wide-ranging conditions.

Because many requirements-influencing factors overlay for manufacturing processes, we can assume for a large proportion of the
manufactured products that the quality-determining characteristics have a
normal distribution. The distribution density of the normal distribution of a
quality characteristic variable X with the parameters μ and σ has the
function:

$$f(x) = \frac{1}{\sqrt{2\pi}\,\sigma} e^{-\frac{(x-\mu)^2}{2\sigma^2}} \qquad (5.24)$$

The following two equations describe the expected value and the
variance:

$$E\{X\} = \mu \qquad\qquad D^2\{X\} = \sigma^2 \qquad (5.25)$$

Figure 5.5 shows the normal distribution with its parameters.

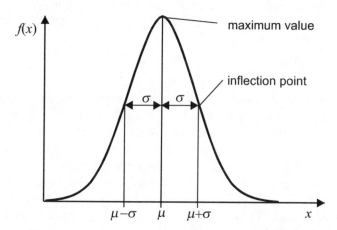

Figure 5.5. Representation of the normal distribution with characteristic values

In words, this is often expressed as:
"The quality characteristic value X has the $N(\mu,\ \sigma^2)$ distribution"
Using the definition, the distribution function $F(x)$ can be written as:

$$F(x) = P(X < x) = \frac{1}{\sqrt{2\pi}\,\sigma} \int_{-\infty}^{x} e^{-\frac{(\xi-\mu)^2}{2\sigma^2}} d\xi \qquad (5.26)$$

We must emphasise that this integral can only be solved numerically. The
popularity of the normal distribution consequently makes it necessary to
have the values of this function available in tabular form. This is not
possible for arbitrary parameters. Rather, the normal distribution is

standardised using a transformation. The quality characteristic value U is produced from the quality characteristic variable X.

$$U = \frac{X - \mu}{\sigma} \tag{5.27}$$

This quality characteristic variable also has a normal distribution with the expected value 0 and the variance 1. The distribution density of this so-called standardised normal distribution is designated by $\varphi(u)$ and the distribution function by $\Phi(u)$. Because the function $\Phi(u)$ is available as a table (see Table A.1), it is easy to determine the sought probabilities.

$$\varphi(u) = \frac{1}{\sqrt{2\pi}} \cdot e^{-\frac{u^2}{2}} \qquad\qquad \Phi(u) = \frac{1}{\sqrt{2\pi}} \int_{-\infty}^{u} e^{-\frac{z^2}{2}} \, dz$$

Example:
Let us assume the interval $[-1,2]$ and a standardised normal distribution. We require the probability with which a realisation falls in the interval.

Solution:

$$
\begin{aligned}
P(-1 \le U < 2) \quad &= \Phi(2) - \Phi(-1) = \Phi(2) - 1 + \Phi(1) \\
&= 0.97725 - 1 + 0.841345 \\
&= 0.818595
\end{aligned}
$$

We can use the fact that $\Phi(-u) = 1 - \Phi(u)$.

In practice, the case can also occur in which a value for $\Phi(u)$ is given and the associated u-value is required. In this case, one can use Table A.2. If the table does not contain the sought value, one can also take this numeric value from Table A.1 by searching for the value $\Phi(u)$ and then reading the associated value u from the column and row headings. If necessary, one must interpolate. The associated notation is:

$$u_q = \Phi^{-1}(q) \tag{5.28}$$

where u_q can also be designated as the *normal distribution quantile*.
Example: u $_{0.7}$ = 0.525

Uses of the Normal Distribution

Quality assurance is widely used in the normal distribution because it can be used to provide a good description of a large proportion of the characteristics for manufactured products. We illustrate this with two examples.

The operating point $\mu = 990\,\Omega$ is known for the manufacturing of resistors. The resistance values obey a normal distribution with a known standard deviation $\sigma = 50\,\Omega$. The proportion of resistors (defect rate p) that lie outside the specified tolerance limits [900 Ω, 1100 Ω] is required.

Solution:
The resistance value is the quality characteristic variable X. The defect rate is the probability with which a realisation of X lies outside the tolerance limits.

$$p = P\big(\big[X < T_u\big] \cup \big[X > T_o\big]\big) = P\big(X < T_u\big) + P\big(X > T_o\big)$$

The operating point can be interpreted as the expected value of the normal distribution. The quality characteristic value X will be standardised to the new quality characteristic value U (5.27). In accordance with Equation (5.26), the first summand is the distribution function of the standardised normal distribution Φ. The second summand can also be calculated using the standardised normal distribution in an intermediate step (reversal of the relation sign and thus interpreted as a complementary event).

$$p = \Phi\left(\frac{T_u - \mu}{\sigma}\right) + 1 - \Phi\left(\frac{T_o - \mu}{\sigma}\right) = \Phi\left(\frac{900 - 990}{50}\right) + 1 - \Phi\left(\frac{1100 - 990}{50}\right)$$

$$= \Phi(-1.8) + 1 - \Phi(2.2) = 1 - 0.96407 + 1 - 0.986097 = 0.049833$$

The defect rate p is consequently 4.98%.

The following example demonstrates the use of quantiles. Resistance values can be compared (*e.g.*, using laser comparison), where, however only an increase of the resistance is possible. A resistor that exceeds the upper tolerance limit (here $=1100\,\Omega$) will be rejected whereas a resistance value that is too small means corrective work. The standard deviation of the resistances is again 50 Ω. The rejection rate p_a should not exceed 1%. What is the maximum of the operating point μ in order to satisfy this requirement? The rejection rate p_a can be interpreted as being the probability with which the resistance value (quality characteristic value X) exceeds the upper tolerance limit.

$$p_a = P\big(X > T_o\big) = 1 - P\left(\frac{X - \mu}{\sigma} < \frac{T_o - \mu}{\sigma}\right) = 1 - P\left(Z < \frac{T_o - \mu}{\sigma}\right)$$

$$= 1 - \Phi\left(\frac{T_o - \mu}{\sigma}\right)$$

After substituting the values, this yields: $0.01 = 1 - \Phi\left(\dfrac{1100\ \Omega - \mu}{50\ \Omega}\right)$

The argument of the standardised normal distribution that yields 0.99 is now sought. This can be solved using the quantile u_q of the normal distribution.

$$u_{0.99} = 2.3263 = \frac{1100\ \Omega - \mu}{50\ \Omega} \qquad \mu = 983.685\ \Omega$$

The operating point must be less than $983.685\ \Omega$ in order that the rejection rate p_a does not exceed the 1% value.

Exponential Distribution

Another important continuous distribution is the exponential distribution. It is used above all in reliability theory (see Chapter 8). Here, the lifetimes of products, components, *etc.*, are considered. Every product fails sometime. The time of failure cannot be predicted with certainty, namely the *lifetime T* is a quality characteristic value and thus also a random value. The time duration *t* when a specified product fails is then a realisation of the quality characteristic value *T*. $P(T < t)$ then specifies the probability with which the product has failed by time *t*. This probability is, however, identical with the distribution function for *T*, $F_T(t)$.

$$P(T < t) = F_T(t)$$

The probability that a product has failed after a time *t* is then:

$$P(T > t) = 1 - F_T(t) \qquad P(t_1 \le t < t_2) = F_T(t_2) - F_T(t_1)$$

The reliability theory is interested in the probability with which a product fails in a time interval d*t* under the condition that it still functions at the beginning of the time interval. This probability is called the failure rate. The exponential distribution assumes that this probability is constant.

$$P(t \le T < t + dt \,|\, T \ge t) = \text{const.} \quad \text{for all } t$$

$$\lambda(t) = \frac{P(t \le T < t + dt \,|\, T \ge t)}{dt} \tag{5.29}$$

These can be used for the special case $\lambda(t) = \lambda = \text{const}$ to determine the distribution function and the distribution density by derivation using a differential equation of the 1st order.

$$F_T(t) = 1 - e^{-\lambda t} \quad \text{for } t \geq 0 \tag{5.30}$$

$$f_T(t) = \lambda \cdot e^{-\lambda t} \quad \text{for } t \geq 0 \tag{5.31}$$

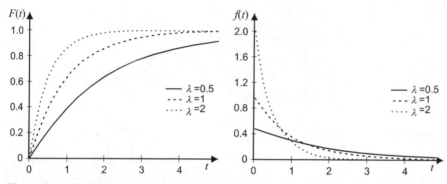

Figure 5.6. Distribution function $F_T(t)$ (left) and density function $f_T(t)$ (right) of the exponential distribution for various values of the parameter λ

These are the density function and distribution function for the exponential distribution. Figure 5.6 shows that the form of the distribution function and the density function depends only on the parameter λ. Neither function exists for $t < 0$. The following two equations specify the expected value and the variance:

$$\mu_T = \frac{1}{\lambda} \quad ; \quad \sigma_T^2 = \frac{1}{\lambda^2} \tag{5.32}$$

The expected value μ_T has particular importance in reliability theory. It specifies the expected average lifetime of a product. If many identical products are used operationally, the average time of failure is μ_T.

5.5 Evaluation of Quality Data and Point Estimates

5.5.1 Obtaining a Sample

For all previous considerations, we have assumed that a large part of the information is already available for the considered quality characteristic variable. The ideal case is knowledge of the probability distribution or the density function. The knowledge of the most important parameters, such as the expected value and variance, also suffices in many circumstances. In practice, however, because such information is normally not available, observations or experiments must first be performed to obtain this information. This is done with the help of one (or more) samples. It is obvious that such a sample cannot be taken from a single product. This is because a value that by chance deviates greatly from the expectations is certainly possible and could lead to totally incorrect conclusions.

A *sample* is characterised by the sample size n and the measured values $x_1,... x_n$. It does not matter here whether the quality characteristic variable is discrete or continuous. Each sample should permit a representative, unaltered statement about the observed quality characteristic. This is largely ensured when the corresponding product is selected totally randomly (*e.g.*, "blind" picking, selection using random numbers). No products may be preferred (taken from the top, removal of apparently defect products, *etc.*).

The result of the sample taking is the original list of sample values. Table 5.2 shows such an original list. Here, 100 values have been measured for the adhesion of printed conductors on the printed-circuit boards (determined using a withdrawal test). Because this original list can scarcely be used to provide information about the observed characteristic, the values must be processed.

Table 5.2. Original list of the adhesion values (in N)

5.70	5.35	5.35	5.50	5.50	7.60	6.65	5.80	7.65	5.90
6.50	6.60	6.05	5.90	6.20	7.30	5.55	6.45	7.35	7.45
6.90	7.00	7.25	6.80	5.90	5.80	6.40	6.20	6.10	6.25
6.25	6.50	6.25	6.80	7.10	5.70	6.25	6.70	6.90	6.80
5.75	6.90	7.10	5.70	7.05	6.05	6.40	5.95	6.30	5.70
6.35	7.30	6.05	7.20	5.85	5.70	7.45	7.10	6.00	7.65
6.65	6.50	6.70	6.50	6.35	6.50	6.50	7.25	6.70	6.30
7.40	7.40	7.00	7.15	6.30	7.00	6.80	7.10	7.05	7.10
6.55	7.00	6.60	6.10	6.25	6.70	6.75	6.65	6.70	6.00
6.10	6.15	6.20	6.35	6.10	6.70	6.80	6.60	6.70	6.65

A first possibility is the *primary distribution table*. The measured values are arranged according to the size and entered in a tally list. Table 5.3 shows a selection from such a primary distribution table.

This shows that the primary distribution table for this example has relatively little expressiveness because only limited (in the extreme case, no) data compression is made. This table is particularly useful for recording discrete quality characteristic variables. An example is an acquisition of the soldering defects per module. A frequency table is useful for continuous quality characteristic variables.

Table 5.3. Primary distribution table (selection) of the measured values from Table 5.2

Measured value in N	Tally list	Absolute frequency h	Measured value in N	Tally list	Absolute frequency h
5.35	//	2	6.60	///	3
5.40		0	6.65	///	3
5.45		0	6.70	//// //	7
5.5	//	2	6.75	/	1

The value range of the measurement results is divided into classes each with the same size (if possible) (classing) and the determined measurement results are assigned to the corresponding classes as a tally list. Several aspects must be considered for the class assignment. The class width d should be as constant as possible. The number of classes K is subject to specific rules. It may be neither too large (excessive evaluation cost) nor too small (excessive information losses because of the classing). The following table provides guidance values:

Table 5.4. Recommendations for the number of classes for a classing

N	50	100	500	1,000	10,000
K	8	10	13	15	20

The range R $(= x_{max} - x_{min})$ can be used to calculate the class width:

$$d \approx \frac{R}{K}$$

The final class assignment should be chosen so that as few values as possible lie on the class boundaries. Even though todays software allows a classing to be performed automatically and fast, a critical consideration of the result is always desirable. The appropriate programs calculate the class limits using specific rules that do not necessarily need to be optimum for the investigated problem. Most programs, however, permit a customisation of the classes. The discussed example has the following values:

$$R = x_{max} - x_{min} = 2.3 \text{ N} \quad n = 100 \Rightarrow K = 10 \quad d \approx \frac{R}{K} = 0.23 \text{ N}$$

The frequency table can be used to produce the histogram as a graphical illustration of the absolute frequencies and the empirical distribution function $\hat{F}(x)$ that results from the relative cumulative frequency ($\sum\%$; column 4) of the frequency table (Figure 5.7).

Table 5.5. Frequency table for the removal force for the printed circuit board

Class limits	Absolute frequency h_k	Relative frequency in %	$\sum\%$
5.35–5.58	5	5	5
5.58–5.81	8	8	13
5.81–6.04	7	7	20
6.04–6.27	16	16	36
6.27–6.50	12	12	48
6.50–6.73	18	18	66
6.73–6.96	9	9	75
6.96–7.19	12	12	87
7.19–7.42	8	8	95
7.42–7.65	5	5	100

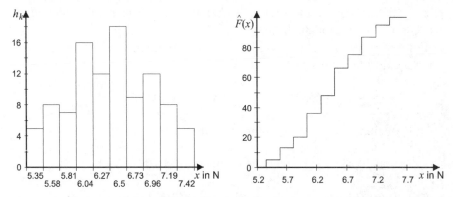

Figure 5.7. Histogram (left) and empirical distribution function $\hat{F}(x)$ (right)

5.5.2 Calculation of Statistical Measured Values

To calculate a statistical measured value, a numeric value is calculated from the individual values x_i $(i = 1...n,)$ of the sample based on a specific calculation rule.

$$\left(x_1, x_2, ...x_n\right) \xrightarrow{\text{calculation rule}} \text{statistical measured value}$$

The following discussion considers the most common statistical measured values.

Mean (arithmetic middle) $\overline{x}$

$$\overline{x} = \frac{1}{n}\sum_{i=1}^{n} x_i \tag{5.33}$$

It characterises the average behaviour of the quality characteristics. Its use requires that an original list is available. If only a classing (K class count, u_k class middles and h_k class frequencies) is present, the following equation should be used.

$$\overline{x} = \frac{1}{n}\sum_{k=1}^{K} h_k \cdot u_k$$

Median (central value) $\tilde{x}$

To determine this measured value, the values of the sample must be ordered according to their size. The median $\tilde{x}$ is then the most central measured value in this ordered series of measured values. Like the mean, it mirrors the average behaviour of the observed characteristic. The median should be used, in particular, for small samples.

For example, the following series of measurements (values in kΩ) was recorded during the resistor production:

9.35; 9.90; 9.65; 10.25; 9.45; 9.85; 10.05; 10.40; 9.55

The ordered series of measurements is then:

9.35; 9.45; 9.55; 9.65; **9.85**; 9.90; 10.05;10.25; 10.40

The fifth value (bold) is the median here. If the sample size is even, the median is calculated using the mean of the two most-middle values. In practice, the median is normally used only for small sample sizes ($n < 10$).

Empirical variance (average squared deviation) s^2

It is calculated using

$$s^2 = \frac{1}{n-1}\sum_{i=1}^{n}(x_i - \overline{x})^2 = \frac{1}{n-1}\left(\sum_{i=1}^{n} x_i^2 - n\overline{x}^2\right) \qquad (5.34)$$

The empirical variance characterises the "spreading" of the measured values around the mean. For a classing, this quantity is:

$$s^2 = \frac{1}{n-1}\sum_{k=1}^{K}(u_k - \overline{x})^2 h_k$$

The calculated empirical variance s^2 can be used to calculate the empirical standard deviation s, which normally is more meaningful for the user (in particular with regard to the numerical values).

$$s = \sqrt{s^2} \tag{5.35}$$

Modified empirical variance s^{*2}

The seldom-used empirical variance is calculated as

$$s^{*2} = \frac{1}{n} \sum_{i=1}^{n} (x_i - \bar{x})^2 \tag{5.36}$$

In practice, the case often arises that several sample results are available (number of samples k; individual sample size n). The means and the standard deviations s_j are calculated for each sample $\bar{x}_j$ $j = 1, \ldots, k$. The mean of all samples $\bar{\bar{x}}$ (the mean of the means) is determined using Equation (5.37).

$$\bar{\bar{x}} = \frac{1}{k} \sum_{j=1}^{k} \bar{x}_j \tag{5.37}$$

There are several possibilities for calculating the standard deviation:

$$s_1 = \sqrt{\bar{s^2}} = \sqrt{\frac{1}{k} \sum_{j=1}^{k} s_j^2} \tag{5.38}$$

$$s_2 = \frac{1}{k} \cdot \sum_{j=1}^{k} s_j \tag{5.39}$$

$$s_3 = \sqrt{\frac{1}{n \cdot k - 1} \sum_{i=1}^{n \cdot k} (x_i - \bar{\bar{x}})^2} \tag{5.40}$$

These three rules can (but need not) return the same results. In particular, differences can be expected when changes in the process occur between taking the samples.

The *range R* is defined as:

$$R = x_{\max} - x_{\min} \tag{5.41}$$

5.5.3 Random Sample and Sample Functions

By taking the sample $(x_1, \ldots x_n)$, a random value, a specific quality characteristic, is observed. Each individual sample value is thus a special realisation of the "observed" quality characteristic variable X. This can be generalised by defining that the sample $(x_1,...x_n)$ is a special realisation of a random vector $(X_1,...X_n)$ designated as a *random sample*. The individual random variables X_i are mutually independent and have the same distribution as the observed random variable X. Measured values, designated here as the *sample function*, can also be calculated from this random sample.

This gives two random variables: a mean $\overline{X}$ and an empirical variance S^2 (note that uppercase letters are used to represent random variables):

$$\overline{X} = \frac{1}{n}\sum_{i=1}^{n} X_i \qquad S^2 = \frac{1}{n-1}\sum_{i=1}^{n}\left(X_i - \overline{X}\right)^2$$

The random variables $\overline{X}$ and S^2 are thus functions of the random sample $(X_1,... X_n)$. Other sample functions can also be defined similarly (*e.g.*, for the median and for the range).

5.5.4 Distributions and Parameters of Sample Functions

The following considerations serve to provide the basis for drawing the appropriate conclusions from the observations that have been made for actual quality characteristics. We initially assume that a normal distribution $N(\mu,\sigma^2)$ exists for the observed quality characteristic X. However, this parameter is not known before each practical investigation (otherwise each observation would be superfluous).

Initially the mean $\overline{X}$ of the sample is considered. If we take a sample of size n from a normal-distributed population, measure the characteristic to be analysed and calculate the arithmetic mean from the determined measured values, this produces a mean that is also a random variable with the following properties:

$$\overline{X} - N\left(\mu, \frac{\sigma^2}{n}\right) \tag{5.42}$$

If we take samples from a normally distributed population (theoretically infinitely often) and calculate the arithmetic mean, this mean is itself a random variable with a normal distribution, where the expected value of this new random variable that corresponds to the expected value of population set and the variance of this random value is n-times smaller. If we now standardise $\overline{X}$ with

$$U = \frac{\overline{X} - \mu}{\sigma}\sqrt{n}$$

then U has a $N(0,1)$ distribution.

We can determine the properties of other sample functions similarly. The median $\tilde{X}$ for large n approximates the following properties

$$\tilde{X} - N\left(\mu, \sigma^2 \cdot \frac{\pi}{2n}\right) \tag{5.43}$$

The range R also has a normal distribution with

$$R - N\left(d_n \cdot \sigma, e_n^2 \cdot \sigma^2\right) \tag{5.44}$$

where d_n and e_n are constants that depend on the sample size (Table A.7).

If we take a sample of size n from a normally distributed population (with a known variance σ^2) that we then use to calculate the sample variance S^2, this dispersion is itself also a random variable. When we use the following transformation

$$Z = \frac{(n-1)}{\sigma^2} S^2 \tag{5.45}$$

then this random value Z has a χ^2 distribution (chi-square) with $m = n - 1$ degrees of freedom.

If we take a sample of size n from a normally distributed population (μ known), calculate from it the sample mean (random value $\overline{X}$) and the sample variance (random value S^2), and then transform it as follows,

$$T = \frac{\overline{X} - \mu}{S} \sqrt{n} \tag{5.46}$$

then T has a t-distribution (*Student distribution*).

The χ^2-distribution and the t-distribution are widely used in mathematical statistics. Their quantiles are normally available as tables (Table A.3 and Table A.4).

If we now consider two normally distributed populations from which the variance σ_1^2 and σ_2^2 are known and where $\sigma_1^2 = \sigma_2^2$ also applies. We take a sample of size n_1 from the population number 1 and a sample n_2 from the second population, and calculate the variance S_1^2 and S_2^2, then

$$F_{m_1; m_2} = \frac{S_1^2}{S_2^2} \tag{5.47}$$

is a F-distributed (Fisher-distributed) random variable with $m_1 = n_1 - 1$ and $m_2 = n_2 - 1$ degrees of freedom. The quantiles of this distribution are shown in Table A.5 and Table A.6. The F-distribution (and others) is required for the comparison of dispersions.

The following transformation is useful for discrete quality characteristic variables. We consider the random value X to be the number of defect products in a set of n products. The random variable Y (relative frequency of defects) is then defined as:

$$Y = \frac{X}{n}$$

If the probability of occurrence of a defect product $= p$, then the random variable Z produced by the transformation

$$Z = \frac{Y - p}{\sqrt{\dfrac{p \cdot (1 - p)}{n}}}$$

(5.48)

has a $N(0,1)$-distribution (approximation for large n). Consequently, the sample size n should not be less than 100.

5.5.5 Point Estimates for Quality Characteristic Variables

From the observed sample values $x_1, x_2,\dots x_n$, a suitable calculation rule is used to determine an *estimate value* $\hat{\lambda}$ for the required quality characteristic value (the character "^" is always an indication that an estimate is involved).

$$(x_1, x_2, \dots x_n) \xrightarrow{\text{calculation rule}} \hat{\lambda}$$

Here, $\hat{\lambda}$ is simply than a statistical measured value. This raises the question as to which statistical measured values can be used to estimate the quality characteristic variable operating point of a process, the variance of a process, the defect rate of a process, *etc.* It appears evident that the arithmetic mean $\bar{x}$ is a suitable estimate value for the operating point. This, however, is not a scientific foundation. Specific requirements are placed on such an estimate.

The most important requirement is an estimate be *unbiased* (absence of distortion), namely the estimate should mirror the true value of the quality characteristic variable to be estimated. If this requirement cannot be met, it should at least be the case for a large sample size (asymptotic unbiased). Furthermore, the estimate should be appropriate (consistent). This means that the probability rises with increasing sample size n and that the calculated estimate agrees with the true, but unknown, value of the quality characteristic variable.

As a last important requirement, the estimate should be sufficient (exhaustive). This means that all information contained in the sample can also be used for the estimate. Sometimes different estimates can be specified for an unknown quality characteristic variable. This raises the possibility of making comparisons of these estimates by comparing the dispersion areas of these estimates with each other. The estimate with the smallest variance is then the most effective estimate.

The following estimates are widely used in quality assurance:

Estimates for the mean

The mean $\bar{x}$ of a sample is an estimate of the operating point (expected value) μ. This estimate is unbiased, consistent and exhaustive. Compared with all other estimates, the mean is also the most effective estimate.

$$\hat{\mu} = \bar{x} \tag{5.49}$$

The size $\bar{\bar{x}}$ (see Equation (5.37)) can also be used here.

The median $\tilde{x}$ of a sample is also an estimate of the operating point, which, however, is only asymptotically unbiased and consistent. This estimate is not exhaustive because only the most central value of a series of numbers (*e.g.*, measured value series) is used for determining the median and all other numeric values (within certain limits) can fluctuate without the median (and thus the estimate) changing. The estimate using the median is also less effective than that of the mean, although the median is less sensitive to stray values. For smaller sample sizes ($n < 20$), however, the median and the mean are generally equally good.

Estimates for variance

The empirical variance s^2 using Equation (5.34) is an unbiased, consistent and exhaustive estimate for the variance σ^2.

$$\hat{\sigma}^2 = s^2 \tag{5.50}$$

The modified empirical variance s^{*2} (5.36) can also be used. This estimate, however, is only asymptotic unbiased.

Generally, the standard deviation σ is of interest. The following equation can be used to determine the standard deviation:

$$\hat{\sigma} = \sqrt{s^2} \tag{5.51}$$

If several samples are used to determine the standard deviation (Equations (5.38) to (5.40)), then:

$$\hat{\sigma} = s_1 \qquad \hat{\sigma} = \frac{1}{a_n} s_2 \qquad \hat{\sigma} = s_3 \qquad\qquad (5.52)$$

Table A.7 lists the constant a_n that depends only on the sample size. It can be omitted for $n > 20$. The range R can also be used to estimate the standard deviation.

$$\hat{\sigma} = \frac{R}{d_n} \qquad\qquad (5.53)$$

Table A.7 lists the constant d_n. This estimate is unbiased and consistent, but not exhaustive (only the smallest and the largest measured value determines the range, all other values can change (within limits)).

The range and the median are currently losing importance. These estimates were constructed in order to save computing effort. However, they can certainly be used for an initial evaluation.

Estimates for the defect rate $\hat{p}$ can be made using the relative frequency y, which is unbiased, consistent and exhaustive.

$$\hat{p} = y = \frac{x}{n} \qquad \hat{a} = 1 - y = \frac{n - x}{n} \qquad\qquad (5.54)$$

The following discussion shows that for the use of discrete quality characteristic variables, the largest possible sample sizes should be used. In the module production, this cannot normally be realised with a lot, so that the requirement for large sample sizes can only be achieved by combining the results of several lots. The simple equation (5.54) must be modified, in particular because of changing lot sizes.

$$\hat{p} = \frac{\sum_{i=1}^{k} x_i}{\sum_{i=1}^{k} n_i}$$

x_i number of defect modules per lot
n_i associated lot size
k number of lots

If one wants to consider the defects per module (general designation μ; warning, do not confuse this with the average value of the normal distribution!), this size can be estimated using the following equation:

$$\hat{\mu} = \frac{1}{k} \sum_{i=1}^{k} x_i$$

k number of modules
x_i number of defects on the ith module

5.6 Statistic Test and Interval Estimate for Quality Characteristics

5.6.1 General Problem Definition

The point estimates give a first instrument to obtain information about the quality characteristic values of a manufacturing process. This information, however, does not suffice for a quality-conform production, as the following example shows.

For the manufacturing of resistors, a sample ($n = 10$) is taken and the arithmetic mean of the sample determined using $\bar{x} = 9.98$ kΩ. The associated nominal value is $\mu_o = 10$ kΩ. This raises the question whether the process is still quality conform (here, this means whether the nominal value is maintained). The point estimate for the operating point of the considered process immediately allows the initial conclusion to be made that the operating point does not agree with the required nominal value. A correction of the process appears unavoidable. This, however, is countered by the fact that the calculated mean was determined using a sample, so that a random effect could cause here an estimate of the operating point that is too small.

This raises the question here whether the observed dispersion can occur because of some random effect or whether a change of the process caused this dispersion. If the process has changed, the cause of the change must be found and rectified. Otherwise, the process can continue to run without intervention.

This question can, however, *never* be answered with absolute certainty, namely incorrect conclusions (errors) cannot be excluded. There are two general types of such errors:

Type I error:
A normally good process (namely the process produces quality-conform products; in the above example, this is a process whose operating point lies exactly at 10 kΩ) will be classified as a bad process. As a consequence of this error, changes will be made to a good process, which normally then cause just the opposite. The probability for the occurrence of such an error is designated in the following discussion as α.

Type II error:
A bad process (operating point deviates from 10 kΩ) is judged as being a quality-conform process. As a consequence, no changes are made to the

process, so that (in the extreme case) rejects can be produced (or corrective work is required). The probability of this error is designated as β.

In general, orders of magnitudes of 0.1%–10% are used to determine α or β. For all the following considerations (including the calculation of confidence intervals), the specification of the type I error is always necessary; the term error probability (or significance level) is generally used in the literature when a type I error is meant. Alternatively, the confidence level $\varepsilon = 1-\alpha$ is used. We emphasise that the mentioned errors (despite such unwelcome consequences) cannot be prevented for statistical tests. Even a complete test (which sometimes is not economically sensible or cannot be performed) cannot prevent errors and incorrect decisions.

5.6.2 Testing the Continuous Quality Characteristic Values

Testing the Operating Points

As a starting point, the example (mean $\bar{x} = 9.98\,\mathrm{k\Omega}$) discussed in Section 5.6.1 is used. It is also known that the manufacturing of the resistors has a normal distribution with the standard deviation $\sigma = 0.03$ kΩ. We now assert that the operating point for the manufacturing of the resistors observes the nominal value 10 kΩ. Such an assertion is called a hypothesis in the following section.

A differentiation is made between two basic types of hypotheses: two-sided and one-sided hypotheses. The two-sided hypothesis H$_0$ is: $\mu = \mu_0$ (the true, but unknown, operating point has the value μ_0). The one-sided hypotheses are $\mu \geq \mu_0$ or $\mu \leq \mu_0$. Initially, we consider the two-sided problem situation.

To test the hypothesis, first a sample (size n) is taken. The arithmetic mean $\bar{x}$ is calculated from the sample. At the beginning we assume that the hypothesis is true (and this is the case for *all* statistical hypotheses), namely the process has just the operating point μ_0. Consequently, the sample was taken from a normally distributed population $N(\mu_0, \sigma^2)$. According to Equation (5.42), the calculated average value is itself a random variable and has a normal distribution. Figure 5.8 illustrates this behaviour. The distribution of the means is also a normal distribution. Because this distribution is defined, at least theoretically, for the range $-\infty$ to $+\infty$, it is possible that an arbitrary mean, even when it also lies far outside the specified nominal value, could originate from quality-conform manufacturing. Thus, the hypothesis must always be confirmed. Although the probability for such a case is extremely small, it is not zero. Consequently, a compromise must be made for the evaluation of the production. It defines when the mean determined from the sample (or in the following discussion, some other statistical measured value or an arbitrary transformation) exceeds or falls below one (or several) limit value(s), the hypothesis is not confirmed and the considered manufacturing process does not produce quality conform. These limit(s) (designated as G_l or G_u in Figure 5.8) are calculated using the definition of the type I error (α).

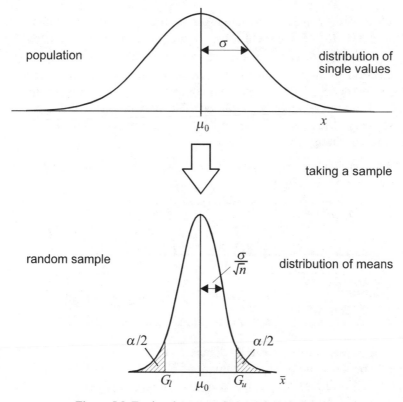

Figure 5.8. Testing the mean of a manufacturing process

We now always assume that the considered sample comes from a manufacturing process that works quality-conform (the operating point here lies exactly at the nominal value), so that the hypothesis would be confirmed and the probability of the rejection of the hypothesis (which would then be unwarranted) does not exceed that of the type I error. Figure 5.8 (top of the population; below the distribution of the means of the taken sample) shows such a production. A two-sided hypothesis now means that a dispersion of the operating point from the nominal value in both directions would cause problems. Consequently, the type I error is distributed equally at both sides of the possible (problematic) deviations ($= \alpha/2$). This allows the limits G_l and G_u to be calculated. They lie exactly where the area under the curve of the normal distribution (Figure 5.8) between $-\infty$ and G_l and between G_u and $+\infty$ each yields the value $\alpha/2$. Using the definition of the quantile for the normal distribution, the following equation results for the limits:

$$G_{l,u} = \mu_0 \mp u_{1-\alpha/2} \frac{\sigma}{\sqrt{n}} \tag{5.55}$$

Table 5.6 shows the calculated limits for the chosen example. It is obvious that different conclusions would result for the measured mean $\bar{x} = 9.98$ kΩ. For example, if the associated process would be accepted for the selection of an error probability $\alpha = 1\%$, whereas for $\alpha = 5\%$ a faulty process is diagnosed. These apparent contradictions, however, are natural as statistical analyses cannot be avoided. The prevention of this error would require very wide limits (theoretically infinitely) that would make an evaluation of the process impossible.

Table 5.6. Limits for the confirmation of the hypothesis $\mu = \mu_0 = 10$ kΩ for different type I error and different sample sizes n

α $(n = 10)$	0.27%	1.00%	5.00%
G_l	9.9715 kΩ	9.9756 kΩ	9.9814 kΩ
G_u	10.0285 kΩ	10.0244 kΩ	10.0186 kΩ

n $(\alpha = 1\%)$	5	10	50
G_l	9.965 kΩ	9.976 kΩ	9.989 kΩ
G_u	10.035 kΩ	10.0244 kΩ	10.011 kΩ

This table shows two basic tendencies. With decreasing type I error, the limits expand, namely, the evaluation of the process becomes "more generous". For increasing sample size, the limits become smaller because of the larger information content of the sample.

In many cases, a single-sided estimate is appropriate, which should be used in particular when the dispersion of the considered quality characteristic is critical only on one side. In this case, the type I error moves only to the critical side. This produces the following calculations for the critical limits:

$$G_l = \mu_0 - u_{1-\alpha} \frac{\sigma}{\sqrt{n}} \quad G_u = \infty \quad \text{for} \quad \mu > \mu_0 \tag{5.56}$$

$$G_u = \mu_0 + u_{1-\alpha} \frac{\sigma}{\sqrt{n}} \quad G_l = -\infty \quad \text{for} \quad \mu < \mu_0 \tag{5.57}$$

Compared with a two-sided estimate (for constant α), the associated limits lie nearer the nominal value.

For a resistor production with subsequent laser matching, for example, resistance values that are too large are critical, whereas resistance values that are too small (within certain limits) are not critical. Thus, Equation (5.57) should be used for this case.

The use of Equations (5.55) to (5.57) requires knowledge of the standard deviation of the process. If this is not so, the standard deviation must be estimated using the sample. In such a case, the sample function (5.46) (t-distribution) must be used to derive the limits.

$$G_{l,u} = \mu_0 \mp t_{n-1;1-\alpha/2} \frac{s}{\sqrt{n}} \tag{5.58}$$

The quantiles of the t-distribution can be taken from Table A.3. In general, the calculated interval for the limits is always somewhat larger than for the use of the normal distribution (see Equation (5.55)). The differences become smaller with increasing sample size and these limits are identical for $n \to \infty$. The causes lie in the increased uncertainty that the estimate of the standard deviation from the sample compared with a given standard deviation. Analogous to Equation (5.56) or (5.57), one-sided estimates can also be specified here. Table 5.7 shows the calculations with the t-distribution under the assumption that $s = 0.03$ kΩ was determined for the standard deviation of the sample. This allows a direct comparison with Table 5.6. Otherwise, the possible conclusions are the same as for a given standard deviation.

Table 5.7. Calculation of the limits for the use of the t-distribution

α $(n = 10)$	0.27%	1.00%	5.00%
G_l	9.961 kΩ	9.969 kΩ	9.978 kΩ
G_u	10.039 kΩ	10.031 kΩ	10.021 kΩ

n $(\alpha = 1\%)$	5	10	50
G_l	9.939 kΩ	9.969 kΩ	9.989 kΩ
G_u	10.061 kΩ	10.031 kΩ	10.011 kΩ

Testing the Variance

The testing of means always takes precedence for the inspection of processes. A change of the mean increases the danger of the production of the faulty products with the resulting rejects and/or correction work. However, changes of the dispersion of the process can also cause quality faults.

The measure for the testing of the dispersion is the specification of a target variance σ_0^2. Initially, a two-sided test is made again, namely the hypothesis H_0 $\sigma^2 = \sigma_0^2$. Equation (5.45) is used to test the variance. This equation can be rearranged to produce:

$$S^2 = \frac{Z}{n-1}\sigma^2$$

In accordance with Equation (5.45) Z has the χ^2-distribution. The quantiles are contained in Table A.4.

As for the mean considerations, the dispersion limits can be derived:

$$G_l = \frac{\chi^2_{n-1;\alpha/2}}{n-1}\sigma_0^2 \qquad\qquad G_u = \frac{\chi^2_{n-1;1-\alpha/2}}{n-1}\sigma_0^2 \qquad (5.59)$$

In contrast to the testing of the mean, these limits are asymmetric to the specified target variance σ_0^2.

The two-sided test of the variance is the exception in practice, because a variance that is too small is not normally damaging. Consequently, the hypothesis $\sigma^2 \leq \sigma_0^2$ normally uses tests whether the variance is smaller than the specified maximum variance σ_0^2,

$$G_u = \frac{\chi^2_{n-1;1-\alpha}}{n-1}\sigma_0^2 \quad G_l = 0 \qquad\qquad (5.60)$$

5.6.3 Comparing Continuous Quality Characteristic Values

Comparing Operating Points (Means)

Often the task involves the comparison of means. Test objects can be, for example, two measuring devices that can be queried as to whether both devices for the same measurement object on average also indicate the same measured values. Or there are two machines that manufacture the same products and it is queried whether there are differences between the two machines with regard to a specific quality characteristic.

The starting point of the test are two samples with n_1 measured values $x_{11} \ldots x_{1n_1}$ from the first machine and n_2 values $x_{21} \ldots x_{2n_2}$ from the second machine. These samples are used to produce the means $\bar{x}_1$ and $\bar{x}_2$ the standard deviations s_1 and s_2. Although the sample sizes are not necessarily equal, this is desirable.

A significant difference of the means between the two machines occurs when the following inequality is true:

$$\frac{|\bar{x}_1 - \bar{x}_2|}{\sqrt{\frac{(n_1-1)s_1^2 + (n_2-1)s_2^2}{n_1 + n_2 - 2} \frac{n_1 + n_2}{n_1 n_2}}} > t_{n_1 + n_2 - 2; 1 - \alpha} \tag{5.61}$$

If the sample sizes are equal, this inequality simplifies to:

$$\frac{|\bar{x}_1 - \bar{x}_2|}{\sqrt{\frac{s_1^2 + s_2^2}{n}}} > t_{2(n-1); 1 - \alpha} \tag{5.62}$$

Example:
An automatic placement machine places chip resistors on a glass plate as a regular matrix (5×4). The displacements of the placed resistors are then measured.

The measurement results are used to make changes to the automatic placement machine after which components are placed on another glass plate and further measurements made. Table 5.8 shows the results of both measurements.

This produces the following equation for the comparison of the means:

$$\frac{|\bar{x}_1 - \bar{x}_2|}{\sqrt{\frac{s_1^2 + s_2^2}{n}}} = \frac{|14.57 - 9.03|}{\sqrt{\frac{9.69^2 + 11.78^2}{20}}} = 1.624 < t_{38; 0.95} = 1.69$$

This leads to the conclusion that there is no significant statistical difference between the two placements.

However, this raises the question whether all available information is used. For example, in the described case, the chip resistor with number 1 for both insertions is set at the same coordinates by the automatic placement machines. This means that the two measurements or the samples are connected here. In such a case it is possible to evaluate the associated differences of the measured values for both samples.

Table 5.8. Measurement results

	Series of measurements 1	Series of measurements 2	Difference
1	26.1	16.7	9.4
2	2.4	−0.6	2.9
3	3.9	7.6	−3.7
4	13.9	5.2	8.7
5	−3.5	−17.1	13.6
6	15.9	18.7	−2.8
7	13.6	13.3	0.3
8	17.3	3.8	13.5
9	4.9	0.9	4.1
10	11.7	6.8	4.9
11	13.5	9.8	3.8
12	26.3	19.7	6.6
13	8.8	12.0	−3.1
14	21.3	11.3	10.0
15	20.4	23.4	−3.0
16	5.4	−7.2	12.6
17	25.5	20.0	5.5
18	6.5	−11.3	17.8
19	25.0	22.1	2.9
20	32.5	25.6	6.9
Mean	$\bar{x}_1 = 14.57$	$\bar{x}_2 = 9.03$	$\bar{d} = 5.54$
Standard deviation	$s_1 = 9.69$	$s_2 = 11.78$	$s_d = 6.18$

For connected samples, a mean value difference is significant when:

$$\frac{\bar{d}}{s_d}\sqrt{n} > t_{n-1;1-\alpha} \tag{5.63}$$

$$\frac{5.54}{6.18}\sqrt{20} = 4.005 > t_{19;0.95} = 2.093$$

This test produces a significant difference. What is the reason for the difference from the previous result? Although there is no difference in the evaluated means ($|\bar{x}_1 - \bar{x}_2| = \bar{d}$), the standard deviation of the differences is less than that for the individual series of measurements. It is significant that the connection of the measured values provides additional information.

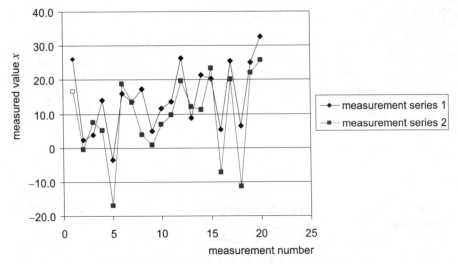

Figure 5.9. Representation of the two measured value series

Figure 5.9 shows the causes. The two series of measurements are not randomly dispersed fully independent of each other, but rather there are specific dependencies, in particular, for the measured values 15 to 19. Furthermore, it shows that measured value series 2 normally has smaller values. The forming of differences represents this much better than through forming just mean values and their comparison. The example shows the importance of supplementary information.

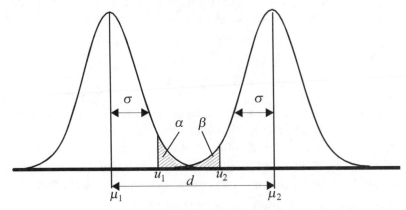

Figure 5.10. Comparison of two tests with regard to their mean

For a mean comparison, one can determine the required sample size n when specific advance knowledge and prerequisites apply. We assume that the standard deviations σ of both tests are known and have a normal distribution. It is to be proved that the means μ_1 and μ_2 differ significantly.

The proof of this difference is made using the means $\bar{x}_1$ und $\bar{x}_2$ from the tests. If we specify that a difference $d\,(=\mu_2-\mu_1)$ of both means with a probability $1-\beta$ (β error probability of the second type) and simultaneously a possible equality of μ_1 and μ_2 with a probability of $1-\alpha$ (α error probability of the first type) is to be recognised, this produces for the two auxiliary quantities u_1 and u_2 in Figure 5.10 for a one-sided problem situation:

$$u_1 = \mu_1 + u_{1-\alpha}\sqrt{\frac{\sigma^2}{n}} \quad ; \quad u_2 = \mu_2 - u_{1-\beta}\sqrt{\frac{\sigma^2}{n}}$$

If u_1 and u_2 now overlap, a difference between μ_2 and μ_1 cannot be proved statistically. The difference d thus can also be considered as being the resolution of the comparison. If we set both quantities equal, we can then calculate the minimum test size for each test that is necessary to prove a difference d of the means:

$$n = \left(\frac{u_{1-\alpha}+u_{1-\beta}}{d}\right)^2 \sigma^2 \tag{5.64}$$

Comparison of Variances

The comparison of the quality of machines and processes also requires the comparison of the variances. The conclusion that process 1 can provide better quality than process 2 requires the proof that the variance of process 1 (σ_1^2) is smaller than the variance of process 2 (σ_2^2). The following hypothesis is made:

$$H_0: \quad \sigma_1^2 = \sigma_2^2$$

If this hypothesis is rejected, the variance behaviour of the two processes is different and consequently the capability to produce quality is also different. As previously, it is assumed that the hypothesis is correct, namely both processes have the same variance. To test this hypothesis, a sample (n_1) is taken from process 1 and the variance s_1^2 calculated. Similarly, a sample (n_2 not necessarily equal to n_1) is taken from process 2 and the variance s_2^2 calculated.

The test quantity $f = \dfrac{s_1^2}{s_2^2}$ is calculated here.

If this quotient is one, the hypothesis H_0 can be accepted, namely both machines have no proven difference in their variances. However, even for a correct hypothesis, random deviations of the quotients in both directions are still possible. These random deviations are handled by the sample function (5.47) (F-distribution). A deviation of the quotients from 1 is random when the following inequalities apply:

$$\frac{s_1^2}{s_2^2} \geq F_{n_1-1;n_2-1;\alpha/2} \qquad ; \qquad \frac{s_1^2}{s_2^2} \leq F_{n_1-1;n_2-1;1-\alpha/2} \qquad (5.65)$$

If an inequality is not satisfied, the hypothesis will be rejected, namely the variances differ significantly from each other. The quantiles of the F-distribution required for the test are contained in Table A.5 and Table A.6 in the Appendix.

This two-sided test is normally seldom used. One-sided tests are usual. The hypothesis $\sigma_1^2 > \sigma_2^2$ would be rejected when the following inequality is satisfied:

$$\frac{s_1^2}{s_2^2} < F_{n_1-1;n_2-1;\alpha} \qquad ; \qquad \left(F_{n_1-1;n_2-1;\alpha} = \frac{1}{F_{n_2-1;n_1-1;1-\alpha}} \right) \qquad (5.66)$$

The hypothesis $\sigma_1^2 < \sigma_2^2$ would be rejected when the following inequality is satisfied:

$$\frac{s_1^2}{s_2^2} > F_{n_1-1;n_2-1;1-\alpha}$$

5.6.4 Test Procedure for Discrete Quality Characteristic Values

The discrete quality characteristic values normally aim at achieving a low defect rate p or a high yield a. What is required is the proof that, for example, a new process does not exceed a specific given maximum defect rate p_{max} or a required first-pass yield (see Section 5.3.6) is exceeded. This can be tested using a trial production run.

Once again, a two-sided problem situation is considered initially. The hypothesis is:

$$H_0: \quad p = p_0$$

To test the hypothesis, a sample of size n is taken from the population. This sample can either be identical with a trial production or it is taken from a running manufacturing process. The number of defects k in the sample is determined. The relative frequency of defects y is used as the test quantity

$$y = \frac{k}{n}$$

The transformation (5.48) is used to calculate the limits.

$$G_{l,u} = p_0 \mp \frac{u_{1-\alpha/2}\sqrt{p_0(1-p_0)}}{\sqrt{n}} \qquad (5.67)$$

If G_l is less than zero, then G_l is set to zero. If the test quantity y lies outside these bounds, the hypothesis is not confirmed. The two-sided problem situation is not normally used. An undershooting of G_l by the determined frequency of faults y means only the rejection of the H_0 hypothesis. This shows that the defect rate of the tested process is better than p_0. This indicates, among other things, that the process has reserves. Because normally for practical uses only exceeding p_0 is "bad", a one-sided problem situation is normally used. The hypothesis is then:

$$H_0: \quad p \le p_0$$

$$G_l = 0 \quad ; \quad G_u = p_0 + \frac{u_{1-\alpha}\sqrt{p_0(1-p_0)}}{\sqrt{n}} \qquad (5.68)$$

5.6.5 Testing of Distributions

All previous tests assumed that the tested quality characteristic obeyed a normal distribution. This, however, is an assertion that must first be proved. A simple test of this assertion is possible using probability paper. In this paper, the ordinate is chosen so that the drawn distribution function $F(x)$ of the normal distribution becomes a straight line.

Figure 5.11 shows the general form of the probability network for the normal distribution. The assignment of the abscissa can be freely chosen and so adapted to the corresponding problem specification. To test a distribution in the network, the empirical distribution function $\hat{F}(x)$ is entered into the network using the ordinate transformation. In Figure 5.11, this is done for the measured removal forces (see Table 5.2). For the evaluation whether this empirical distribution function corresponds to a normal distribution, the theoretical distribution function $F(x)$ is also entered. This distribution function is a straight line in the probability network. The position of the straight line can be determined using the expected value and the standard deviation of the theoretical distribution function. These parameters are determined using the point estimate from the available data. For the removal forces, this produces:

$$\hat{\mu} = \bar{x} = 6.51 \text{ N} \qquad ; \qquad \hat{\sigma} = s = 0.55 \text{ N}$$

This allows the following entries to be made for the straight line:

$$\text{point 1: } [\bar{x}, \ 50\%]$$

$$\text{point 2: } [\bar{x} - s, \ 15.87\%] = [5.96 \text{ N}, \ 15.87\%]$$

Other possibilities exist for the entry of the point 2. The corresponding auxiliary points on the left-hand ordinate of the probability network provide orientation here.

Finally, a visual comparison between the theoretical (straight line) and the empirical (step function, the dispersion between the straight line and the associated upper step value is decisive) distribution function is then made. If both functions show excessive deviations, it can be assumed that a normal distribution does not exist. In particular, distinctive bends of the staircase curve indicate deviations from the normal distribution. For the removal forces example, it can be assumed that a normal distribution exists. Note that larger deviations can occur (and should be tolerated) for small sample sizes.

The visual comparison is obviously not an exact mathematical-statistical criterion. However, appropriate methods are available. Examples for such methods are the χ^2 adaptation test and the Kolmogorov–Smirnov test.

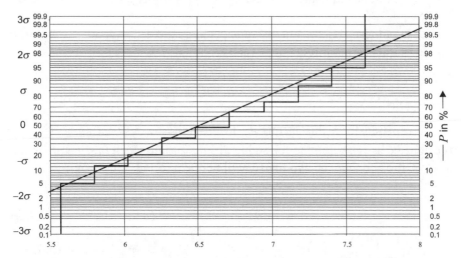

Figure 5.11. The empirical distribution function $\hat{F}(x)$ in the probability network

5.6.6 Confidence Estimates

The test for individual quality characteristic values and the performed point estimates normally do not provide any statement about the accuracy of the estimate. Generally, the test of the quality characteristic value provides only an acceptance or a rejection of the hypothesis with the associated conclusions for the considered process. Confidence estimates can be used to correct this deficiency. In general, a confidence estimate is based on the following problem definition. A point estimate is used to estimate a quality characteristic value (*e.g.*, operating point, expected value μ). It is known that random factors mean this point estimate has a certain inaccuracy. The task of the confidence estimate is now to specify an interval in which the actual value of the quality characteristic value is present with a specific probability.

Confidence Estimates for Operating Points

Again, we consider the example of the resistor production. The mean was estimated $\hat{\mu} = 9.98$ kΩ using the sample as a basis. The standard deviation was given by $\sigma = 0.03$ kΩ and a normal distribution of the population is assumed. We require the range $[g_l, g_u]$ in which the unknown operating point μ of the manufacturing process has a probability ε (in general 95% or $1-\alpha$).

The hypothesis $H_0 : \mu = \mu_0$ is considered again. This hypothesis will not be rejected when the following inequality applies:

$$\left| \frac{\overline{x} - \mu_0}{\sigma} \sqrt{n} \right| < u_{1-\alpha/2}$$

We can transform this inequality to provide as a result a range for μ_0 (designated as $[g_l, g_u]$ in the following discussion) in which this inequality is always true and thus the hypothesis will not be rejected. This provides:

$$\overline{x} - u_{1-\alpha/2} \frac{\sigma}{\sqrt{n}} < \mu < \overline{x} + u_{1-\alpha/2} \frac{\sigma}{\sqrt{n}}$$

$$g_l = \overline{x} - u_{1-\alpha/2} \frac{\sigma}{\sqrt{n}} \qquad ; \qquad g_u = \overline{x} + u_{1-\alpha/2} \frac{\sigma}{\sqrt{n}} \qquad (5.69)$$

This is a calculation of a two-sided confidence interval. Similarly, we can also calculate one-sided confidence intervals.

$$g_l = \overline{x} - u_{1-\alpha}\frac{\sigma}{\sqrt{n}} \quad ; \quad g_u = \infty \tag{5.70}$$

$$g_l = 0 \quad (\text{resp. } -\infty) \quad ; \quad g_u = \overline{x} + u_{1-\alpha}\frac{\sigma}{\sqrt{n}}$$

If the standard deviation is not known, the confidence limits can be calculated using the t-distribution.

$$g_l = \overline{x} - t_{n-1;1-\alpha/2}\frac{s}{\sqrt{n}} \qquad g_u = \overline{x} + t_{n-1;1-\alpha/2}\frac{s}{\sqrt{n}} \tag{5.71}$$

For the considered example, the actual limits (for $n=10$ and $\alpha=1\%$) are:

	Normal distribution	t-distribution
g_l	9.955 kΩ	9.949 kΩ
g_u	10.004 kΩ	10.011 kΩ

This means that the unknown operating point with a probability of 99% lies in the [9.955 kΩ; 10.004 kΩ] range. There is a malfunction in the process if the sought nominal value does not lie in this range.

In addition to the conclusion whether a fault is present in the process, which is identical to the conclusion of an equivalent hypothesis, in particular, the width of the confidence intervals, in particular, provides a statement about the precision of the statistical test. It is also possible to use a specified size d for the confidence interval to calculate the required sample size for this proof. We can transform Equation (5.69) to provide:

$$d = g_u - g_l \qquad n = 4 \cdot \left(\frac{(u_{1-\alpha/2}) \cdot \sigma}{d}\right)^2 \tag{5.72}$$

If the standard deviation is not known, this equation can also be used, where, however, it must be considered that the quantile of the t-distribution also depends on the sample size. This means that the solution is not possible by the simple transformation of the equation based on the sample size, but can be solved by iteration.

Confidence Estimates for Variance/Standard Deviations

Similar to the interval estimates for the operating point, an interval estimate can also be derived for the standard deviation using the χ^2-distribution. The following equations can be used to determine the limits of the standard deviation for a two-sided interval estimate:

$$g_l = \sqrt{\frac{(n-1)}{\chi^2_{n-1,1-\alpha/2}}}\, s \quad ; \quad g_u = \sqrt{\frac{(n-1)}{\chi^2_{n-1,\alpha/2}}}\, s \qquad (5.73)$$

One-sided estimates can also be used here, where, depending on the purpose, two possibilities exist:

$$g_l = 0 \quad ; \quad g_u = \sqrt{\frac{(n-1)}{\chi^2_{n-1,\alpha}}}\, s \qquad (5.74)$$

This estimate is suitable for making conclusions about the upper limit of the standard deviation, namely g_u specifies how large the standard deviation in the population set can be in the worst case. This is equivalent to a pessimistic estimate of the standard deviation.

$$g_l = \sqrt{\frac{(n-1)}{\chi^2_{n-1,1-\alpha}}}\, s \quad ; \quad g_u = \infty \qquad (5.75)$$

An optimistic estimate of the standard deviation is performed here, namely, g_l specifies the minimum size of the standard deviation in the population.

Confidence Estimates for Discrete Quality Characteristic Values

The discrete quality characteristic values centre on the *defect rate*. The starting point is a sample in which k defect products were determined. The interval of the defect rate from which this sample could originate is required. The interval estimate uses the sample function (5.48). For the two-sided hypothesis $p = p_0$, the limit for the defect rate can be derived by transforming the equation to provide p_0.

$$g_{l,u} = \frac{1}{n+u^2_{1-\alpha/2}}\left(k + \frac{u^2_{1-\alpha/2}}{2} \mp u_{1-\alpha/2}\sqrt{\frac{k\cdot(n-k)}{n} + \frac{u^2_{1-\alpha/2}}{4}}\, \right) \qquad (5.76)$$

These limits are only an approximation and are true for $n > 100$. One-sided estimates, similar to the standard deviation, can be derived and interpreted appropriately here.

Table 5.9 shows a sample calculation depending on the lot size. It again shows for typical lot sizes in the module production that conclusions from the confidence intervals are very uncertain. The main reason is the significantly smaller information content of a good-bad statement compared with a measured value. This again strengthens the necessity for searching for measurable characteristics in a process.

Table 5.9. Confidence limits for attributive characteristics from the sample for an estimated average defect rate of 2% and an error probability of $\alpha = 5\%$

n	k	g_l	g_u
100	2	0.6%	7.1%
200	4	0.8%	5.1%
500	10	1.1%	3.7%
1,000	20	1.3%	3.1%
2,000	40	1.5%	2.7%
5,000	100	1.6%	2.4%
10,000	200	1.7%	2.3%

5.7 Quality–Control Charts

5.7.1 Introduction

Quality–control charts are an aid in quality control for processes, where a process is considered to be a single machine or also several manufacturing steps that can be influenced by the 5 causes man, machine, material, method and human environment. It should ensure that the characteristic values of a product and/or the process data lie within specified limits. It is assumed here that the investigated process would generally be capable (even without testing) of producing quality. Because, however, malfunctions cannot be excluded, the use of such charts allows the effects of such defects to be detected as deviations in time and their causes rectified.

Deviations are divided into systematic and random deviations. The systematic deviations include tool changes, material changes, incorrect operation, changes of the ambient effects, *etc*. Actual examples are the systematic displacement of components during the placement and the systematic displacement of the solder paste deposits during the solder paste printing. The causes of these malfunctions are normally faulty machine settings (in particular for the solder paste printing due to the incorrect setup) or a "creeping" drift of the machine. This occurs, in particular, for automatic placement machines.

We assume that systematic deviations can generally be prevented. If a process is free from systematic deviations, it is designated as being not faulty, so that only random deviations occur. If one is capable of correcting detected systematic deviations, the process can be said to be mastered. This is done, for example, by calibrating the machine (*e.g.*, automatic placement machine) or by setting additional offsets for the solder paste printer.

The random deviations include limited machine accuracy, environmental effects, material variations, *etc.* These deviations are a "natural" property of processes. If these deviations do not exceed specific limits, they are normally also not avoidable. Consequently, in contrast to the correction of systematic deviations, these random deviations can be rectified only to a specific limit. If these deviations are so small that the probability of randomly producing a defect product is nearly zero (in practice $\ll 0.1\%$), this is called a (quality-)conform process (see Section 5.8).

Quality–control charts have the task of detecting and showing systematic and excessive random deviations. The causes for determined deviations, however, cannot be determined directly in the charts because they are specific for the monitored process. In addition, such charts also have a recording function for the quality and reflect the quality level of the manufacturing. The recording function is particularly important for quality certificates (audits, certification, possible customer problems).

5.7.2 Process Types

Before a process can be controlled and monitored, it is necessary to determine its most important properties so that optimum monitoring can be achieved. These properties include:

- process type
- operating point μ_0 (process mean)
- information about the standard deviation σ.

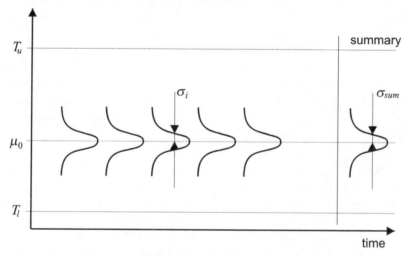

Figure 5.12. Process type A

The process type A is the "desired type". The operating point of the process is independent of the time. The standard deviation σ_i (internal standard deviation) of the process is constant, namely the standard deviation determined over an extended period by making observations (total standard deviation σ_{sum}) corresponds to this internal standard deviation.

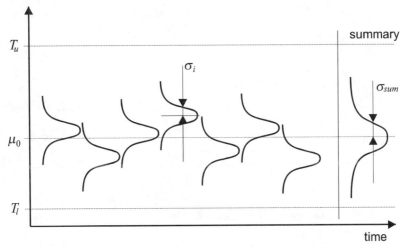

Figure 5.13. Process type B

The current mean fluctuates for process type B. The internal standard deviation is constant. The fluctuations of the mean produce a large total standard deviation. The fluctuations of the mean are described by the external standard deviation σ_a. The following equation describes the relationship between these standard deviations:

$$\sigma_{sum} = \sqrt{\sigma_i^2 + \sigma_a^2} \tag{5.77}$$

These standard deviations are determined using the variance analysis (see Section 7.4).

Process type C changes systematically with the process mean, where there is a linear change as a function of time here. The internal standard deviation remains constant. This is also called a trend process.

We have mentioned only the most important process types. Reference [1] describes other types.

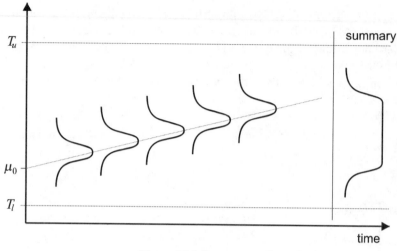

Figure 5.14. Process type C

5.7.3 Shewhart Quality–Control Charts

Basic Principle of Quality–Control Charts

We use a quality–control chart to monitor an arbitrary quality characteristic G (*e.g.*, a resistance value) that is comprehended as being a random variable. Figure 5.15 shows the general form of such a quality–control chart.

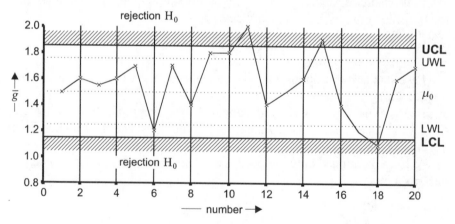

Figure 5.15. Basic form of a quality–control chart

At specified times, a sample (size n) is taken, a test quantity calculated and entered in the chart. Provided this test quantity lies within an interval

[*LCL, UCL*] (lower or upper control limit, also designated as the control range), the observed process is under control, namely there is no reason to make any changes to the process. However, for example, this is not the case for the 11th control time, namely the technological process is no longer under control with regard to the observance of the nominal value (here the observance of a mean μ_0). This requires a regulating intervention in the process. Furthermore, the control chart also contains a centre line M (here μ_0). This normally corresponds to the monitored nominal value.

To calculate the control limits, the following considerations assume that the size G has a normal distribution, namely it is also possible to write:

$$G - N\left(\mu_G, \sigma_G^2\right)$$

G should now observe a specific nominal value μ_0, namely the hypothesis is initially:

$$H_0: \quad \mu_G = \mu_0$$

This hypothesis is tested using a sample of size n ($g_1, \ldots g_n$). This is used to form a test quantity (*e.g.*, the mean $\bar{g}$) This test quantity has (when the mean is used) the following theoretical probability properties:

$$\bar{G} - N\left(\mu_G, \frac{\sigma_G^2}{n}\right)$$

We now consider a process that *exactly* observes the nominal value μ_0 and for which the process standard deviation σ_G is known. This assumes that a nondefect (and thus good) process is involved. When a sample is taken from this process, the probability that this process (which is in fact "good") is considered as being faulty will not be larger than α (type I error probability). The control limits are then calculated as:

$$LCL = \mu_o - u_{1-\alpha/2}\frac{\sigma_G}{\sqrt{n}} \quad \text{and} \quad UCL = \mu_o + u_{1-\alpha/2}\frac{\sigma_G}{\sqrt{n}} \tag{5.78}$$

The significance level α specifies the probability with which one is prepared to correct a well-running process, which normally involves a useless or even damaging effort. Currently 1% is normally used for α. However, the technological or economic situation may mean that other values should be used (*e.g.*, 0.27%, 5%, 10%). The smaller this probability is selected, the larger the control range becomes. However, this also increases the danger of not detecting a process that has really got out of control (here a changed process mean; for the evaluation, also refer to

Section 5.7.5). Quality–control charts for which the control limits are calculated using this principle are also called *Shewhart control charts*.

The sample size n is small, and normally lies between 1 and 13. It is usually selected as an odd number because the median then can be read directly. The preferred value is $n = 5$. The larger the sample size, the smaller the control range becomes with a corresponding increase in the control level. This advantage, however, is countered by the effort (test costs) that a large sample size causes.

The samples should be taken at regular intervals. The interval size depends on the actual process, where, on the one hand, the test costs play a role, but, on the other hand, also the resulting costs when a fault occurs in this time interval. Typical time intervals are 30 min, 1 h, 2 h.

In addition to the previously discussed control chart that detects deviations of a quality characteristic value (the operating point here) at both sides of μ_0 and thus also designated as a two-sided chart, appropriately modified one-sided quality–control charts each with only one intervention limit exist for the case that μ_0 should only not undershoot or exceed the nominal value. Whereas, for a two-sided chart, the type I error α is "distributed" equally on both sides, it can be placed on the critical side for one-sided charts. Consequently, the corresponding control limits are now:

$$LCL = \mu_0 - u_{1-\alpha}\frac{\sigma_G}{\sqrt{n}} \; ; \quad \text{for H}_0: \; \mu_G > \mu_0 \; \text{ or} \tag{5.79}$$

$$UCL = \mu_0 + u_{1-\alpha}\frac{\sigma_G}{\sqrt{n}} \; ; \quad \text{for H}_0: \; \mu_G < \mu_0 \tag{5.80}$$

If the use of one-sided control charts is possible (namely the monitored quality characteristic value G can deviate (almost) arbitrarily far from μ_0 at the noncritical side), it should be used in preference to the two-sided chart, because deviations from the nominal value can be detected faster for the one-sided chart (the control limits are nearer to μ_0 than for a comparable two-sided chart).

In addition to the control limits, warning limits (*LCL* and *UCL*) are usually drawn on the charts. An overshoot or undershoot of these limits indicates a certain danger that the process could get out of control. Although measures are normally not yet necessary, increased attention should be paid to the process. Another possibility is the taking of a further sample in order to provide certainly about a possible fault. These limits can be calculated with Equations (5.78) to (5.80), although 5% rather than 1% should then be used for α.

In addition to exceeding of the control limits, control charts can also indicate approaching malfunctions. These involve a *run* and a *trend*. A run has occurred when the entries in the chart lie seven times *in succession* below or above the centre line (in this case μ_0). If such a case occurs, the probability $(= 0.5^7 \approx 0.8\%)$ is extremely small that the process can meet the requirement $\mu = \mu_0$. It must be assumed that the process deviates systematically from μ_0. A correction is then desirable, even when no control limits have been exceeded.

A trend is present when the entries in the chart reduce or increase seven times *in succession*. The probability is then very large that the process mean moves systematically in one direction. This can be caused, for example, by tool wear. A correction is also desirable here.

We must emphasise that this approach functions for the construction of a quality control only when process type A is present. Otherwise, an increased proportion of incorrect decisions can be expected.

Charts for Monitoring the Process Mean

The equations for the discussions contained in the last section correspond to a *mean chart* (or $\bar{x}$ chart). This chart is primarily used for the monitoring of process means, where the process standard deviation σ must be known (or at least an estimate, see Section 5.7.6). For practical uses, not Equations (5.78) to (5.80), but rather constants (here A_E for the control limits or A_W for the warning limits), are used for calculating the control limits. These constants depending on the sample size are available in tables (Table A.8) and can be used only for the significance level for $\alpha = 1\%$ and for two-sided charts.

Another chart for the monitoring of the process mean is the *median chart*. To allow the median to be determined quickly and rationally, the sample size should be odd. The following equations can be used to calculate the control limits:

$$LCL = \mu_0 - u_{1-\alpha/2} \cdot c_n \cdot \frac{\sigma}{\sqrt{n}} \quad \text{and} \quad UCL = \mu_0 + u_{1-\alpha/2} \cdot c_n \frac{\sigma}{\sqrt{n}} \quad (5.81)$$

c_n is a constant that depends only on the sample size. It can be obtained from Table A.7. More practical, however, is the use of the C_E or C_W constants (Table A.8).

The *single–value chart* or *original–value chart* can also be used for monitoring the process mean. In this case, a test quantity is not formed, but rather all sample characteristics are entered in the chart. The process must be corrected if one (or several) value lies outside the control limits. The following assumption is used for calculating the control limits. The probability P_{sum} that *all n* sample characteristics for a nonfaulty process lie within the intervention limits should be $1-\alpha$. P_{sum} is the product of the probabilities from P_1 to P_n. Where $P_1 = P_2 = \ldots = P_n = P$ are the associated probabilities that the individual sample characteristic lies within the control limits. Thus:

$$P_{sum} = P^n = 1 - \alpha \quad \text{and so:} \quad P = \sqrt[n]{1-\alpha}$$

(5.82)

$$LCL = \mu_0 - u_{(1+\sqrt[n]{1-\alpha})/2} \cdot \sigma \quad ; \quad UCL = \mu_0 - u_{(1+\sqrt[n]{1-\alpha})/2} \cdot \sigma$$

The mean chart responds best for possible faults. Although the median chart is somewhat more insensitive for the same sample size, it is easier to perform manually (namely without computer support). The original value chart is the least–sensitive chart, but has the advantage that, in addition to the mean monitoring, dispersion deviations can also be detected. Section 5.7.5 has a detailed comparison of these charts. The other equations required for calculating the median chart or the original value chart are contained in Table 5.10.

Charts for Monitoring the Process Variance

The process mean contains the most significant quality statement, because any deviations here lead most quickly to quality failures. The observance of the mean by itself, however, does not suffice. At the second place here is the *process standard deviation* σ. The starting point for an appropriate control chart (also called s-chart) is the fact that the variance of a sample originating from a normally distributed population has the χ^2 distribution. If a process standard deviation of σ_0 should not be exceeded, this yields, for example, for the upper control limit:

$$UCL = \sqrt{\frac{\chi^2_{n-1;1-\alpha/2}}{n-1}} \sigma_0$$

The other equations are contained in Table 5.10, which, however, always assume two-sided charts. The chart is maintained similarly to the mean chart. The standard deviation of the sample is calculated as the test quantity.

Only those process standard deviations that are too large reduce the quality. This means a lower control limit is not normally necessary, because the regulating of a standard deviation that is too small is not useful. For practical uses, consequently a one-sided chart should be preferred. (The necessary change of the equations is made similar to the mean chart; $1-\alpha$ is used instead of $1-\alpha/2$). A two-sided chart for testing the process standard deviation is useful when:

- Reserves in the process are to be detected (indicated by an undershooting of *LCL.*)
- Measurement inaccuracies or operator faults lead to small standard deviations and so can be displayed.

Another chart for monitoring the dispersion is the *range chart* (R-chart) in which the range of the sample is used as the test quantity. The range chart is less effective than the s-chart and is used in particular when a simple manual (without computer) realisation is needed.

Table 5.10. Calculation of the intervention or warning limits for all common chart types (the constants $A_E \ldots D_{LWL}$ in Table A.8 can be taken for $\alpha = 1\%$ and two-sided charts)

Chart type	LCL LWL	Mean	UWL UCL
Single-value chart	$\mu - u_{(1+\sqrt[n]{1-\alpha})/2} \cdot \sigma$ $\mu - u_{(1+\sqrt[n]{0.95})/2} \cdot \sigma$	μ	$\mu + u_{(1+\sqrt[n]{0.95})/2} \cdot \sigma$ $\mu + u_{(1+\sqrt[n]{1-\alpha})/2} \cdot \sigma$
Mean chart	$\mu - u_{1-\alpha/2} \cdot \dfrac{\sigma}{\sqrt{n}} = \mu - A_C \cdot \sigma$ $\mu - u_{0.975}\dfrac{\sigma}{\sqrt{n}} = \mu - A_W \cdot \sigma$	μ	$\mu + u_{0.975}\dfrac{\sigma}{\sqrt{n}} = \mu + A_W \cdot \sigma$ $\mu + u_{1-\alpha/2} \cdot \dfrac{\sigma}{\sqrt{n}} = \mu + A_C \cdot \sigma$
Median chart	$\mu - u_{1-\alpha/2} \cdot c_n \dfrac{\sigma}{\sqrt{n}} = \mu - C_C \cdot \sigma$ $\mu - u_{0.975} \cdot c_n \dfrac{\sigma}{\sqrt{n}} = \mu - C_W \cdot \sigma$	μ	$\mu + u_{0.975} \cdot c_n \dfrac{\sigma}{\sqrt{n}} = \mu + C_W \cdot \sigma$ $\mu + u_{1-\alpha/2} \cdot c_n \dfrac{\sigma}{\sqrt{n}} = \mu + C_C \cdot \sigma$
Disper-sion chart	$\sqrt{\dfrac{\chi^2_{n-1;\alpha/2}}{n-1}}\,\sigma = B_{LCL} \cdot \sigma$ $\sqrt{\dfrac{\chi^2_{n-1;0.025}}{n-1}} \cdot \sigma = B_{LWL} \cdot \sigma$	$a_n \cdot \sigma$	$\sqrt{\dfrac{\chi^2_{n-1;0.975}}{n-1}} \cdot \sigma = B_{UWL} \cdot \sigma$ $\sqrt{\dfrac{\chi^2_{n-1;1-\alpha/2}}{n-1}} \cdot \sigma = B_{UCL} \cdot \sigma$
Range chart	$D_{LCL} \cdot \sigma$ $D_{LWL} \cdot \sigma$	$d_n \cdot \sigma$	$D_{UWL} \cdot \sigma$ $D_{UCL} \cdot \sigma$

5.7.4 Control Charts

All previous charts can only be used for measurable quality characteristics (variable, continuous). Different charts must be developed for unmeasure-able (attributive) characteristics, which, however, use the same principle. A significant difference is, however, that control charts for attributive characteristics show something when defects are already present, whereas for measurable characteristics at least warnings are present before faults occur. Conversely, no degree of dispersion can be monitored for attributive characteristics. Because present defects are used for maintaining the charts and an incorrect evaluation of a module or a defect can lead to wrong decisions, great care must be paid to the problems associated with the precise definition of a defect. For example, different test results for the same test object are possible for the evaluation of solder joints.

Before control charts can be produced for attributive characteristics, it must first be defined what is to be judged. This requires an initial differentiation between defects per module and defect modules per lot. In analogy to the measurable characteristics, attributive quality control charts also have two *control limits*. The difference, however, is the response for the control–limit violation. The overshooting of the upper limit unambiguously indicates quality problems in the test process. An under-shooting of the lower limit is not really desired. There are two reasons for this. Firstly, this indicates a process improvement, and, secondly, there is the possibility that incorrect data was used to produce the chart. If a process improvement is the cause, the chart should always be recalculated.

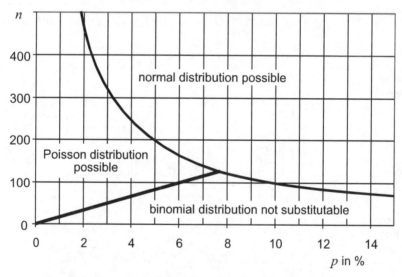

Figure 5.16. Approximation possibilities for attributive control charts

A large problem for the calculation of attributive control charts is the *distribution model* used as basis. In general, the binomial distribution (for defect modules per lot) or the Poisson distribution (for defects per module) is used. For both distributions, the control limits can only be calculated using computer programs, tables or special nomograms (the Larsson diagram for the binomial distribution and the Thorndike diagram for the Poisson distribution [5]). Under certain conditions ($n\hat{p} \cdot (1 - \hat{p}) > 9$), an approximation is possible using the normal distribution. Figure 5.16 shows these limits.

If the prerequisites for an approximation using the normal distribution are satisfied, the following consideration applies to the test quantity defect rate (or the relative frequency of defects, random variable Y) of the sample. If a sample of size n is taken from a population (with a defect rate $p = p_0$), then the relative frequency of defects Y approximates the normal distribution with:

$$Y = N\left(\mu_Y, \sigma_Y^2\right) \quad \text{with} \quad \mu_Y = p_0 \quad \text{and} \quad \sigma_Y^2 = \frac{p_0 \cdot (1 - p_0)}{n}$$

Another problem to be solved for these charts is the specification of the *nominal values* of the chart. This can be done either using preliminary information obtained from the investigated process or by the specification of a maximum defect rate. The nominal value of the charts for the defect rate is designated with p_0 in the following discussion. As an alternative, the specification of a required first-pass yield is also possible.

Similar to the control charts of the measurable characteristics, the *type I error* α must also be specified here. Typical are 0.27% for the use of an approximation with the normal distribution or 1% when the calculation is performed using the binomial or Poisson distribution.

The final problem is the size of the *sample or the test n*. It should, in general, be constant and as large as possible. This requirement, however, cannot always be realised. In particular, the monitoring of the quality for lots must cope with a fluctuating lot size, which then can also have an affect on the intervention limits.

Using the p-Chart to Monitor the Proportion of Defect Modules in the Lot

If the prerequisites for an approximation using the normal distribution apply, the control limits for the *p-chart* (for $\alpha = 0.27\%$) are calculated as follows:

$$LCL/UCL = p_0 \mp 3 \cdot \sqrt{\frac{p_0 \cdot (1 - p_0)}{\overline{n}}} \tag{5.83}$$

$\overline{n}$ average lot size

The average lot size can also be replaced by the actual lot size. The lower control limit can also be negative for small lot sizes and/or small defect rates. This limit must be set to zero in such a case.

Example:
A new module requires a first–pass yield of 90%. The average lot size is 200. The approximation condition for the normal distribution is thus satisfied ($np_0 \cdot (1 - p_0) = 18 > 9$). The following equations specify the control limits (for $\alpha = 0.27\%$; the details for $\alpha = 1\%$ are specified within parentheses):

$$LCL = 0.1 - 3 \times \sqrt{\frac{0.1 \times (1 - 0.1)}{200}} = 3.6\% \quad (= 5.1\%)$$

$$UCL = 0.1 + 3 \times \sqrt{\frac{0.1 \times (1 - 0.1)}{200}} = 16.4\% \quad (= 14.9\%)$$

This also clearly shows the problems of the attributive characteristics because of the size of the interval between the two control limits. The defect rates $p_i (= x_i/n_i)$ of the individual lots are entered in the chart. A modified form of the p-chart is the *np-chart*. Rather than the defect rates of the individual lots, the number of defect modules are entered in the chart. This requires, however, a constant lot size. The following equation provides the control limits:

$$LCL/UCL = np_0 \mp u_{1-\alpha/2} \sqrt{np_0 \cdot (1 - p_0)} \tag{5.84}$$

$$(= np_0 \mp 3 \cdot \sqrt{np_0 \cdot (1 - p_0)}) \quad \text{for} \quad \alpha = 0.27\%$$

Charts for the Number of Defects per Unit

x-chart, c-chart and u-chart for the number of defects per unit
A large disadvantage of the previously discussed attributive charts is the relatively low inspection level, caused by the limitation on the maximum lot sizes normally present. A limited escape can be derived from the fact that a module consists of solder joints, components, *etc.* Because a defect-free module eventuates only when all solder points, components, *etc.*, are defect free, the subassembly consideration ignores the information whether a defect module has just one or several defects. This disadvantage can be countered to some extent by using charts that monitor the number of defects per unit. A unit here can be a single module, but also a number of modules (namely also a lot). The starting point is the specification of the mean

number of defects μ_0 as the quality level to be achieved or determined for the investigated process.

$$\mu_0 = \hat{\mu} = \frac{1}{k}\sum_{i=1}^{k}x_i \quad \text{with} \quad \begin{array}{ll} k & \text{number of modules (constant)} \\ x_i & \text{number of defects on module number } i \end{array}$$

$$LCL/UCL = \mu_0 \mp 2.576 \cdot \sqrt{\mu_0} \quad \text{for } \alpha = 1\%$$

$$LCL/UCL = \mu_0 \mp 3 \cdot \sqrt{\mu_0} \qquad \text{for } \alpha = 0.27\%$$

If $\alpha = 0.27\%$ is selected for the type I error, it is designated as c-chart, otherwise it receives the designation x-chart.

Example:

We consider modules with a total of 2,500 solder joints. Preliminary investigations have shown that the defect rate for a solder joint is 75 DPM. Each sample size is 50 modules. The value 1% is specified for the type I error α.

$$\mu_0 = 50 \times 2500 \times \frac{75 \, \text{DPM}}{1,000,000} = 9.375$$

$$LCL = 9.375 - 2.57 \cdot \sqrt{9.375} = 1.49$$
$$UCL = 9.375 + 2.57 \cdot \sqrt{9.375} = 17.26$$

Another variant is the u-chart. It is similar to the x-chart (or c-chart), although the sample size (number of modules) can fluctuate here. For this reason, the average number of defects per unit must be specified as a nominal value $\overline{x}(F)$. This number, for example, could originate from the preliminary investigations.

$$LCL/UCL = \overline{x}(F) \mp 3 \cdot \sqrt{\frac{\overline{x}(F)}{\overline{n}}}$$

$$\text{with} \quad \overline{x}(F) = \frac{x_1 + x_2 + ... + x_k}{n_1 + n_2 + ... + n_k} \qquad \overline{n} = \frac{1}{k}\sum_{i=1}^{k}n_i$$

Inspection Charts

A large number of various defects can occur in the module production. For the previous charts, all these defects have been combined and considered together in the charts. This summation can mean that under some circumstances an accumulation of a specific individual defect type does not become noticeable. This occurs, in particular, when the other defects behave as expected. Inspection charts can be used to handle this situation.

Table 5.11 shows an example of the data acquired for such a chart. The control limits are calculated similarly to a u-chart; the average number of defects is used here $\bar{x}(F)$ (here $= 56$). The following table shows the control limits:

$$LCL = \bar{x}(F) - 3 \cdot \sqrt{\frac{\bar{x}(F)}{\bar{n}}} = 0.1288 - 3 \cdot \sqrt{\frac{0.1288}{15.2}} = -0.1468$$

$$UCL = \bar{x}(F) + 3 \cdot \sqrt{\frac{\bar{x}(F)}{\bar{n}}} = 0.1288 + 3 \cdot \sqrt{\frac{0.1288}{15.2}} = 0.3924$$

with $\bar{x}(F) = \dfrac{x_1 + x_2 + ... + x_k}{n_1 + n_2 + ... + n_k} = \dfrac{56}{456} = 0.1288$

$$\bar{n} = \frac{1}{k} \sum_{i=1}^{k} n_i = \frac{456}{30} = 15.2$$

Table 5.11. Data for an inspection chart (selection)

Sample no.	1	2	3	4	...	27	28	29	30	Total	Relative defect frequency
Defect type											
Wetting defects	1	0	0	3		4	0	0	0	24	5.26
Tombstone defects	0	0	0	0		0	0	0	1	3	0.66
Offset	0	0	0	0		0	0	0	0	8	1.75
Voids	0	0	1	1		0	1	0	0	3	0.66
Solder balls	0	0	0	0		3	0	0	0	18	3.95
Number of modules	15	10	20	15		12	15	12	15	456	
Total defects	1	0	1	4		7	1	0	1	56	
Defect each 100 modules as %	6.7	0.0	5.0	26.7		58.3	6.7	0.0	6.7	13.02	

The lower control limit vanishes. Sample 27 shows a violation of the upper control limit. The individual defect types are evaluated in the last column. This can be used to construct a Pareto diagram from which the major defects can be derived.

Other Quality–Control Charts

For industrial applications, charts are frequently used for monitoring the process mean and the process standard deviation combined together.

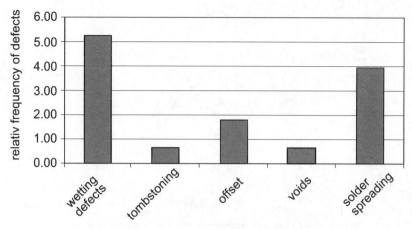

Figure 5.17. Pareto diagram for the inspection chart

This permits the simultaneous monitoring of both quantities. Typical combinations are:

- $\bar{x}$-s-chart (mean-standard deviation chart, currently most used)
- $\bar{x}$-R-chart (mean-range chart)
- $\tilde{x}$-R-chart (median-range chart).

5.7.5 Operations Characteristics of Quality–Control Charts

The operations characteristics (also called working characteristics) are used to evaluate the effectiveness of quality control charts compared with the systematic deviations of the observed process. We initially consider the mean chart. A fault here is a deviation of the process mean μ by $\Delta\mu$ from the nominal value. The response of a control chart to such a fault must be determined, namely, the probability P_E (intervention probability) with which a control–limit violation occurs and thus an intervention is made in the process.

If the process mean shifts, the distribution of the means also shifts (see Figure 5.18). An intervention is made in the process when a mean lies outside the control limits. The following equation specifies the intervention probability P_E:

$$P_E = P_{El} + P_{Eu} = 1 - \Phi\left(u_{up}\right) + \Phi\left(u_{lo}\right)$$

$$\text{with: } u_{up} = \frac{UCL - (\mu + \Delta\mu)}{\sigma/\sqrt{n}}, \quad \text{resp. } u_{lo} = \frac{LCL - (\mu + \Delta\mu)}{\sigma/\sqrt{n}}$$

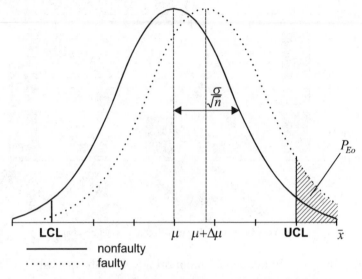

Figure 5.18. Comparison of faulty and nonfaulty processes

After calculating the control limits (Equations (5.78)), this yields:

$$P_E = 1 - \Phi\left(u_{1-\alpha/2} - \sqrt{n}\,\frac{\Delta\mu}{\sigma}\right) + \Phi\left(-u_{1-\alpha/2} - \sqrt{n}\,\frac{\Delta\mu}{\sigma}\right)$$

(5.85)

Figure 5.19 shows typical graphs for the intervention probability of a mean chart for different sample sizes depending on the deviation of the process mean, normalised to the process standard deviation.

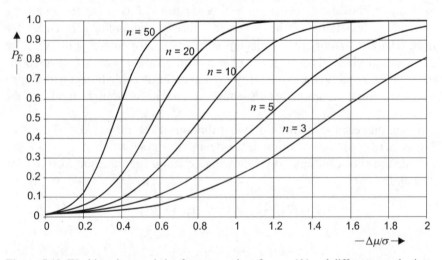

Figure 5.19. Working characteristics for a mean chart for $\alpha = 1\%$ and different sample sizes

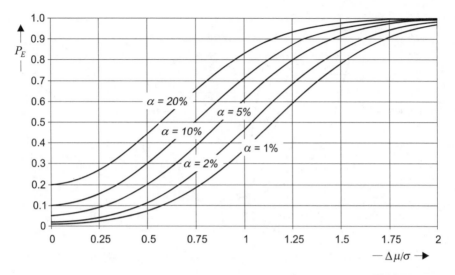

Figure 5.20. Working characteristics for a mean chart ($n = 5$) for different α

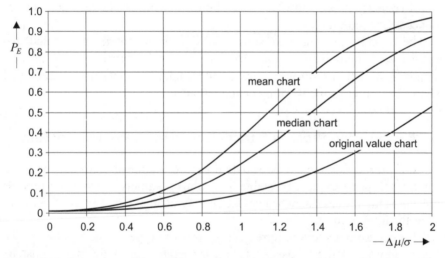

Figure 5.21. Comparison of the working characteristics for mean chart, median chart and the original value chart (all charts for $n = 5$ and $\alpha = 1\%$)

This shows that for a specified standard deviation of $\sigma = 1$ μm, a mean chart with $n = 5$ shows a shift of the process mean by 1 μm only with a probability of 35%. This deviation is shown better when the sample size increases. Thus, when you specify the sample size, you should calculate these characteristic curves and estimate the consequences that occur for a possible fault (if they are not detected) and compare these consequences with the higher test effort. Similarly, the representation in Figure 5.20 shows how the significance level α varies for a constant sample size ($n = 5$).

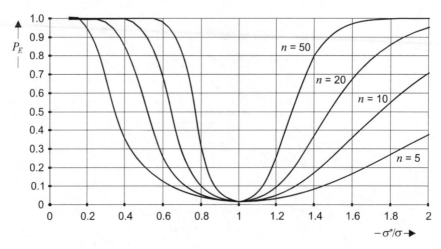

Figure 5.22. Working characteristics of an s-chart for a variable sample size ($\alpha = 1\%$)

An increase of α yields a better recognition of mean deviations. This improvement must, however, be bought with a larger number of unnecessary interventions.

The operations characteristics for deviation charts are calculated similarly. A fault here is considered to be the difference of the process standard deviation σ^* from the required standard deviation σ. Figure 5.22 shows examples of the working characteristics.

5.7.6 Estimating Process Parameters Using a Preliminary Run

Certain prerequisites must be satisfied before a control chart can be used. Firstly, the process must be stable and robust. Namely, neither the process mean nor the process standard deviation may change either in the short term or in the long term.

This corresponds to a requirement for process type A (see Section 5.7.2). Furthermore, a largest possible resistance against possible faults must be present. The second important requirement is that the observed process should be capable of producing quality (see Section 5.8). A preliminary run is performed to test whether these requirements are satisfied. This preliminary run also provides the data needed for calculating the control limits.

This preliminary run should cover at least 25 samples (in regular intervals). Ensure that similar conditions to those expected later in the planned manufacturing monitoring apply here. The sample size should at least be as large as that for the planned control chart.

Testing the process dispersion
This test must always be performed first. The process dispersion or the process standard deviation σ must be known in order to construct control charts. k samples should be taken.

Initially, the sample variances s_i^2 are calculated from the n characteristic values for each sample. The mean is determined from these k variance values s_i^2.

$$\overline{s^2} = \frac{1}{k} \sum_{i=1}^{k} s_i^2$$

This yields the following estimate for the standard deviation:

$$\hat{\sigma} = \sqrt{\overline{s^2}} \quad \text{with the degree of freedom} \quad f = k \cdot (n-1)$$

A control chart in accordance with Shewhart using $\hat{\sigma}$ is constructed and the results of the preliminary run entered. The entries are compared with the control limits. If some (maximum 1–2) sample characteristics lie outside the control limits, these values must be eliminated, a new estimate for $\hat{\sigma}$ performed and the control chart rebuilt.

If values again lie outside the control limits, they must once again be eliminated. When individual samples are eliminated, this reduces the number of available samples from k to k' (necessary for the further calculation of the degrees of freedom f).

If more than two values lie outside the control limits, it is probable that the considered process is unstable. This then requires a more detailed analysis. The use of a control chart is not permitted.

The remaining sample characteristics must be compared with the theoretical distribution function (this is the χ^2-distribution here). The sample characteristics s_i^2 must be entered into an appropriate probability network (similar to Section 6.4). The ith value here corresponds to a frequency sum of $i/(k+1)$.

For small sample sizes ($n < 10$), the χ^2 network (available for degrees of freedom $f < 10$) must be used as a probability network. The χ^2-distribution with the parameter $\hat{\sigma}$ used as the theoretical distribution is represented as a straight line in this network.

A normal probability network can be used for larger sample sizes. If systematic deviations from the theoretical distribution function are present, the process then should be considered as being unstable. If this is not the case, then $\hat{\sigma}$ is an estimate of the process standard deviation σ and can be used for calculating control charts.

Evaluation of the process mean

Another prerequisite for the use of a quality–control chart for monitoring the mean is the mastery of the process, namely systematic influences may not be present. The mean is stationary for a mastered process, namely μ does not change with time. The variance of the means of the sample $\bar{x}_i$ $(\sigma_{\bar{x}}^2)$ of the taken samples is caused only by the current process standard deviation σ. This means that the following dependency must be true:

$$\sigma_{\bar{x}} = \frac{\sigma}{\sqrt{n}} \text{ or } \sigma_{\bar{x}}^2 = \frac{\sigma^2}{n}$$

Thus, the quantity $\dfrac{n\hat{\sigma}_{\bar{x}}^2}{\hat{\sigma}^2}$ may only randomly vary from 1. The current process standard deviation σ is estimated from the preliminary run. The remaining k' samples are used to determine the overall mean $\hat{\mu}$ and the dispersion of the means:

$$\hat{\mu} = \bar{\bar{x}} = \frac{1}{k'}\sum_{i=1}^{k'}\bar{x}_i$$

$$\hat{\sigma}_{\bar{x}}^2 = s_{\bar{x}}^2 = \frac{1}{k'-1}\sum_{i=1}^{k'}\left(\bar{x}_i - \bar{\bar{x}}\right)^2 \; ; \; f' = k'-1$$

The $LCL/UCL = \hat{\mu} \mp 2.576 \cdot \hat{\sigma}_{\bar{x}}$ is used to build a control chart ($\alpha = 1\%$). This control chart is specially designed for the analysis of the preliminary run and so is not directly comparable with the control charts for the running manufacturing process. The means of the sample $\bar{x}$ are entered in the control chart. Similar to the procedure for estimating the variance, values are eliminated. If more than three values lie outside the control limits, the process should be considered as being unstable. The number of samples that now remain is designated as k''. An analysis of the means in the probability network is also performed.

Finally, the variances are compared. This requires the determination of the test quantity $F = \dfrac{n\hat{\sigma}_{\bar{x}}^2}{\hat{\sigma}^2}$ that is then compared with the critical value

$F_{crit} = F_{f1;f2;0.99}$,
   with
   $f_1 = k''-1$ (k'' number of the remaining samples)
   $f_2 = k'(n-1)$ (k' number of samples before the elimination)

If $F \le F_{crit}$ is true, the process can be considered as being mastered. This allows the use of a quality–control chart. If, however, $F > F_{crit}$ is true, the process is not mastered and a more detailed analysis of the process must be made.

Example:

Sample no.	1	2	3	4	5	6	7	8	9
$\bar{x}$ in mA	46.40	45.20	44.60	49.00	53.80	56.20	50.00	53.20	53.00
s in mA	14.84	3.90	13.90	7.35	9.83	6.91	6.93	9.98	11.20
Sample no.	10	11	12	13	14	15	16	17	18
$\bar{x}$ in mA	46.40	43.60	58.00	52.20	47.40	53.00	46.40	50.00	49.60
s in mA	7.64	6.19	17.44	5.72	14.15	10.84	6.02	14.32	6.58
Sample no.	19	20	21	22	23	24	25		
$\bar{x}$ in mA	45.40	46.80	52.40	49.00	55.00	51.20	54.60		
s in mA	7.44	8.90	11.61	9.30	9.35	5.81	23.82		

The average dispersion is calculated as

$$\overline{s^2} = 119.22 \text{ mA}^2 \quad \text{and} \quad \hat{\sigma} = 10.92 \text{ mA}$$

Figure 5.23 shows the built s-chart. Because the last sample lies outside the control limits, the rules specify that it will be eliminated. The new estimate for the standard deviation yields $\hat{\sigma} = 10.03$ mA. The recalculated control chart (not shown) does not show any violations of the control limits.

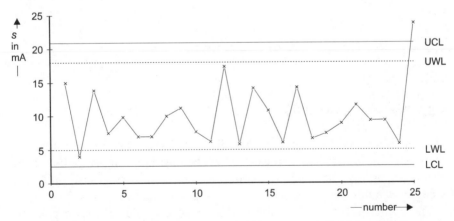

Figure 5.23. s-chart in accordance with Shewhart for the analysis of the preliminary run

The remaining deviations are ordered according to the size and entered in a χ^2-network. The data for the upper and lower intervention limits are used to estimate the theoretical distribution function. A significant deviation from the theoretical distribution function is not apparent. This allows an s-chart to be used with the following values:

UCL = 19.33 mA   ; LWL = 3.49 mA
UWL = 16.74 mA   ; LCL = 2.28 mA

The remaining samples are used to test the mean.

$$\bar{\mu} = \bar{\bar{x}} = 49.91 \text{ mA} \quad \text{and} \quad \hat{\sigma}_{\bar{x}} = s_{\bar{x}} = 3.925 \text{ mA}$$

A control chart built using these values does not show any violations of the intervention limits. The entry in the probability network also does not have any complaints.

Comparison of the variants:

$$F = \frac{n \cdot \hat{\sigma}_{\bar{x}}^2}{\hat{\sigma}^2} = 0.766 \quad ; \quad F_{crit} = F_{23.96;0.99} = 2.01 \quad ; \quad F < F_{crit}$$

The process can be considered to be mastered and so $\bar{\mu}$ and $\hat{\sigma}_{\bar{x}}$ can be used as estimates for a mean chart.

UCL = 61.46 mA   ; UWL = 58.70 mA
M = 49.91 mA
LCL = 38.36 mA   ; LWL = 41.12 mA

The corresponding tolerance centre can also be used instead of $\bar{\mu}$.

5.8 Quality Capability of Processes

5.8.1 Process Capability

The capability coefficients were initially used by the automobile industry. The goal was to provide a means of evaluating the comparability of the quality capability for machines, processes and suppliers. These ratios are used, for example, to make decisions about component suppliers.

These coefficients include the process potential, the process capability, the machine capability, the critical process capability, *etc.*

The process potential is described by a quality–capability ratio that expresses the potential capability of a process to consistently produce a specific characteristic within the given specification limits (tolerance limits). The process potential judges the quality performance of a process using a comparison of the process spread with the tolerance width.

In general, these quality–capability ratios for given tolerance limits (T_l or T_u) are calculated in industry (with the automobile industry as initiator) using the following classic rules that are independent of the actual product and its production:

$$C_p = \frac{T_u - T_l}{6\sigma} \qquad (5.86)$$

$$C_{pk} = \text{Min}\left\{ \frac{\overline{x} - T_l}{3\sigma} ; \frac{T_u - \overline{x}}{3\sigma} \right\} \qquad (5.87)$$

C_p process capability
C_{pk} critical process capability
$\overline{x}$ process mean (determined using the characteristics of the product)
σ measure for the process standard deviation (similar)

The process capability C_p specifies the capability that the controlled process can attain its maximum. The critical process capability C_{pk} also takes into account that, in general, the process mean does not ideally lie in the tolerance centre. Consequently, C_{pk} is (almost) always less than C_p.

Before these quality–capability ratios can be calculated, some specifications and clarifications *must* be made:

- Specification of the characteristic (or also the characteristics)
- Unambiguous definition of the tolerance limits (or specification limits)
- How can the characteristics be measured/tested
- Statistical properties of the characteristics (*e.g.*, distribution type, process type).

Although these demands sound simple, an orderly comparability of capability ratios requires that they are strictly observed, which, however, is not always immediately possible.

The validity range of Equations (5.86) and (5.87) is restricted. The following prerequisites must be satisfied:

- The considered quality characteristic has a normal distribution and corresponds to process type A (see Section 5.7.2).
- The considered quality characteristic can be measured (namely, a variable characteristic but *no* attributive characteristic (good/bad)).

Both prerequisites normally (but not always) apply. In particular, the process type A occurs only very seldom (in its "pure form").

In industrial practice, these ratios are generally used for the evaluation and the comparison of the quality capability for processes. If a normal distribution applies, it is (theoretically) possible to estimate the expected defect rate for the given process capability C_p of a process. If high demands are placed on the process capability ($C_p > 1.67$), it can be assumed that defects will scarcely occur by chance.

However, the numerical values of the theoretical defect rates (in particular, for comparisons) should not be used directly because these numbers are correct only for an exact validity of the normal distribution (namely, including the boundary areas). Such a proof is, however, practically almost impossible to perform because it requires a test size of more than 1 million measurements and the usual distribution tests would immediately cause a rejection for even small deviations of the theoretical values from the practical values. Consequently, such proofs are not practicable.

Table 5.12. General evaluation of processes and *theoretically* expected defect rates for a given process capability C_p; the theoretical defect rates for given C_{pk} values are shown in parentheses

Process capabilities C_p, $C_{pk}=$	General evaluation of the process	Theoretical defect rate in DPM	Designation
1	Not quality conform	$2,700 = 0.27\%$ (1,350)	3 sigma
1.33	Conditionally quality conform	63.34 (31.67)	4 sigma
$C_p = 2.0$ $C_{pk} = 1.5$	Quality conform	3.4	6 sigma (Motorola)
1.67	Quality conform	0.5734 (0.287)	5 sigma
2.00	Quality conform	0.0019 (0.00095)	6 sigma

However, it remains an important conclusion that irrespective of the distribution type, a high numerical value of the process capability is equivalent to a very small probability for the occurrence of defects. Simultaneously, a high process capability means that defects that affect the mean and/or the process dispersion do not immediately cause a rapid increase in the defect rate.

The practical demands placed on C_{pk} are somewhat lower than the demands placed on C_p. Thus, the process mean can fluctuate by approximately $\pm 1\sigma$ from the tolerance centre.

The 6-sigma requirement as defined by Motorola takes into account that it is practically impossible to keep the operating point of a process exactly at the tolerance centre, which would correspond to the optimum setting of a process with regard to the defect rate. This definition allows the operating point a certain freedom of $\pm 1.5\sigma$.

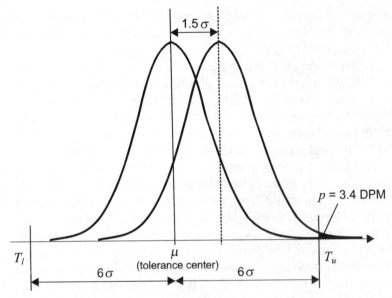

Figure 5.24. Illustration of 6-sigma using the Motorola definition

This behaviour is illustrated in Figure 5.24.

The theoretical defect rate p based on a given C_{pk} value is estimated using the following equation:

$$p \approx 1 - \Phi\left(3C_{pk}\right) \tag{5.88}$$

The process capability characterises simply and relatively precisely (even when not complete) the capability of a complete process to produce long term a specific product with good quality (namely with a defect rate (almost) zero) and with good observance of the specification (= nominal value) while taking account of material, man, machine, environment, *etc*. A specific product should be used to provide the data required for the numeric determination of these coefficients. For automatic placement machines, this is the manufactured module. Although this emphasis may appear to be obvious, it cannot always be realised in practice.

5.8.2 Machine Capability

A prerequisite for capable processes are capable machines, namely, a capable process can never be expected when incapable machines are used. This means that before determining the process capability, it is desirable to determine the machine capability of the machines and equipment required for manufacturing the product. The calculation rules are similar to those for the process capabilities, namely Equation (5.86) and Equation (5.87) are used to calculate the *machine capability* C_m and for the critical machine capability C_{mk}, respectively.

In contrast to the process capability, the *critical machine capability* C_{mk} characterises the capability of manufacturing a specific product with good quality (namely with a defect rate near zero) and in conformance with the specification (= nominal value), where the investigation of the machine capability takes place only over a short time period. This means effects caused by material changes, changing environmental conditions, *etc.*, have little affect here. A specific product should also be used here to provide the data required for the numeric determination of these coefficients. The short interval of the investigations means that the formulation short-term capability is also used instead of the term machine capability.

The process capability is then the sum of several machine capabilities and the fluctuating effects of material, man, *etc.* Thus, the process capability can never be better than the smallest (= worst) machine capability of the associated machines. Consequently, higher numerical demands are placed on the machine capability than for the process capability. Namely, in general, a capability value of $C_{mk} > 1.67$ is demanded from a machine in order for it to be classified as being quality conform. Reference [7] contains detailed information for mathematical handling; special adaptations for the module assembly are contained in [8].

5.9 Acceptance Sample Test

5.9.1 Introduction

In many situations, the production quality of items or lots must be tested. Typical tasks are the receiving inspection (incoming-goods inspection), intermediate inspections, final inspections and quality control. The inspection is made here using samples. Generally, this inspection should decide whether the associated lot conforms to the specified quality requirements, namely it can be added to a further processing or to a delivery. If this is not the case, the associated lot will be rejected.

Possible further processing for rejected lots are:

- Complaint to the manufacturer
- Complete sorting
- Scrap
- Sell at reduced price.

For rejected lots, it is specified for all following calculations that a complete inspection of the remaining lot (complete sorting) is performed and the determined defect units (a unit is the standard conform designation for a product) removed and replaced by defect-free units.

To make the evaluation, the qualitative (attributive) characteristics are counted, such as the number of faulty units x (products) or the number of defects x in a sample. Another possibility is the measurement of quantitative (variable) characteristics (see Section 5.9.7). The ratio of units that lie outside the specified limits (*e.g.*, tolerance limits) is determined. A characteristic value that does not satisfy the specified required demand is designated as being a defect. Thus, a defect unit has at least one defect. Defects are classified. ISO 3951 [6] specifies the following classification:

Critical defect
A critical defect is a defect for which it can be assumed or is known that it will probably cause dangerous or unsafe situations for persons that use, maintain, *etc.*, the affected unit. Or it is a defect for which it can be assumed or is known that it will probably hinder the fulfilment of the function of a large plant, such as a ship, an aircraft, a computer system, *etc.*

For critical defects, however, one should carefully consider whether the use of sample tests is appropriate, because this immediately means that defects that result from chance (when a defect product is not in the sample) will be accepted.

Main defect
A main defect is a noncritical defect that probably will lead to a failure or will significantly reduce the usability for the envisaged purpose.

Secondary defect
A secondary defect is a defect that probably will not significantly reduce the usability for the envisaged purpose or is a deviation from the applicable standards that affects the use or the operation of the unit only to a limited extent. When a sample test is made, it is assumed that some proportion of defective units will not be detected, because only a subset of the lot is tested.

The proportion of not-detected defective units is designated as average outgoing quality. This implicit disadvantage can be compensated only through the use of a 100% inspection (complete test). However, testing errors mean that even a complete test must expect an average outgoing quality (in particular for a manual inspection).

In addition to the mentioned disadvantage of the sample test, it also has significant advantages compared to the complete test:

- Low test costs
- Use of destructive inspection possible
- The lots are available faster
- Fewer (but trained) testing personnel necessary
- The more-skilled work is less monotonous with a subsequent reduction of test errors.

The following prerequisites must be satisfied while performing a sample test:

- Homogeneous lots, namely the random distribution of defective units within the lot
- Random removal of the sample from the lot (each unit has the same "chance" of being taken). It must not be the case, for example, because of ease, that the sample is only taken from the top of a lot.

5.9.2 Single Sampling Plans

Initially, attribute sampling plans are considered, namely the characteristics are evaluated as being "good" (defect free) or "bad" (defect).

The simplest form of the sample test is the *single sampling plan*. The following procedure should be adopted when such a plan is used:

1. Take a sample of size n from the lot (lot size N).
2. Test all n units and count the number of defective units determined in the sample (or the number of defects) x.
3. If $x <= c$ (acceptance number), the lot will be accepted.
4. If $x > c$, the lot will be rejected.

For the practical use of such sampling plans, the following notation has been agreed and is designated as a sampling instruction:

$$n - c$$

The acceptance or rejection of a lot depends heavily on the proportion of defective units (or defect rate) in the lot p. This proportion is designated in the following discussion as the quality level of the lot. Random influences must be expected.

Characteristic curves are calculated to evaluate the effectiveness of such a sampling plan. The most important characteristic curve is the operation characteristic. It represents the dependency on the acceptance probability P_a of the quality level p.

The hypergeometric distribution is used to calculate the exact acceptance probability for the proportion of defective units.

$$P_a = \sum_{k=0}^{c} \frac{\binom{N-M}{n-k} \cdot \binom{M}{k}}{\binom{N}{n}} \qquad (5.89)$$

For practical calculations, the hypergeometric distribution can be replaced by the binomial distribution (see Section 5.3.5) provided the prerequisites are satisfied $(n < 0.1 \cdot N)$.

$$P_a = G(c; n; p)$$

Figure 5.25 shows a typical curve for the operation characteristic. A percentage (%) is normally used as the dimension for the proportion of defective units. Characteristic points are:

$p_{1-\alpha}$ Quality level (proportion of defective units) for the supplier's risk α

α Supplier's risk (namely the probability that a lot with a $p_{1-\alpha}$ proportion of defective units will be rejected). The designation producer's risk is also common. 10% (more seldom 5%) is normally used as size for α.

p_{β} Limiting quality level, other possible designations:
     LQ   Limiting quality
     LTPD   Lot tolerance percent defective
     RQL   Rejectable quality level

β Customer's risk (preferably 10%) (the probability that a lot with a p_{β} proportion of defective units will be accepted)

$p_{0.5}$ Indifferent quality level, 50%-point; at this quality level, the corresponding lot will be accepted with 50% probability (also called *IQL:* indifference quality level)

The operation characteristic can be divided into three ranges:

$$0 \le p \le p_{1-\alpha}$$

Range of high acceptance probability. The vendor must try to supply only such lots, because there is only a small probability that they will be rejected.

$$p \ge p_{\beta}$$

Range of small acceptance probability. The customer must ensure that such lots with such a high proportion of defective units, which would cause larger faults, will be rejected with a high probability $(>1-\beta)$.

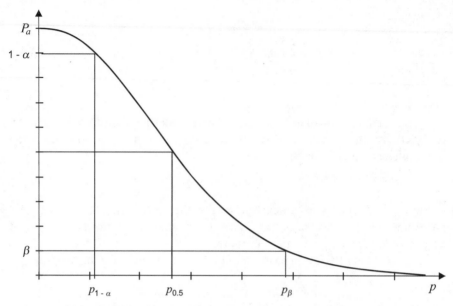

Figure 5.25. Operation characteristic curve with the characteristic points

$$p_{1-\alpha} < p < p_\beta \quad \text{(uncertainty range)}$$

Average acceptance probability. The acceptance or rejection will be largely determined by chance. This is the price that must be paid for the low cost of the sample test.

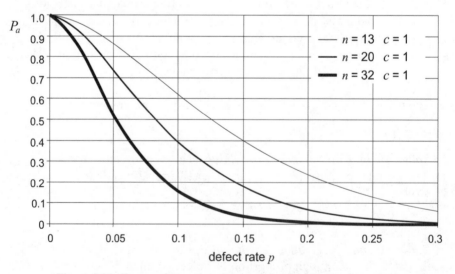

Figure 5.26. Change of operation characteristic with change of sample size

Supplier's risk (α) or the customer's risk (β) are undesirable properties of a sampling plan. A lot with the quality level $p_{1-\alpha}$ will be considered as being a (just) good lot and should consequently really always be accepted (will, however, be rejected with the probability of the supplier's risk!). Similarly, a lot with the quality level p_β is a bad lot and should really always be rejected.

If the sampling instruction is changed, this yields the quantitative changes graphs for the operation characteristics shown in Figures 5.26 to 5.28.

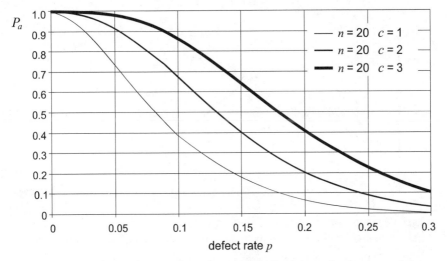

Figure 5.27. Changes of operation characteristic for change of acceptance number

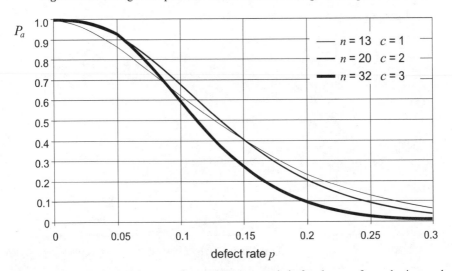

Figure 5.28. Quantitative changes of operation characteristic for change of sample sizes and acceptance number

If a series of lots with the same size N with constant quality level are tested using a given sampling instruction, the proportion P_a of the lots will be accepted. The not-tested part of the lots ($N-n$ units) contains defective units. As specified, rejected lots will be completely tested and so contain no defective units (assuming an ideal inspection). This yields the *average outgoing quality* (*AOQ*, proportion of the defective units in the lots after the sample test):

$$AOQ = \frac{1}{N}\sum_{k=0}^{c}(M-k)\cdot P(X=k) \approx P_a \cdot p \qquad (5.90)$$

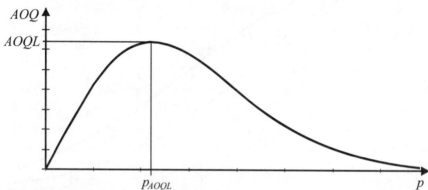

AOQL   Average outgoing quality limit (maximum average outgoing quality)
p_{AOQL} Quality level for which the maximum average outgoing quality occurs

Figure 5.29. General form of the *AOQ* average outgoing quality

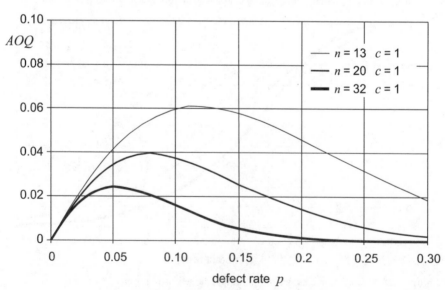

Figure 5.30. Average outgoing quality characteristic curves for changes of sample size

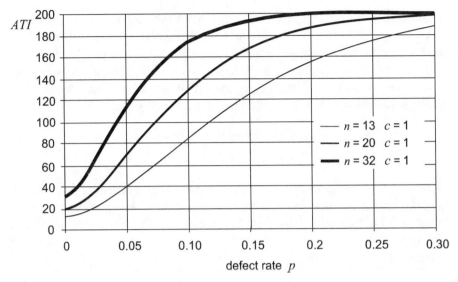

Figure 5.31. Graph for the characteristic curve of the average test size ($N = 200$)

The average outgoing quality characteristic curve in Figure 5.29 has a characteristic point.

The maximum average outgoing quality is the proportion of defective units, which, in the least favourable case (when the quality level of the incoming lots = p_{AOQL}), occurs *on average* (never for a single lot) after the inspection. Figure 5.30 shows the quantitative changes of the average outgoing quality characteristic curve for changes of sample size. It shows that the average outgoing quality sinks with increasing sample size.

The average test size *ATI* (average total inspection) is a measure of the test cost of the sample test including a possible complete test).

$$ATI = n + (1 - P_a) \cdot (N - n) \tag{5.91}$$

5.9.3 Determination of Single Sampling Plans

The previous section showed how for a given sampling instruction, the characteristic curves, including the characteristic points, can be determined. For practical use, however, this does not suffice, because normally the sampling instruction $(n - c)$ must be determined on the basis of given requirements.

A vendor wants, if possible, all of the delivered lots to be accepted. Because a total absence of defects cannot normally be guaranteed, the quality level agreed with the customer specifies the level at which a lot is still considered as being good. This is normally $p_{1-\alpha}$. In the most unfavourable cases, such lots are rejected with the probability α (supplier's risk). Conversely, the user wants to protect himself against bad lots. This means that the user specifies a quality level p_β. Such lots may be accepted with a maximum probability β (customer's risk, 10%). A sampling instruction should now be determined that satisfies both needs. The hypergeometric distribution would yield the following inequality system:

$$P_a \geq 1 - \alpha = G\left(N, c, n, p_{1-\alpha}\right)$$
$$P_a < \beta \qquad = G\left(N, c, n, p_\beta\right)$$

The exact solution of this inequality system is possible only with search methods and is complicated (only software solutions are appropriate). If the hypergeometric distribution is approximated using the binomial distribution and this with the normal distribution, the sampling instruction can be approximated as follows:

$$n = \left(\frac{u_{1-\alpha}\sqrt{p_{1-\alpha}(1-p_{1-\alpha})} - u_\beta\sqrt{p_\beta(1-p_\beta)}}{p_\beta - p_{1-\alpha}}\right)^2 \qquad (5.92)$$

$$c = n \cdot \left(p_{1-\alpha} + u_{1-\alpha}\sqrt{\frac{p_{1-\alpha}(1-p_{1-\alpha})}{n}}\right) \qquad (5.93)$$

The solutions must be rounded appropriately, where the sample size must always be rounded up. The use of the binomial distribution removes the dependency on the lot size. Because the sample size is not known before making the calculation, the reliability of the determined solution (replacement for the hypergeometric distribution) must be tested after determining the sampling instruction. Namely, if the sample size is less than 1/10 of the lot size, the determined sampling instruction is usable. The solution must be rejected if the sample size is larger than 30% of the lot size. If it lies between these limits, its usability depends on how it is used. The use of the Larsson diagram is another possible way of approximating the problem definition.

5.9.4 Double and Multiple Sampling Plans

When a single sampling plan is used, the complete sample must be tested. However, there can be situations in which the result of inspection becomes apparent beforehand (*e.g.*, overshooting the acceptance number). To save testing effort, double sampling (maximum two samples) and multiple sampling plans (more than two samples, normally seven) have been constructed.

A double sampling instruction has the following form:

Identification: $n - c_1/d_1 - c_2$

n Sample size of the first and the second sample

c_1 Acceptance number for the first sample

d_1 Rejection number for the first sample

c_2 Cumulative acceptance number for the first and the second sample

For the inspection with a double sampling plan, initially the first sample is taken and the number of defective units (or defects) x_1 determined. If $x_1 <= c_1$, the inspection has ended; the lot will be accepted. If $x_1 >= d_1$, the lot will be rejected. If neither condition is satisfied, a second sample will be taken and tested; x_2 defective units are determined. If $x_1 + x_2 > c_2$, it will be rejected, otherwise it will be accepted.

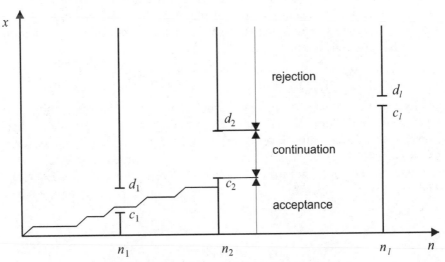

Figure 5.32. Screen diagram for multiple sampling plans

n_i Sample size for the ith sample ($i = 1...l$)

l Number of samples

c_i Acceptance number on the ith sample

d_i Rejection number on the ith sample

A multiple sampling plan functions similarly, a maximum of 7 samples are taken. Figure 5.32 uses a screen diagram in which actual test curves can be entered to graphically illustrate a multiple sampling plan.

Double and multiple sampling plans have the same characteristic curves as for single sampling plans. The calculation is, however, correspondingly more difficult. Equation (5.94) can be used to approximate the calculation of the operation characteristic of a double sampling plan:

$$P_a = G(c_1, n, p) + \sum_{i=c_1+1}^{d_1-1} g(i; n, p)\ G(c_2 - i; n, p) \tag{5.94}$$

Equation (5.90) can be used to approximate the average outgoing quality.

Using a calculated operation characteristic of a double or multiple sampling plan as a basis, an equivalent single sampling plan (namely with an approximately equal quantitative of operations characteristic curve) can be calculated by reading the characteristic points. For example, the single plan 129-5 is equivalent to the double plan 80-2/5-6. This means, that for the same operations characteristic, 49 fewer units need to be tested than for the double plan, if it is possible to cancel after the inspection of the first sample. If, however, a second sample must be tested, 31 more units must be tested. To prove that the use of double and multiple plans has advantages, an additional characteristic curve is necessary, because the required size of the sample is not known at the start of the inspection. The average sample size *ASN* (average sample number) provides details. The following equation is used for a double sampling plan:

$$ASN = n \cdot \left(1 + G(d_2 - 1; n; p) - G(c_1; n; p)\right) \tag{5.95}$$

Figure 5.33 shows the general form of this characteristic curve for equivalent single, double and multiple sampling plans. It is obvious that, irrespective of the quality level of the tested lot, the multiple plan requires the least test effort and the simple plan has the largest test effort. This statement is, however, not true for a single lot (refer to the above example). The test effort savings are particularly large for very good or very bad quality levels.

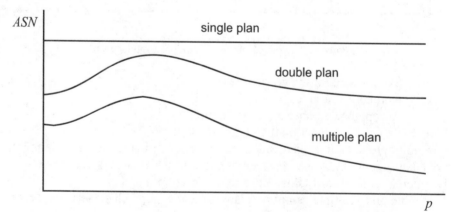

Figure 5.33. Graph of *ASN* characteristic curves for various plan types

5.9.5 Sequential Plans

When we consistently continue the consideration for the introduction of multiple plans, we reach the limit case in which the size of a sample is 1, namely, after each tested unit, a decision is possible concerning the acceptance, continuation of the inspection or rejection. Such sampling plans are called sequential plans.

The calculation of such problems assumes the specification of two characteristic points of the operations characteristic $(p_{1-\alpha}, 1-\alpha; p_{\beta}, \beta)$. Two hypotheses are made:

H_0 : $p < p_{1-\alpha}$ (for confirmation acceptance of the lot)

H_1 : $p > p_{\beta}$ (for confirmation rejection of the lot)

To test the hypotheses, samples are taken from the lot. After each taken unit, a possible confirmation of the hypotheses is tested using a probability ratio. The following possibilities occur:

$$\frac{P\left(p_{\beta}, k, x\right)}{P\left(p_{1-\alpha}, k, x\right)} \leq \frac{\beta}{1-\alpha} \quad \text{acceptance } H_0 \quad \text{rejection } H_1$$

$$\frac{P\left(p_{\beta}, k, x\right)}{P\left(p_{1-\alpha}, k, x\right)} \leq \frac{1-\beta}{\alpha} \quad \text{acceptance } H_1 \quad \text{rejection } H_0$$

The inspection continues when none of the inequalities is satisfied. k is the number of tested units and x the number of determined defective units. $P(p, k, x)$ is the probability that the previously tested total sample (in the determined sequence of the defective units) comes from a population with the quality level p. It is approximated (ignoring the lot sizes) as follows:

$$P(p,k,x) = p^x \cdot (1-p)^{k-x}$$

When we set the quality levels $p_{1-\alpha}$ and p_{β} in the above inequalities, this yields:

$$Y_A(k) = s \cdot k - h_0 \quad \text{acceptance line}$$
$$Y_R(k) = s \cdot k + h_1 \quad \text{rejection line}$$

with

$$h_0 = \frac{1}{A} \ln\left(\frac{1-\alpha}{\beta}\right) \quad ; \quad h_1 = \frac{1}{A} \ln\left(\frac{1-\alpha}{\beta}\right)$$

$$s = \frac{1}{A} \ln\left(\frac{1-p_{1-\alpha}}{1-p_{\beta}}\right) \quad ; \quad A = \ln\left(\frac{p_{\beta} \cdot (1-p_{1-\alpha})}{p_{1-\alpha} \cdot (1-p_{\beta})}\right)$$

Figure 5.34 shows these as straight lines. This diagram (comparable with the screen diagram for multiple plans) can be used for inspection. The testing continues until the random path of the inspection intersects one of the straight lines.

k_{min} is the smallest sample size necessary for the acceptance of the lot.

$$k_{min} = \frac{h_0}{s}$$

The specified points can be used to calculate the sequential plan so that it is equivalent to other plans. For the double plan (80-2/5-6) used in Section 5.9.4, this would yield the following sequential plan:

$h_0 = 1.9849;$ $h_1 = 1.9874;$ $s = 0.04388;$ $k_{min} = 46$

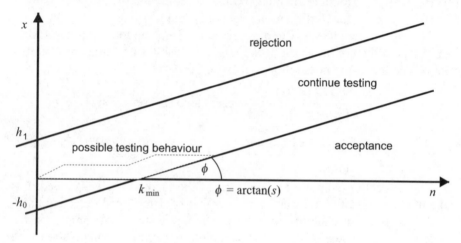

Figure 5.34. General form of a sequential plan

Table 5.13. Characteristic parameters for sequential plans

p	P_a	ASN
0	1	h_0 / s
$p_{1-\alpha}$	$1-\alpha$	$\dfrac{(1-\alpha)h_0 - \alpha h_1}{s - p_{1-\alpha}}$
s	$\dfrac{h_1}{h_0 + h_1}$	$\dfrac{h_0 \cdot h_1}{s(1-s)}$
p_β	β	$\dfrac{(1-\beta)h_1 - \beta h_0}{p_\beta - s}$
1	0	$h_1 /(1-s)$

Because the two straight lines are parallel, there is the danger that under some circumstances it "never" (the maximum is the lot size) comes to an end of the inspection. The probability for this case is, however, extremely small. To exclude this case, normally a maximum sample size is also specified (approximately 1.5 times the sample size of an equivalent simple plan). For the calculation and representation of the operation characteristic and the average sample size, it suffices when Table 5.13 is used.

Compared with all previously considered sampling plans, a sequential plan guarantees the smallest average sample size. This advantage is, however, partially offset by the complicated handling of the plans. Furthermore, the impossibility of an exact prediction of the actual necessary test size becomes more important.

5.9.6 Sample Test for the Number of Defects per Unit

For many products it is not primarily important to test whether or not a unit is defective, but rather the number of defects per unit is required (for example, the number of soldering defects on a circuit board, the number of blisters; scratches, *etc.* on a surface, *etc.*).

To evaluate such lots, a sample test similar to the previously known procedures can be performed, where, instead of the defective units, the defects are counted. The acceptance probability for a single sampling plan then approximates the value of the distribution function of the Poisson distribution.

p – average number of defects 100 units in the lot

$$P_a = \sum_{k=0}^{c} \frac{\mu^k}{k!} \cdot e^{-\mu} \qquad \mu = n \cdot p$$

There are no changes for the other characteristic curves. Double and multiple instructions can also be constructed. It is only necessary that the quality level p is used as dimension "defects per 100 units". Although the specification as a percentage is not permitted, it is sometimes used.

5.9.7 Variable Sampling Plans

Attribute sampling plans can in principle be used to test all possible characteristics. In practice, however, many characteristics cannot be measured. The attribute test only classifies such characteristics whether they lie inside or outside the tolerance limits. This means that much information present in a determined measured value is not used. Variable sampling plans can compensate for this deficit. An important prerequisite for the use of such plans is that the distribution of the tested characteristic corresponds to a normal distribution. Furthermore, it is initially assumed that the standard deviation σ of the distribution is known. This standard deviation must be sufficiently small so that a significant overshooting proportion p (reject or corrective work) can occur only on one side of the tolerance range.

A sample of size n_σ is taken (σ indicates that this is a plan that uses a specified standard deviation). The units are measured and the sample mean $\bar{x}$ calculated.

Decision:

When $\qquad \bar{x} + k_\sigma \sigma \leq T_o \quad$ and $\quad \bar{x} - k_\sigma \sigma \geq T_u \Rightarrow$ acceptance

k_σ is the acceptance constant (comparable with the acceptance number for attribute plans).

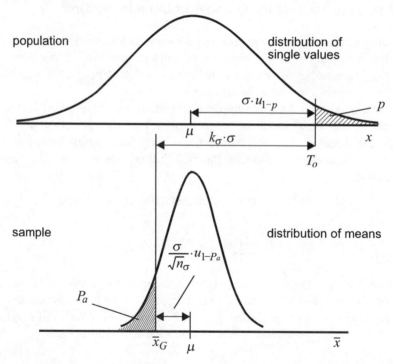

Figure 5.35. Illustration of the operation of a variable sampling plan; $\bar{x}_G$ is the mean of the sample where the decision between acceptance and rejection tips

The operation characteristic of such a sampling plan is calculated as

$$P_a = \Phi\left(\left(u_{1-p} - k_\sigma\right) \cdot \sqrt{n_\sigma}\right) \qquad (5.96)$$

Equation (5.90) can be used for the average outgoing quality.

Similar to the attribute sampling plans, a variable sampling plan can be calculated when two points of the operation characteristic are specified:

$$n_\sigma = \left(\frac{u_{1-\alpha} - u_\beta}{u_{1-p_{1-\alpha}} - u_{1-p_\beta}} \right)^2 \quad ; \quad k_\sigma = \frac{u_{1-\alpha} \cdot u_{1-p_\beta} - u_\beta \cdot u_{1-p_{1-\alpha}}}{u_{1-\alpha} - u_\beta}$$

The sample size n_σ must always be *rounded up*.

It is also possible to determine equivalent sampling plans. This gives for the double plan (80-2/5-6):

$$n_\sigma = 27 \quad ; \quad k_\sigma = 1.718$$

This shows that for the same decision reliability, the variable plan requires significantly less effort for the inspection. This saving results to some extent from the better use of the information contained in the measured values (instead of "good" or "bad" of the specific measured value). Another saving, however, is achieved by using the available preliminary information (normal distribution and known standard deviation). If this preliminary information is incorrect, this causes additional risks, the effect of which can only be predicted with difficultly.

If the standard deviation is not known, the necessary sample size n_s is approximated (valid for $n_s > 5$) as follows:

$$n_s = (1 + \frac{k_\sigma}{2}) \cdot n_\sigma$$

The acceptance constant k_σ (namely $k_s = k_\sigma$) does not change. The equivalent sample size n_s would be 67 here. This provides for the operation characteristic:

$$P_a = \Phi\left(\frac{(u_{1-p} - k_s) \cdot \sqrt{n_s}}{1 + k_s / 2} \right)$$

Variable plans guarantee significantly smaller sample sizes than attribute sampling plans. This advantage, however, is partially countered by the following disadvantages:

- normal distribution is a prerequisite
- incorrect preliminary information (in particular, the standard deviation) means wrong decisions are possible
- only one characteristic can be tested (per plan).

If several quality characteristics for a product are to be tested, an appropriate variable plan *must* be prepared for each characteristic. For attribute plans, individual characteristics *can* be combined to form groups. An attribute plan is then assigned to each group.

5.9.8 Sampling Systems

A specific sampling instruction can be defined for each problem (the basis is the specification of two characteristic points). This procedure, however, has the following disadvantages:

- no systematic present
- a large number of different sampling instructions results
- the lot sizes are not normally considered
- a consideration of preliminary information is scarcely possible
- only limited cooperation between different partners is possible.

Sampling systems avoid such disadvantages. Each sampling system focuses on a specific test strategy. Such test strategies can be:

- keep the probability of the rejection of good lots small (when preliminary information about the quality level of the supplied lots is known)
- keep the probability of the acceptance of bad lots small
- keep the average outgoing quality small
- minimise the costs.

The preliminary information can include factors such as:

- the test results of the previously tested lots of the same product
- the average quality level of the supplied lots
- the distribution of the quality level of the supplied lots
- cost factors for inspection, rejection, defect consequence, *etc.*

The AQL Sampling System in Accordance with ISO 2859

AQL sampling systems are the most common sampling systems. The systems are based on the American military standards (MIL 105 D) developed during the Second World War. These standards were used as basis for the international standard ISO 2859 [6]. The tables in the individual standards are identical. The following preliminary information must be specified or known for the selection of a sampling instruction:

- agreed acceptable quality level (*AQL)*
- lot size N
- test level
- preliminary information from the results of previous sample inspections
- specification of the number of the samples to be selected (choice of plan type).

It is assumed that a series of lots is tested. The most important characteristic of this sampling system is the AQL value.

AQL =acceptable quality level

The AQL value is defined as the maximum proportion of defective units as a percentage (or the maximum number of defects per 100 units) that is accepted for the sample test as (maximum) acceptable manufacturer quality. The AQL value defines what is (just) accepted as good quality. Simultaneously, the AQL value serves as identification for the sampling plans. The sampling system guarantees that lots with a quality level of the AQL value will be accepted with a high probability (ISO 2859 specifies between 87% and 99%). The AQL value is comparable with the quality level $p_{1-\alpha}$. Because, however, this sampling system does not have any fixed producer's risk, a direct equivalence of both terms is not appropriate.

Sampling instructions of the normal, intensified and reduced inspection (also designated as inspection types) are specified for each AQL value. These inspection types serve as the response to quality fluctuations. The single sample instruction in the known form $(n - c)$ is specified for the normal and the intensified inspection. The reduced inspection used as sampling instruction:

$$n - c/d$$

where d is the rejection coefficient. The following cases must be differentiated for the evaluation of the specific inspection result:

$x \leq c$ acceptance of the lot, continue with the reduced inspection
$c < x < d$ acceptance of the lot, continue with the normal inspection
$x \geq d$ rejection of the lot, continue with the normal inspection

These inspection types are used depending on the results of the preceding inspections. Figure 5.36 shows the general relationships.

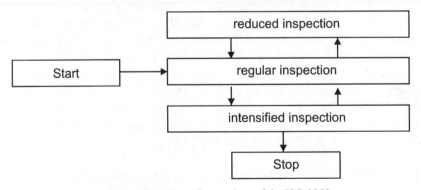

Figure 5.36. Jump instructions of the ISO 2859

The test levels are used to specify the inspection level, where the inspection level is defined as being the capability of differentiating between a good and a bad lot quality. A sampling instruction with a high inspection level has a small uncertainty area and thus a relatively steep operation characteristic. The inspection level is primarily determined by the sample size.

The ISO 2859 has seven test levels:

S1 to S4 special test levels

I, II, III normal test levels

The special test level S1 and the normal test level III give the lowest and the highest inspection level, respectively. Normally, test level II is chosen, namely, this level should be selected when nothing is specified. Special test levels should be chosen only for destructive or very expensive inspections.

The lot size is a fixed condition that can scarcely be affected. In the ISO 2859, the lot sizes are divided into 15 groups. This means that a small change of the lot size does not necessary cause a change of the sampling instruction. The larger the lot size, the smaller is the relative sample size.

The choice of the plan type is scarcely subject to fixed rules. It is possible to choose between single, double or multiple sampling plans (7 samples). The individual plans are so constructed that the operation characteristic scarcely depends on the choice of the plan type.

The sample system is constructed so that a high acceptance probability ($> 90\%$) is guaranteed for lots with the quality level $p < AQL$. Because each manufacturer must expect fluctuations in its own quality level, its average quality level should satisfy the following condition:

$$p \leq 0.5 \cdot AQL$$

In such a case, rejections are seldom expected. This condition, which is normally used to define the AQL value, represents, primarily, the interests of the manufacturer. It is important, however, that the user must agree to the manufacturer's suggested AQL value. The user will largely be guided by the consequence that undetected defective units can cause. Indirectly, such conclusions can be derived by the specification of an LQ value (its meaning is analogous to p_β) or an $AOQL$ value. For this purpose, the ISO 2859 contains control tables that provide these characteristic values for the selected plans. If such a value does not meet the requirements, the test level should first be varied. Only when this is also not successful, must the AQL value be changed (however, in agreement with the manufacturer).

AQL Sampling System for Measurable Characteristics

Similar to the ISO 2859 for the inspection of attributive characteristics, an AQL sampling system also exists for variable characteristics. ISO 3951 [4] contains the appropriate sampling instructions, where the required specifications primarily conform to ISO 2859. The following changes apply:

- The special test levels S-1 and S-2 are not present.
- Only single-sampling plans are present.

A prerequisite is that the tested characteristic has the normal distribution. Depending on the preliminary information, s-plans (the standard deviation is determined from the sample) or σ-plans are taken from the standard. It is used similar to Section 5.9.7. The plans are calculated so that for the same starting sizes, the operation characteristics have a generally similar form as ISO 2859. When 3951 is used, particular care must be taken to ensure that the mentioned prerequisites are observed (normal distribution, accuracy of the specified standard deviation).

References

[1]   DIETRICH, E.; SCHULZE, A.: *Statistische Verfahren zur Maschinen und Prozessqualifikation.* 4th revised edition. Hanser Verlag, Munich/Vienna 2002

[2]   *DIN 55350 Begriffe der Qualitätssicherung und Statistik.* Edition:1995-08 Beuth Verlag Berlin

[3]   *DIN EN ISO 9001*, Qualitätsmanagementsysteme - Anforderungen (ISO 9001:2000-09). Edition: 2000-12, Beuth Verlag Berlin

[4]   *DIN ISO 3951*, Verfahren und Tabellen für Stichprobenprüfung auf den Anteil fehlerhafter Einheiten in Prozent anhand quantitativer Merkmale (Variablenprüfung). Edition:1992-08, Beuth Verlag Berlin

[5]   GRAF, U.; HENNING, H.-J.; STANGE, K.; WILRICH, P.-T.: *Formeln und Tabellen der angewandten mathematischen Statistik.* 3rd edition. Springer-Verlag Berlin 1987

[6]   *ISO 2859-1*; Annahmestichprobenprüfung anhand der Anzahl fehlerhafter Einheiten oder Fehler (Attributprüfung). Edition:1993-04, Beuth Verlag Berlin

[7]   RINNE, H.; MITTAG, H.-J.: *Prozeßfähigkeitsmessung für die industrielle Praxis.* Hanser Verlag, Munich/Vienna 1999

[8]   WOHLRABE, H.: *Maschinen- und Prozessfähigkeit von Bestückausrüstungen der SMT;* Verlag Dr. Markus A. Detert Templin, 2nd edition 2001

6 Process Cost Optimisation

6.1 Process Chains in Electronics Manufacturing

6.1.1 Principal Technologies of Electronics Production

Modern electronics production is characterised by the assembly of electronics modules, on which a wide range of electronic, electrical and mechanical components must be connected to form operational modules or devices.

The very different technological levels ranging from the placement of flip-chip components, including the assembly of resistances in through-hole technology (THT), through to the fastening of heat sinks, under some circumstances must be realised on a single printed-circuit board. This means automatic and manual work steps can coexist in modern electronics manufacturing. This coexistence of a wide range of technologies can be found especially in the manufacturing of capital goods. Figure 6.1 shows the construction hierarchy of electronic devices in accordance with [16].

In general, the trend in electronics is to the processing of *surface-mounted devices* (SMD). The first industrially used method for the assembly of electronics components on printed-circuit boards was, however, the through hole technology (THT). For this technology, wired components with their connections are inserted in the holes on the printed-circuit board and, after any necessary cutting of the connection wires on the opposite side, soldered with the conducting structures of the printed-circuit board. The disadvantages of the through-hole technology are the high surface requirement of the component connection and the long leads to the actual component (in particular for ICs). The decisive restriction for through-hole assembly, however, is the connection mounting grid of the components.

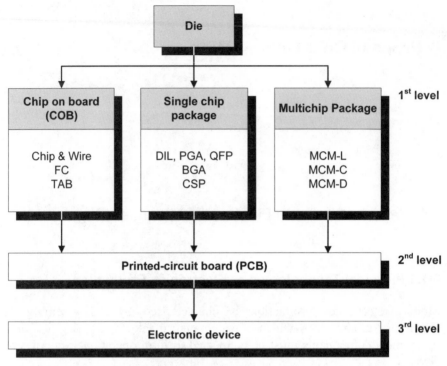

Figure 6.1. Construction hierarchy of electronic devices

The implementation of the automatic placement means this size cannot be reduced significantly below 2.54 mm (100 mil). The resulting reduction of the packing density led to the development of the surface-mounted technology (SMT). The surface-mounted technology is currently the most important technology for the assembly of electronic modules. For the surface-mounted technology, the component connections are placed on the corresponding connection surfaces of the printed-circuit board and soldered there. This allows the connection mounting grid of the components to be significantly reduced (the current minimum is approximately 0.3 mm (also see [21])), which permits a higher number of connections per component and a higher packing density than for the through-hole technology.

The resulting continuing reduction of the connection leads improves the HF and switching behaviour of the modules. References [11] and [16] provide comprehensive representations of the current technologies and the technological sequences in electronics production.

Mounting variations of PCBs (according to IPC-CM-770D):

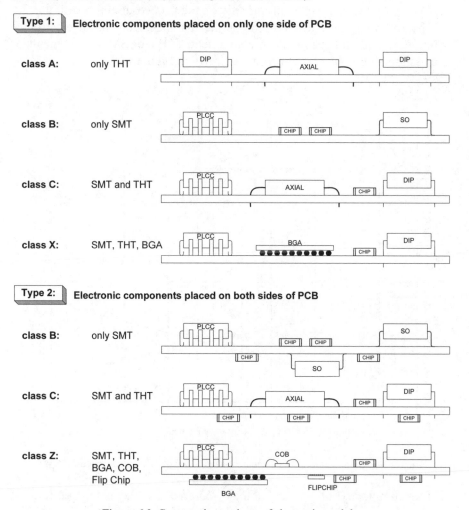

Figure 6.2. Construction variants of electronic modules

Despite the rapid development of the SMT, the THT is still widely used in electronics manufacturing – particularly in the area of "low-cost" electronics (home electronics, toys) and in the capital goods industry (machine controllers, power-electronics assemblies, modules with a small number of units). Often, especially in the capital goods area, components whose modules are assembled using both technologies (THT and SMT) (so-called mixed technologies) are also found. Figure 6.2 shows the currently usual construction variants of electronic modules and their classification in accordance with the standard IPC-CM-770 [10]. Another technology has developed from the classic SMT, the assembly of nonhoused components (DCA – direct chip attachment) directly on the printed-circuit board.

The technologies actually used for the manufacturing of an electronics module depend on the associated components (THT or SMT components or the assignment in the classification as shown in Figure 6.2).

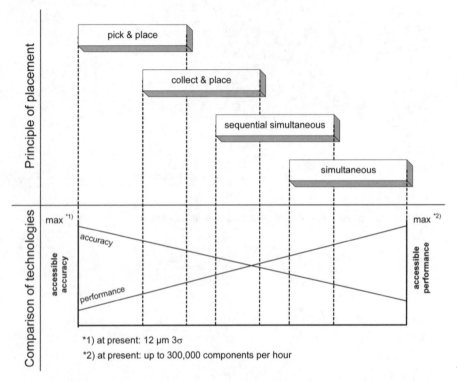

Figure 6.3. Comparison of the placement accuracy with placement speed for SMT

The complexity of the technology required for the manufacturing of electronics modules depends on the base technologies and on the components to be processed. Here, a component is also considered to be the module carrier (normally printed-circuit boards). Economic aspects, such as the technology loading, placement speed and flexibility, also play an important role in the selection of the populating technology. However, the demands for a high placement speed and a high placement accuracy are contradictory. This relationship is shown in Figure 6.3 for the various populating principles of the SMD automatic placement machines.

Other currently common technologies cannot be assigned to the basic technologies and can occur for each type of module. These are technologies used for the assembly of mechanical or electromechanical components. These include, for example, the pressing in of plug-in connectors, the riveting, screwing or gluing of heat sinks, the assembly of components that cannot be soldered (*e.g.*, LC displays and programmed components) and the fastening of large-volume or heavy components (*e.g.*, electrolytic capacitors, relays, power resistors) by gluing, screwing or fastening using the appropriate equipment.

6.1.2 Quality Assurance in Modern Electronics Production

The large number of possible, and sometimes very complex, technological steps in modern electronics production presented in the previous section imply high demands on the process monitoring and the quality assurance. The end product in electronics manufacturing is the module that is then itself assembled into devices. A significant quality characteristic of this module is its function capability under defined operating and environmental conditions. Maintaining this function capability over, if possible, the complete planned lifetime of the module, designated as *reliability* (see also Chapter 8), has equal importance. A special feature of the inspection of electronics modules is the almost complete absence of measurable quality characteristic values in the manufacturing processes before the electrical operability of the modules is achieved. Most of the characteristics have an attributive character. Their size must be determined by counting (*e.g.*, the number of soldering defects). To achieve a stable high product quality in production, this requires decisions to be made already in the planning and transition phase of a product. Tools that permit an evaluation of the required processes and procedure, include the design of experiments (DoE), the failure modes and effects analysis (FMEA) and the quality function deployment (QFD) method for the systematic and complete product and quality planning. All three methods are described in detail and in practical terms in [12].

Another significant factor for achieving the required product quality is the guaranteeing of the capabilities of the processes and the associated equipment that must be performed on the manufacturing tasks with the specified parameters and tolerances. To prove these capabilities, the so-called *process-capability coefficients* were successfully introduced in the automobile industry at the start of the 1980s (also see Section 1.4.5, 5.8 and [12]). Since the beginning of the 1990s, the proof of these coefficients is also being increasingly demanded from electronics manufacturers. Investigations have been initiated as to whether and how the regulations from the automobile industry can be transferred to the electronics industry. Initial investigations were performed on automatic placement machines and the corresponding calculation rules for determining the capability coefficients prepared. The inclusion of solder paste printing machines and soldering machines followed. Reference [20] describes the fundamental principles for the determination of the machine and process capabilities in the electronics industry with focus on the consideration of the placement technology. Generally applicable statements concerning the machine and process capability about the processes beyond the electronics manufacturing are contained in [12].

To ensure the process and product quality in electronics manufacturing, quality processes must be coupled to the technological processes. A *quality process* is defined as being the unit from the inspection process and the rejection process (also see Section 6.2.3 and Figure 6.5). The rejection process handles all products identified as being defective by the inspection process. This handling can involve the repair, the marking (mapping, inking) or the separation (scrapping). In the repair case, the subsequent reinspection of the repaired modules also belongs to the rejection process. For the execution of inspection processes, various nondestructive test procedures have established themselves in electronics manufacturing depending on the technological processes to be tested and their quality characteristics. The most important representatives here are automatic optical inspection (AOI), X-ray inspection (AXI), in-circuit test (ICT) and functional test (FCT). Other important test technologies and their application areas are described in [14].

Each necessary rejection process (in particular for repair processes) must be equipped appropriately depending on the technology and the defects involved (Table 6.1).

In all cases, all inspection and rejection processes should record the quality data (*e.g.*, lot, module, defect type, frequency, affected components or networks, defect location, repair measure – "traceability") as completely as possible.

An important aspect for the evaluation of the inspection processes is the *measuring-equipment capability*. "All inspection and measurement systems have a certain degree of uncertainty and error of dimensions, which normally cannot be quantified exactly" ([12]: 323). This is the cause of the occurrence of random deviations and systematic deviations for the determination of measured values. To estimate this effect on the inspection process, a measuring-equipment capability investigation is performed. It provides a statement whether a specific inspection process (or the selected test equipment) is at all suitable for a specific inspection task. The necessity for such an investigation can be legal or internal and external customer requirements. Discussions on the necessity, the theory, the measuring-equipment capability coefficients and their derivation and for the execution of this method are contained in [12].

Table 6.1. Selection of inspection and repair technologies

Defect-causing technology	Possible inspection processes	Repair process / equipment
Solder paste printing / dispensing	Manual optical inspect. AOI Profile inspection	Washing the incorrectly printed-circuit board Inclusion the washed circuit board in the printing/dispensing process Reinspection
Placement SMT placement THT placement Manual placement	Manual optical inspect. AOI	Component removal and placement or position correction of the incorrectly placed components using handling technology (*e.g.*, Fineplacer with microscope) and manual tools
Soldering Reflow soldering Wave soldering Manual soldering Selective soldering	Manual optical inspect. AOI AXI ICT flying-probe test FCT	Opening or remelting of defective solder joints with technology-dependent technology (*e.g.*, Fineplacer with above and below heating, minisolder wave, solder immersion bath, manual soldering iron) Possibly removal and placement of the affected components Reinspection
Final assembly	FCT System test	Removal of defective components and replacement with fault-free components (see placement and soldering) Correction of contact faults (*e.g.*, short-circuit, open connection) Reinspection

Detailed considerations not only for the fault classes and fault scenarios, but also for inspection and rejection processes that concern the SMT are contained in [11]. The IPC-A-610 standard [8] provides a comprehensive fault catalogue for electronics production (with regard to SMT, THT, mechanical process steps). The IPC-7912 standard [9] defines standard methods for the categorising of defects in electronics manufacturing.

6.2 Quality Cost Optimisation to Reduce Process Costs

6.2.1 Goal of Quality Cost Optimisation

Electronics manufacturing, in particular printed-circuit board production, involves the successive execution of a number of technological steps whose quality characteristics have a largely attributive character. Figure 6.4 shows a typical placement of machines and equipment for the production of modules with surface-mounted components (SMT).

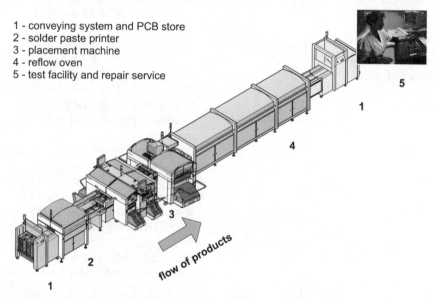

1 - conveying system and PCB store
2 - solder paste printer
3 - placement machine
4 - reflow oven
5 - test facility and repair service

Figure 6.4. Typical manufacturing line for SMT modules

The practice shows there is a dependency between the properties of the module to be produced (*e.g.*, number, type and complexity of the used components, packing density of the components, design of the module) and the produced lot size on the one hand, and the quality behaviour of associated technologies on the other hand. This dependency is described by the defect rate p or the yield y of the lot. A proven size is thus required with which the technological processes can be compared in their quality behaviour with each other and depending on the manufactured product. Such quantities that can be used here are the costs caused by the technologies and the quality processes. Namely, the costs can be used to compare the quality behaviour of completely different technologies and products for which no relationship can be made using physical or technological parameters. Once the costs of the technological processes and the quality process have been determined and related to each other, these cost relationships can be analysed and optimised. The goal of the quality costs analysis is to minimise the quality costs depending on the defect rate p of the considered module or the product. Thus, the costs can be used indirectly to develop an optimum process and test strategy whose goal is a cost-effective product production coupled with the highest product quality. The basic model of the quality cost optimisation provides a fundamental description of the behaviour of the costs in dependency on the product, the process and the required general conditions.

6.2.2 Yield y and Defect Rate p

As mentioned in the previous section, the defect rate p (also called the process defect rate) is a significant characteristic value for describing the quality behaviour of technological processes. The *defect rate p* specifies the proportion of defective modules related to a population (usually the lot). Its complementary size is the *yield y* of a technological process. The sum of the defect rate and the yield is always one. In the manufacturing of printed-circuit boards, the specification of the defect rate has established itself for describing the quality behaviour of the individual processes. In the semiconductor industry, in contrast, the yield is preferred.

If the task is to plan and optimise the quality processes for a complete manufacturing line (for a product or a group of very similar products), statements about the individual defect rates of the processes and about the overall defect rate p_{total} for this process chain are necessary. The defect rate of a process is generally calculated using the following formula:

$$p = \frac{k}{N} \tag{6.1}$$

(k = number of defective modules, N = number of tested modules).
   The yield of a process is defined as

$$y = \frac{N - k}{N} = 1 - p \tag{6.2}$$

Using [7], the overall yield y_{total} of a process chain consisting of n technological processes is calculated as being the product of the individual yields of these processes:

$$y_{total} = y_1 \cdot y_2 \cdot \ldots \cdot y_n = \prod_{k=1}^{n} y_k \tag{6.3}$$

If the yield values $y_1...y_n$ in Equation (6.3) are replaced by the defect rate p (using Equation (6.2)), this gives

$$y_{total} = 1 - p_{total} = (1 - p_1) \cdot (1 - p_2) \cdot ... \cdot (1 - p_n) = \prod_{k=1}^{n} (1 - p_k) \tag{6.4}$$

By reforming, this produces the equation for determining the overall defect rate of a process chain:

$$p_{total} = 1 - y_{total} = 1 - (1 - p_1) \cdot (1 - p_2) \cdot ... \cdot (1 - p_n) = 1 - \prod_{k=1}^{n} (1 - p_k) \tag{6.5}$$

For small defect rates and for few technological processes in a chain, the overall defect rate p_{total} can be approximated by the addition of the individual defect rates of the processes:

$$p_{total} \approx p_1 + p_2 + ... + p_n \tag{6.6}$$

Table 6.2. Comparison of computed and estimated values for determining p_{total}

Defect rate for each process No. of process steps	$p_k = 1\%$		$p_k = 2\%$		$p_k = 5\%$		$p_k = 10\%$	
	p_{total} Estimate	p_{total} Calculation	p_{total} Estimate	p_{total} Calculation	p_{total} Estimate	p_{total} Calculation	p_{total} Estimate	p_{total} Calculation
2	2%	1.99%	4%	3.96%	10%	9.75%	20%	19.00%
5	5%	4.90%	10%	9.61%	25%	22.62%	50%	40.95%
10	10%	9.56%	20%	18.29%	50%	40.13%	100%	65.13%
15	15%	13.99%	30%	26.14%	75%	53.67%	150% (!)	79.41%

Table 6.2 shows the result for the estimate-forming sums and for the exact calculation for various number of process steps and various defect

rates for each process. For simplification, it is assumed that all technological processes in the associated chain have the same defect rate p_k. The grey fields in the table specify the combinations for which the estimate for Equation (6.6) produces unacceptable or even incorrect results (*e.g.*, "150%"). The table shows that an acceptably precise estimate of the overall defect rate is certainly possible and also practicable.

6.2.3 Basic Quality Cost Model and Quality Distribution

In electronics manufacturing, the technological processes and quality processes are coupled with each other to not only produce printed-circuit boards but also to prove their functional capability. Because many parameters and disturbances act on the technological processes and on the quality processes, a pure process monitoring for achieving the end quality of the products normally does not suffice (in particular with regard to the production of many different modules each with only a small number of unit counts – "high mix/low volume"). Consequently, inspection processes, normally with directly following rejection processes, are added to the technological sequence in order to keep the defect rate p small. A rejection process can mean either the repair or the separation of the module. These processes result in costs: The inspection process causes the inspection costs and the rejection process the rejection costs.

These considerations raise the question as to when, from a cost point-of-view, is it appropriate for a quality process (unit from the inspection and rejection process – see Figure 6.5) to directly follow a technological process and when is it more cost effective to perform the quality inspection of the products after a later technological process? An abstraction of the manufacturing and quality processes is introduced to answer this question. Figure 6.5 also shows such an abstraction for an SMT manufacturing line.

A technological process T is characterised by its defect rate p. As mentioned previously, this defect rate p has a value range that lies between 0 and 1. An ideal technological process has the defect rate $p = 0$ and so produces no defective products. The quality process Q, which consists of an inspection process I and a rejection process R, has the task of finding the defective products from the product flow by using some suitable inspection process I. Such products are then removed and either repaired in the rejection process R or discarded (*e.g.*, scrapped). The repaired modules can either be returned to the inspection process I or directly passed to the subsequent technology.

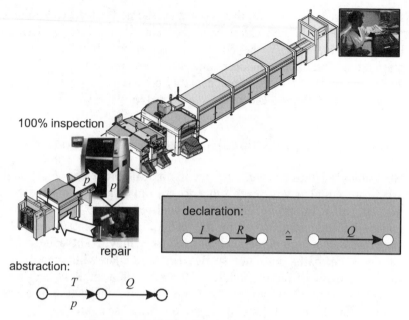

Figure 6.5. SMT line with abstraction for a process

The rejection process in Figure 6.5 is unusual. For the special case of solder paste printing, it consists of washing off the incorrectly printed solder paste, reprinting using process T and reinspection in the AOI.

As mentioned several times previously, the definition has been made that each quality process Q consists of the unit from the inspection process I and the rejection process R (Figure 6.5). This definition largely represents the reality in electronics manufacturing. Exceptions in the semiconductor industry, where the rejection process does not directly follow the inspection process, are known as inking or mapping. For this procedure, the semiconductor components detected as being defective are marked with a colour point (inking) on the wafer or electronically in a file (mapping). The defective chips are fed to rejection (separation) only after the components have been separated. Similar procedures are used in electronics manufacturing for the processing of panels (unit of similar modules that are separated from each other only after processing).

There are two ways of performing the quality strategy after a technological process T (see Figure 6.6):

1. A quality process Q is performed directly after the technological process T.
2. *No* quality process Q is performed after the technological process T, but rather the next technological step follows directly.

Technological process with following quality process:

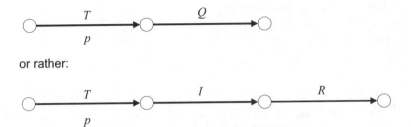

or rather:

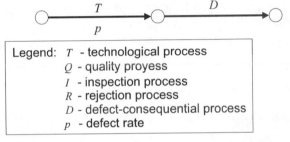

Technological process without following quality process:

Legend: T - technological process
Q - quality proyess
I - inspection process
R - rejection process
D - defect-consequential process
p - defect rate

Figure 6.6. Representation of the possible test strategies for the basic model

The description of the cost functions in these cases depending on the defect rate p is designated as the basic model of the quality cost optimisation (also see [14, 15]). A similar concept is also chosen and discussed in [2].

If no quality process Q follows a technological process T and $p \neq 0$ is true for the defect rate p of the process, the defective products caused by technology T will be further processed and thus also their defects. These defects must be discovered by a quality process after a subsequent technological process and corrected to prevent defective products from being delivered. The defect-consequential process D is defined to describe this mechanism. The process is, in effect, attached to a technological process that is not immediately followed by a quality process. The defect-consequential process D is thus a virtual process that directly follows the technological process T. The process D is characterised in that it causes real costs, the defect-consequential costs C_D. These defect-consequential costs C_D are directly proportional to the defect rate p of the preceding technology T that caused the defect. The following definitions represent the various cost elements (the index X indicates the cost element T, D, R, I or Q):

- complete costs related to a lot: C_X
- costs per product/module: c_X

- $C_X = N \cdot c_X$ with N – number of products (universal set, *e.g.*, lot, job).

Furthermore, for the basic model, the rejection costs C_R and the defect-consequential costs C_D depend on the defect rate p of the directly preceding technological process:

$$C_D = N \cdot c_D \cdot p \text{ and } C_R = N \cdot c_R \cdot p$$

The product of N and p describes the number of defective modules in the considered lot of size N. If the defect rate $p = 0$, then $C_R = 0$ and $C_D = 0$.

The following linear equations can be used to calculate the complete costs of a technological process T with subsequent defect-consequential process D (costs C_0 – index 0: no directly following quality process) or a technological process with subsequent quality process Q (costs C_1 – index 1: directly following quality process):

Costs with defect-consequential process F:

$$C_0 = C_T + C_D = N \cdot c_0 \tag{6.7}$$
$$c_0 = c_T + p \cdot c_D$$

Costs with quality process Q:

$$C_1 = C_T + C_I + C_R = N \cdot c_1 \tag{6.8}$$
$$c_1 = c_T + c_I + p \cdot c_R$$

where:

C_T technological costs per lot
C_D defect-consequential costs per lot
C_I test costs per lot
C_R rejection costs per lot
c_T technological costs per module
c_I test costs per module
c_D defect-consequential costs per defective module
c_R rejection costs per defective module
p defect rate of the technological process T.

To allow the comparison of the costs for different processes and products, the model considerations are specified as costs per module (unit, for example, €/unit). Furthermore, the following prerequisites must be satisfied: the equipment for the inspection processes is available (no investment), the last process in the manufacturing chain is a quality process (*e.g.*, ICT or FCT) and the technological costs C_T are independent of the chosen test strategy. The latter condition specifies that the technological costs are used only as constant summands in the calculations and do not affect the optimisation of the quality costs. It is defined that these constant-cost components are not represented in the graphical representations, but only the quality-determining costs.

Figure 6.7 shows the two possible cases for the form of the cost functions c_0 and c_1 in the interval $0 \le p \le 1$:

1. The graphs of the functions do not have *any* intersection point within the interval $0 \le p \le 1$; the function value of c_0 is always is less than that for c_1.
2. The graphs of the functions have *one* intersection point within the interval $0 \le p \le 1$; the function value of c_0 at $p = 0$ is smaller than the function value for c_1 at $p = 0$; one point (p^*, c^*) exists that describes the position of the intersection point between both functions.

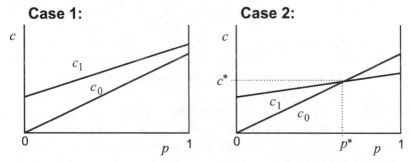

Figure 6.7. Two cases of the cost graphs of the basic model

In the first case, this means that for each value of the defect rate p the addition of a quality process is more expensive to perform than the inspection and rejection after a later technological step. In the second case, an intersection point exists between the cost straight lines at point p^*. This point is designated as being the separation defect rate and is calculated as follows:

$$ p^* = \frac{c_I}{c_D - c_R} \quad \text{for} \quad \begin{matrix} c_0 = c_1 = c^* \\ p = p^* \end{matrix} \quad \text{when} \quad c_D > c_I + c_R \tag{6.9} $$

If the real value of the defect rate p (the "measured" value in the manufacturing process) is less than p^*, then, for cost reasons, it is not necessary to add a quality process directly after the technological process. If the value of p is larger than p^*, economic considerations mean that a quality process should be kept directly after the technological process sequences (see Figure 6.8). In a small value range around the separation defect rate p^*, which depends on the process and quality costs, the costs for testing and not-testing are almost identical. This means that a change of the test strategy is not necessary in such a case.

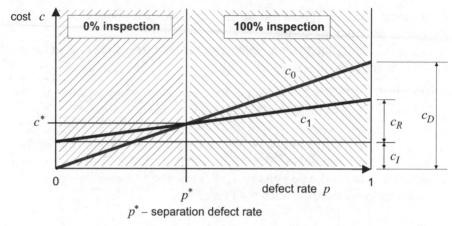

Figure 6.8. The test strategy depending on the defect rate

However, the cost considerations alone do not motivate the use of a quality process after a technological process, but other factors, such as customer requirements, accountability and process control.

In the previous considerations for the basic model, the defect rate p was considered as being a rational number with a value range $0 \le p \le 1$. The decision concerning the introduction of a quality process was only made dependent on the size of the numeric value for p in the ratio to the separation defect rate p^*:

$p < p^*$ no quality process Q after T
$p > p^*$ quality process Q after T
$p = p^*$ no change of the previous test strategy.

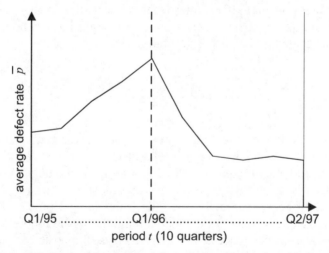

Figure 6.9. Graph of the average defect rate for a module over 10 quarters

It was always assumed that the real defect rate of the considered process is known. This is not the case in practice. The only value that can be assumed as being known is the defect rates of the previous lots of the same product. These defect rates normally have different values. The fluctuations are caused by systematic and random effects on the technological process and on the processed materials. Figure 6.9 shows the quality graph for a real product for the determined defect rates of the lots over an interval of 10 quarters.

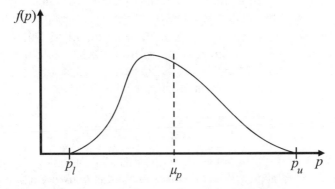

Figure 6.10. Example of a continuous defect-rate distribution

Each actual mean of the defect rates for all lots of a product per quarter is the result of the effect of systematic and random effects and is consequently designated as being a random event. This is also the case for the defect rates p of the individual lots. If the assumption is made that the defect rate p fluctuates and that each defect rate p of a lot represents a realisation from a (theoretically) infinitely large number of considered lots, a continuous random distribution with a distribution density $f(p)$ can also be assumed for the defect rate Figure 6.10.

Thus, if previously a numeric value was assumed for the defect rate and assuming a continuous distribution of the defect rate p and constant costs, the value for the defect rate can now be replaced by the expected value μ_p of the distribution density $f(p)$. The rejection costs C_R of a lot then can be calculated using the follow equation:

$$C_R = N \cdot c_R \cdot \mu_p = N \cdot c_R \cdot \int_{p_l}^{p_u} p \cdot f(p)\,\mathrm{d}p \qquad (6.10)$$

Similarly, the defect rate p for the calculation of the defect-consequential costs C_D of a lot can be replaced by the expected value μ_p of its distribution density. Here, the integral part of the expected value μ_p represents the distribution density $f(p)$ of the defect rate p. The expected value represents a strong characterisation of the distribution density and "can be indicated as the centre of gravity of those masses currently described by the "mass density" *f(x)* [= here *f(p)*, M.O.] " ([5]: 705). This allows the use of the expected value μ_p of the distribution density *f(p)* as the decision criterion for selecting the most cost effective test strategy in accordance with the basic model of the quality cost optimisation.

Generally, the distribution density of the defect rate *p* is not known. However, the different realisations of this function are known. These can be either the defect rates of a known number of lots of a product or the average defect rates $\overline{p}$ of a product over a specific interval *t* (example Figure 6.9). If we consider these realisations as a sample from the distribution density of the defect rate *p*, the arithmetic mean from these realisations can be calculated as a good estimate for the expected value μ_p of the density function *f(p)*. The arithmetic mean is calculated as follows:

$$\hat{\mu} = \overline{x} = \frac{1}{n}\sum_{i=1}^{n} x_i \tag{6.11}$$

where *n* represents the number of individual realisations x_i (*e.g.*, sample size). The mean $\overline{x}$ is an *unbiased, consistent* and *exhaustive* estimate for the expected value μ represented by the symbol $\hat{\mu}$. In accordance with [5] and [18], the following definitions apply:

- *Unbiased:* An estimate is called *unbiased* when its expected value is the same as the parameter to be estimated.
- *Consistent:* An estimate is *consistent* when with increasing sample size *n* the estimated value converges to the true unknown parameter.
- *Exhaustive:* An estimate is *exhaustive* when all information contained in the sample is used for the unknown parameter and no other estimate provides additional information about the true unknown parameter.

When used for the defect rates *p* of the individual lots, the arithmetic mean $\overline{p}$ is calculated as

$$\hat{\mu}_p = \overline{p} = \frac{1}{n}\sum_{i=1}^{n} p_i \tag{6.12}$$

When defect rates are considered over specific fixed time intervals (*e.g.*, quarters), this normally concerns the determination of an arithmetic mean using data already determined using this method (*e.g.*, the grouping of the defect rates of the lots for a module type produced in a quarter to produce the defect rate for the quarter $\overline{p}$. This arithmetic mean from arithmetic means is designated as $\overline{\overline{p}}$. Equation (6.12) is used to calculate $\overline{\overline{p}}$. Note that the lots used for the calculation of $\bar{\bar{p}}$ must have the same lot size (the same population). If this is not the case, $\bar{p}$ can also be calculated as follows:

$$\bar{p} = \frac{\text{number of defect modules of one type during a time period}}{\text{number of produced modules of the same type during a time period}}$$

The link to the lot size no longer applies in this case. A comparison with the defect rates from other production intervals is again only possible when the numbers of produced modules are equal (or, at least, almost equal) in the compared time intervals. Other discussions concerning the calculation of the arithmetic mean are contained in [18].

In addition, the spreads provide another parameter used to describe a distribution. They represent a measure for the dispersion from the expected value of the distribution. The most important spread for continuous distributions is the dispersion σ^2 or its *consistent*, *unbiased* and *exhaustive* estimate, the variance s^2. The variance for the defect rate p is calculated as follows:

$$s_p^2 = \frac{1}{n-1}\sum_{i=1}^{n}(p_i - \bar{p})^2 \tag{6.13}$$

If the defect rates of a product (*e.g.*, of a module) are now recorded over a specific interval or over a number of lots, and the average defect rate $\bar{p}$ and the variance s_p^2 can be calculated to produce unbiased estimates for the parameters of the distribution density of the defect rate p for this considered product. The more realisations of the distribution density that are known, the better is the estimate of the parameters and thus the reliability of finding the correct test strategy (0% inspection or 100% inspection) by using the basic model of the quality cost optimisation.

6.2.4 Extensions to the Basic Quality Cost Model

The basic quality cost model describes the optimisation of the test strategy based only on a single technological process. This cannot be used in practice, but rather represents the fundamental considerations and relationships. In a real technological chain, normally several technological processes follow each other before a product is completed. This means model extensions are necessary to represent real manufacturing chains. As an extension of Figure 6.5, Figure 6.11 shows an SMT line with the abstraction all associated manufacturing and quality processes.

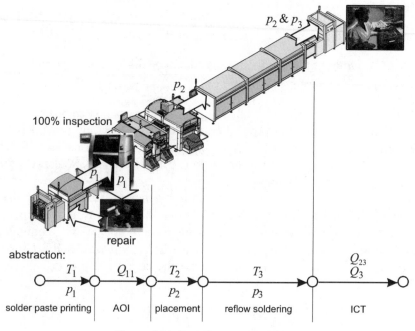

Figure 6.11. SMT line with abstraction

Two technological processes with quality process after each process:

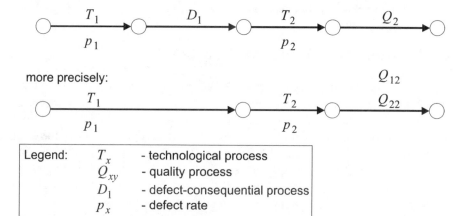

Two technological processes with quality process only after T_2:

Two technological processes with quality process only after T_2:

more precisely:

Legend: T_x - technological process
         Q_{xy} - quality process
         D_1 - defect-consequential process
         p_x - defect rate

Figure 6.12. Abstraction for two processes

Initially, the model extension is considered for a technological line with two processes. If two technological steps follow each other, there is the possibility to test after each technology or to provide an inspection step only after the second technology but which is qualified to test for defects resulting from the two technologies. Figure 6.12 shows the abstraction of such a chain with the associated quality process.

The possible assignments of the quality process with regard to the technological process T_1 are the following:

- A quality process Q_1 follows directly after the T_1
  $\Rightarrow$ all defects caused by T_1 will be found by I_1 immediately after T_1, fed to the rejection process R_1 and corrected there
- The quality process that follows the technology T_2 is characterised by being able to detect and assign the defects caused by the technologies T_1 and T_2
  $\Rightarrow$ all defects caused by T_1 will be found and corrected by the quality process after T_2 – representation of this property: Q_{12}
  $\Rightarrow$ all defects caused by T_2 will also be found and corrected by the quality process after T_2 – representation of this property: Q_{22}.

Figure 6.12 shows that the assignment of the defect-causing technologies to the associated quality processes (expressed by the indexing) completely replaces the (virtual) defect-consequential process (introduced with the basic model) with real quality processes with the corresponding cost factors. This leads to a simplification of the use of the cost model because the costs of real processes are already present or can be determined in the manufacturing.

As for the basic model, the cost functions for this model extension are described by two linear equations:

Costs without quality process after T_1:

$$c_0 = c_{T1} + c_{T2} + c_{I12} + c_{P22} + p_1 \cdot c_{R12} + p_2 \cdot c_{R22} \qquad (6.14)$$

Costs with quality process after T_1:

$$c_1 = c_{T1} + c_{I11} + p_1 \cdot c_{R11} + c_{T2} + c_{I22} + p_2 \cdot c_{R22} \qquad (6.15)$$

where:

c_{T1}, c_{T2}	technological costs per module
c_{I11}, c_{I12}, c_{I22}	inspection costs per module
c_{R11}, c_{R12}, c_{R22}	rejection costs per defective module
p_1, p_2	defect rates of the T_1 and T_2 processes.

It is also assumed here that the associated test strategy does not have any effect on the technological costs c_{T1} and c_{T2}. These costs are assumed to be constant for both quality process assignments and so do not need to be considered for an optimisation.

For the case that the total of the rejection and test costs of the technology T_1 with inspection after T_2 is larger than the total of the rejection and test costs with inspection directly after T_1, the two linear equations c_0 and c_1 can be set equal. The intersection point of the two straight lines that describes the equations is defined by the separation defect rate p_1^* and the cost value c^* (Figure 6.13).

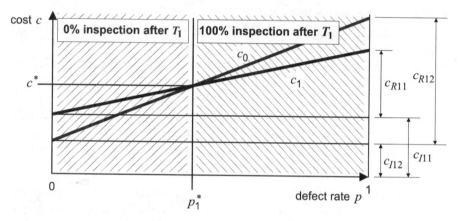

p_1^* - separation defect rate regarding inspection or no inspection after T_1

Figure 6.13. Cost graph for two processes

The difference compared with the basic model of the quality costs analysis is the cost proportion c_{I12} (see Figure 6.13), which leads to a fixed cost proportion also for the case of no inspection after T_1. The fixed cost proportion results because the quality process after T_2 must have the ability to find and correct defects that were caused by the technology T_1. These costs can be, for example, equipment costs, time costs and personnel costs caused by the adaptation of the complete quality process to the extended task.

The separation defect rate p_1^* can be calculated as the intersection point of the two cost straight lines as follows:

$$p_1^* = \frac{c_{I11} - c_{I12}}{c_{R12} - c_{R11}} \quad \text{for} \quad \begin{matrix} c_0 = c_1 = c^* \\ p_1 = p_1^* \end{matrix} \quad \text{when } c_{R12} + c_{I12} > c_{R11} + c_{I11} \qquad (6.16)$$

The case can also be envisaged for the extended basic model. It can be proved in reality that the noninspection after technology T_1 is always cheaper (namely for each defect rate) than the introduction of a quality process after T_1. Such an example is described in [14].

By modifying the basic model for a chain formed from two technological processes, a general representation of the model for any number of technological processes in a chain (n units) can be derived. For the production of electronics modules, a number of technological processes follow each other to which quality processes can also be assigned. This gives 2^N possibilities for assigning quality processes for the technologies T_1 ... T_N. Assuming that at the end of the technological passage, just one quality process is assigned, the remaining possibilities reduce to 2^{N-1} assignments (Figure 6.14).

Number of possible quality process arrangements: 2^{N-1}

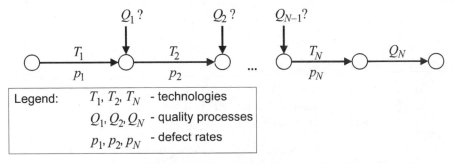

Figure 6.14. Theoretical number of possible assignments of quality processes

It is also assumed that there is the possibility to test directly after a technological process or to assign the quality process after a subsequent technological process. The representation is generalised using indices (see Figure 6.15).

The costs are calculated analogously to those in the extended basic model for two technologies. The linear equations are then:

Costs without quality process after T_m:

$$c_0 = c_{Tm} + c_{Tn} + c_{Imn} + c_{Inn} + p_m \cdot c_{Rmn} + p_n \cdot c_{Rnn} \qquad (6.17)$$

Costs with quality process after T_m:

$$c_1 = c_{Tm} + c_{Imm} + p_m \cdot c_{Rmm} + c_{Tn} + c_{Inn} + p_n \cdot c_{Rnn} \qquad (6.18)$$

where:

c_{Tm}, c_{Tn}	technological costs per module
$c_{Imm}, c_{Inn}, c_{Imn}$	test costs per module
$c_{Rmm}, c_{Rnn}, c_{Rmn}$	rejection costs per defective module
p_m, p_n	defect rates of the processes T_n and T_m.

Technological process with following quality process:

Technological process with quality process after T_n :

more precisely:

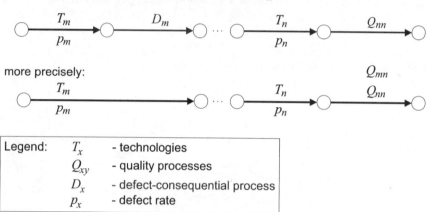

Figure 6.15. Abstraction for n processes

Because the technological costs c_{Tm} and c_{Tn} are assumed to be constant, they can be omitted in the subsequent considerations. The case should also be considered in which there is an intersection point between the two straight lines described by the separation defect rate p^*_m and the costs c^* (Figure 6.16).

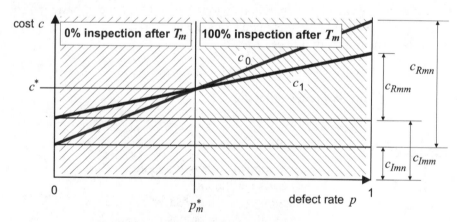

p^*_m - separation defect rate regarding inspection or no inspection after T_m

Figure 6.16. General representation of the cost curve

The separation defect rate $p_m{}^*$ for the generalised model can be calculated as follows:

$$p_m{}^* = \frac{c_{Imm} - c_{Imn}}{c_{Rmn} - c_{Rmm}} \text{ for } \begin{matrix} c_0 = c_1 = c^* \\ p_m = p{}^*{}_m \end{matrix} \text{ when } c_{Rmn} + c_{Imn} > c_{Rmm} + c_{Imm} \qquad (6.19)$$

Obviously, this model must also consider the case that no inspection (considered in the interval $0 \leq p \leq 1$) is always more cost effective than the use of a quality process directly after the technology T_m (the straight lines c_0 and c_1 do not intersect).

The product life cycle:

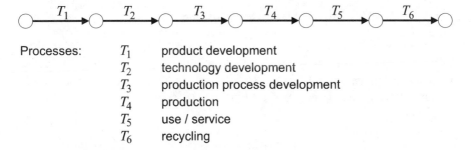

Processes: T_1 product development
             T_2 technology development
             T_3 production process development
             T_4 production
             T_5 use / service
             T_6 recycling

Use of the generalized basic model:

With inspection after T_m :

With inspection not until after T_n :

Figure 6.17. Abstraction of the product life cycle

In the discussed model extensions for two and for n processes (as shown previously), the "virtual" defect-consequential process D introduced in the basic model is represented by the cost components c_{Imn} and c_{Rmn}. The cost factors can actually be determined in the manufacturing and so permit the realisation of the extended models in the real manufacturing world. The applicability of the generalised basic models is not limited just to the technological processes of the module placement. The model can be used globally for considering the complete product life cycle, even outside electronics production (Figure 6.17). However, there can be problems with obtaining the necessary data from the real processes.

This allows the abstract definition: a quality process forms the last step in the product life cycle before recycling, namely, the use of the product by the customer or user until the product's end of life or its failure.

6.2.5 Features of the Inspection Process

The technological processes, being the cause of the defect rate p, played the decisive role in the previous considerations. The inspection processes themselves appear merely as a "black box" in which only the yield y is separated from the defect rate p. Figure 6.18 shows this ideal inspection process.

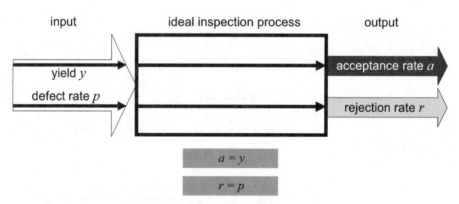

Figure 6.18. The ideal inspection process

The reality is, however, different. Like technological processes, inspection processes can also have faults. The duality of the input flow (faulty and fault-free modules) means two fault types can also occur in the inspection process:

- **Test blurring 1st type α** (also known as pseudodefects):
  Modules that actually meet specification are declared as being defective by the inspection process and are fed to the rejection process.

- **Test blurring 2nd type** β (also known as defect slippage):
  Modules that are actually defective are declared as being fault free by the inspection process and are passed to the next technological step.

This mechanism leads to a change of the acceptance rate a or the rejection rate r by the performed real inspection process (Figure 6.19).

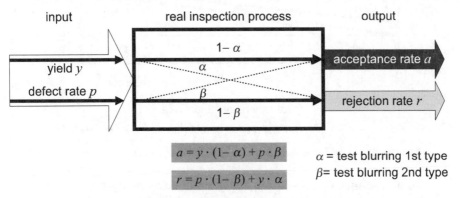

$$a = y \cdot (1 - \alpha) + p \cdot \beta$$

$$r = p \cdot (1 - \beta) + y \cdot \alpha$$

α = test blurring 1st type
β = test blurring 2nd type

Figure 6.19. The real inspection process

Some reasons for the causes of the occurrence of test blurring of type 1 and 2 lie in the fluctuations of the parameters of the technological processes and in the finite accuracy of the inspection processes. Other discussions for these problems are contained in Section 1.5 and in Section 8.7.4.

If test blurrings of type 1 and 2 are included in the basic model, this produces the following equations:

Costs without quality process after T (Equation (6.7)):

$$c_0 = c_T + p \cdot c_D$$

Costs with quality process T for the real inspection process:

$$c_1 = c_T + c_I + p \cdot (1 - \beta) \cdot c_R + (1 - p) \cdot \alpha \cdot c_R + p \cdot \beta \cdot c_D \qquad (6.20)$$

where:

c_T technological costs per module
c_I inspection costs per module
c_R rejection costs per defective module
c_D defect-consequential costs per defective module
p defect rate of the process T
α test blurring type 1 (pseudo defect)
β test blurring type 2 (defect slippage).

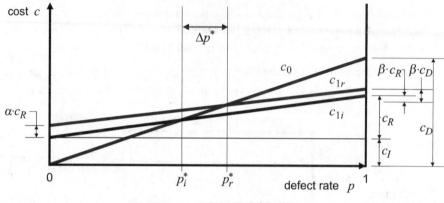

p_i^* - ideal separation defect rate

p_r^* - real separation defect rate

Figure 6.20. Costs of the various strategies for the ideal and real inspection process

The technological costs c_T are again assumed as being constant for both cases of the quality process form and so will be omitted from further consideration. As previously, it is assumed that an intersection point of the two straight lines c_0 and c_1 exists. Figure 6.20 shows the graph for the cost lines. The figure also shows the ratio of the costs for the ideal inspection process (straight line c_{1i}) in the real inspection process (straight line c_{1r}).

The *separation defect rate* p^*_r for a real inspection process can be calculated as follows:

$$p^*_r = \frac{c_I + \alpha \cdot c_R}{(1-\beta) \cdot (c_D - c_R) + \alpha \cdot c_R} \tag{6.21}$$

The recognisable shift of the separation defect rate p^* in Figure 6.20 to higher defect rates for the consideration of the real inspection process compared with the ideal inspection process is interesting. This implies that the more uncertain an inspection process is ($\alpha \neq 0$, $\beta \neq 0$), the higher may be the defect rate p of a technology T before a quality process Q used directly the technology T is desirable from the cost viewpoint.

The considerations for the real inspection process used for the generalisation of the extended basic model show that test blurrings of type 1 and 2 must also be expected for a subsequent inspection process. Because the inspection processes run mutually independent, the values for α and β are also different and must be differentiated using the appropriate indexing. The starting point for these considerations are the relationships shown in Section 6.2.4. The discussions of the linear equations and the equation for the calculation of the separation defect rate can be found in [14].

6.2.6 Cost Model for the Sampling Test

The previously described basic model with its extensions is used for quality cost analysis and optimisation for products that have been manufactured for some time and for which the appropriate quality data are available. Provided specific conditions are observed, the model can also be used for a new product, namely when the new product in its complexity (component density, used technologies) corresponds to the used component spectrum (associated pitch and pin count) and the module size (number of solder joints) of a previously manufactured module so that its quality data can be adapted to the new product. However, even the use of different design rules for the circuit-board design can permanently change the quality behaviour of the new product (despite the same complexity and the same component spectrum), so that the adaptation of the quality data to new products must be used with care. The only alternative is the general 100% inspection of new products after critical processes or the use of statistical process control methods (SPC). These methods were developed in mechanical engineering in the first half of the 20th century and belong since then to the standard techniques of quality assurance in mechanical engineering. They permit the conclusion drawn from a few observation values of a sample to be applied to a total set of products produced under the same conditions and provide important information about the subsequent production sequence. The SPC methods permit statements to be made about the observance of the quality standards prescribed for the product. Here, in accordance with [18], two basic types of SPC can be differentiated:

- statistical process control using control charts, namely the continuous monitoring of the manufacturing process with mathematical-statistical methods
- the mathematical-statistical sampling method for testing previously manufactured products.

While statistical process control permits the production of defective parts to be prevented with the immediate intervention in the processes, the mathematical-statistical sampling method has no direct effect on the production. It is used for the inspection of the products and thus the running processes.

Continuous quality characteristics can be used for statistical process control. Their property of being measurable, when used with the graphical representation on control charts, leads to a graph that, given an appropriate representation, permits a fast estimate of the current quality level and possible trends. A classic example from mechanical engineering is the monitoring of the shaft diameter using control charts. Control charts have also been developed for discrete characteristics. Statements concerning the types, function, form and use of control charts for continuous and discrete quality characteristics are contained in [12, 18] and in Section 5.7 of this book.

The mathematical-statistical sampling methods in their action mechanism with regard to the quality cost model can be placed between the 0% inspection and the 100% inspection. More precisely, the 0% inspection and the 100% inspection are special cases of the sampling test. Figure 6.21 shows the abstraction of a technological process with a subsequent sampling test.

Technological process with sampling test:

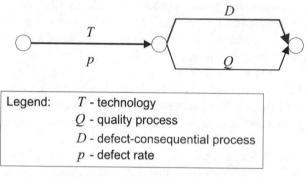

Legend: T - technology
          Q - quality process
          D - defect-consequential process
          p - defect rate

Figure 6.21. Abstraction of the sample test

If a sampling test (represented by the quality process Q) is performed after a technological process T, then only the sample, taken from a lot (also designated as the population), passes through this quality process. If the inspection process I finds defective modules in this sample, they will be fed to the appropriate rejection process R. If the lot is accepted, namely the acceptance number Ac is not exceeded, the remaining modules of this lot (the population less the sample) will not be subjected to any quality process directly after the technology T. They will be fed to a later quality process, represented by the defect-consequential process D. Figure 6.21 shows a "parallel switching" of the quality process Q and the defect-consequential process D. The sample size here should be small compared with the lot size in order to achieve the largest possible cost advantage from the use of the sample test. If the lot is not accepted, namely the acceptance number Ac is exceeded in the inspection of the sample, the complete lot will be fed to the quality process Q. The further procedure for subsequent lots of the same product depends on the associated sampling system (*e.g.*, in accordance with ISO 2859-1 [6]).

Thus, for the cost consideration of the sample test, two cases must be differentiated:

1. The lot is accepted: the modules in the sample will be subjected to the quality process Q; the defect-consequential process D acts on the nontested modules.
2. The lot is rejected: all modules will be subjected to the quality process Q.

Whether a lot is accepted depends on the acceptance probability P_a, which itself depends on the actual defect rate p in the complete lot (also see [1]). For the determination of the cost curve of the sample test depending on the defect rate p, the following quantities must be present or can be determined in the production process:

N lot size (the number of modules in the lot)

n sample size (the number of modules in the sample)

v relative sample size $v = \dfrac{n}{N}$

c_I inspection costs per module

c_R rejection costs per defective module

c_D defect-consequential costs per defective module

p average defect rate of the lot

p_{Sa} average defect rate of the sample of the accepted lot

ρ reduction coefficient of the defect rate $p_{Sa} = p \cdot \rho$ $(\rho \leq 1)$

P_a acceptance probability of the lot.

The cost factors c_I, c_R and c_D are known or given. Also known are the lot size N, the sample size n and the acceptance number Ac in accordance with the chosen sampling rule. An average defect rate p for the lot for the individual process steps has been determined in the series preliminary run. Unknown are the acceptance probability of the lot P_a and the average defect rate of the sample p_{Sa} for an accepted lot.

The complete costs for the sample test consist of the subcosts for the accepted and the rejected lot together. The accepted lot has the following subcosts:

- test costs of the sample for the acceptance of the lot

$$C_{PSa} = c_I \cdot N \cdot v = c_I \cdot n \tag{6.22}$$

- rejection costs of the sample for the acceptance of the lot

$$C_{RSa} = c_R \cdot N \cdot v \cdot p \cdot \rho = c_R \cdot n \cdot p_{Sa} \tag{6.23}$$

- defect-consequential costs for the acceptance of the lot

$$C_{FSa} = c_D \cdot (N \cdot p - N \cdot p \cdot \rho \cdot v) = c_D \cdot N \cdot p \cdot (1 - \rho \cdot v) \qquad (6.24)$$

where:

$N \cdot p$ $\qquad$ total number of defective modules in the lot

$N \cdot p \cdot \rho \cdot v$ $\quad$ number of repaired modules in the sample.

This yields, using the acceptance probability P_a, the following complete costs for the accepted lot:

$$C_{TotalSa} = P_a \cdot (C_{ISa} + C_{RSa} + C_{DSa}) \qquad (6.25)$$
$$= N \cdot P_a \cdot (c_I \cdot v + c_R \cdot p \cdot \rho \cdot v + c_D \cdot p \cdot (1 - \rho \cdot v))$$

In the case of the rejection of the lot, a transition will be made to 100% inspection. The associated costs are then:

$$C_{TotalSna} = (1 - P_a) \cdot N \cdot (c_I + p \cdot c_R) \qquad (6.26)$$

The complete cost of the sample test per lot, considered over a large number of lots, then results from the partial sums:

$$C_{TotalS} = C_{TotalSa} + C_{TotalSna} \qquad (6.27)$$
$$= N \cdot [c_I + p \cdot c_R + P_a \cdot (p \cdot (c_D - c_R) \cdot (1 - \rho \cdot v) - c_I \cdot (1 - v))]$$

After division by the lot size N, this gives the costs for the sample test per module:

$$c_{TotalS} = c_I + p \cdot c_R + P_a \cdot (p \cdot (c_D - c_R) \cdot (1 - \rho \cdot v) - c_I \cdot (1 - v)) \qquad (6.28)$$

To use this equation to calculate the actual costs for the sampling test, the acceptance probability P_a of the lot and the average defect rate p_{Sa} in the sample of the accepted lot (and thus also ρ) must be determined. The starting point here is the average defect rate p of the considered process determined from the statistical values of the previously manufactured lots of this product or from the series preliminary run. This average defect rate and the lot size N are used to derive a sampling plan. This is usually a sampling plan for attributive characteristics. In general, the standard ISO 2859-1 [6] can be used for the selection of a sampling plan for attributive characteristics. A further discussion for the determination and the use of sampling plans is contained in Section 5.9.

If, now in accordance with the restrictions, a sampling plan is selected, it will be characterised by the quantities, sample size n and acceptance number Ac. To determine the the acceptance probability P_a, the individual occurrence probabilities for defects up to acceptance number Ac (including Ac) must be determined (probability for zero defects in the sample, probability for one defect in the sample, ..., probability for Ac unit defects in the sample):

$$P_a = p_0 + p_1 + ... + p_c = \sum_{k=0}^{A_c} p_k \qquad (6.29)$$

The distribution models for attributive (discrete) characteristics must be used to determine the individual probabilities p_0 to p_{Ac}. In general, the hypergeometric distribution should be used for the calculation. If the lot size is high (in the ideal case, infinite) and $v \leq 0.1$ is true for the relative sample size, the binomial distribution can be used as a good approximation (also refer to Section 5.3).

The average defect rate p_{Sa} in the sample of the accepted lot represents a partial expected value from the occurrence probabilities of the defects permitted in the sample (maximum c units) related to the sample size:

$$p_{Sa} = \frac{1}{n} \cdot \frac{\sum_{k=0}^{c} k \cdot p_k}{\sum_{k=0}^{c} p_k} \qquad (6.30)$$

It is now possible to calculate the reduction coefficients ρ. The complete costs per unit c_{TotalS} have a nonlinear curve over the defect rate p, because with changing p, the acceptance probability P_a, and thus also the average defect rate p_{Sa} in the sample, also change. A rising defect rate causes a reduction of the acceptance probability and an increase of the average defect rate of the sample until the acceptance probability $P_a = 0$ is true and thus each lot will be rejected. A 100% inspection is the consequence. If, however, the defect rate falls, the acceptance probability P_a tends to 1, and only the test costs of the sample result as costs. Figure 6.22 shows a possible cost curve for a 0% inspection (c_0), 100% inspection (c_1) and sampling test (c_2).

The separation defect rate between 100% inspection and a sampling test ($p^*_{SPC;100\%}$) can only be determined numerically.

The curve of the graphs in Figure 6.22 shows that the sampling test never represents the lowest-cost solution. The sampling test has the advantage that it is cheaper than 100% inspection for small defect rates while at the same time provides data for the process monitoring. It permits a fast detection of process problems and can also be used for new products for which either no or only little quality data is available. The sample test is particularly cost effective for fluctuating defect rates. The sample test can always be used when 100% inspection is already planned or will be performed because it requires the same technical equipment.

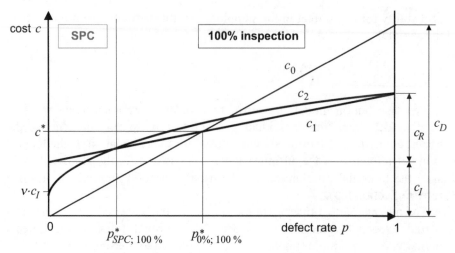

Figure 6.22. Possible cost curve for sample test (k_2)

Considered mathematically, the cost function for the sample test represents the general case of the cost analysis. The equations for 0% and 100% inspection represent special cases of this cost function:

- *Derivation of the 0% inspection:*

given:

$$n = 0 \rightarrow v = \frac{0}{N} = 0 \rightarrow \rho = 0 \rightarrow P_a = 1$$

$$c_I = 0$$

$$c_R = 0$$

$$\begin{aligned}
c_{TotalS} &= c_I + p \cdot c_R + P_a \cdot (p \cdot (c_D - c_R) \cdot (1 - \rho \cdot v) - c_I \cdot (1 - v)) \\
&= 0 + p \cdot 0 + 1 \cdot (p \cdot (c_D - 0) \cdot (1 - 0 \cdot 0) - 0 \cdot (1 - 0)) \\
&= \underline{\underline{p \cdot c_D}}
\end{aligned}$$

- *Derivation of the 100% inspection (for the ideal inspection process):*

given:

$$n = N \rightarrow v = \frac{N}{N} = 1 \rightarrow \rho = 1 \rightarrow 0 \leq P_a \leq 1$$

$$\begin{aligned}
c_{TotalS} &= c_I + p \cdot c_R + P_a \cdot (p \cdot (c_D - c_R) \cdot (1 - \rho \cdot v) - c_I \cdot (1 - v)) \\
&= c_I + p \cdot c_R + P_a \cdot (p \cdot (c_D - c_R) \cdot (1 - 1 \cdot 1) - c_I \cdot (1 - 1)) \\
&= \underline{\underline{c_I + p \cdot c_R}}
\end{aligned}$$

These derivations show that after the implementation of the model for the sample test in a system for the quality cost optimisation they also represent and can evaluate the special cases of 0% inspection and 100% inspection. Practical computed examples, not only for the model of the sample test, but also for the basic model and its extensions, are provided in [13] and [14]. These sources also contain a Microsoft Excel file on a CD that can be used to illustrate and test the quality cost model. References [12] and [13] also contain the software used to calculate sampling plans.

6.3 Effects of Process Optimisation

In the previous discussions, the fluctuations of the defect rate did not have any affect on the costs C_T or c_T of the technological process T because the fluctuation is caused by random processes. If, however, technological, control-technical, organisational and other measures are intentionally used in the technological process with the goal of systematically affecting the defect rate, this causes costs when the defect rate is reduced.

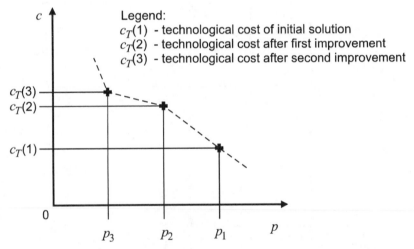

Figure 6.23. General dependency of the technological costs on the goal defect rate

The specific process costs increase with reduced goal defect rate (envisaged defect rate). This results from the investment that must be made in order to achieve this goal. Figure 6.23 shows an example of this dependency. The lower the envisaged defect rate, the higher are the technological costs. Theoretically, the technological costs for the goal of a 0% defect rate approach infinity.

For the following consideration, the costs of the technological initial solution are designated as $c_T(1)$, the costs after the first process improvement and second process improvement are specified with $c_T(2)$ and

$c_T(3)$, respectively. No linear relationship between the reduction of the defect rate and the associated costs, expressed by the technological costs, can be assumed here.

The costs for the quality strategy remain unaffected by the change of the technological costs. The effect of the increase of the technological costs on the complete costs of the process (sum of the technological costs and quality costs for the envisaged defect rate) must now be evaluated. The cost development is shown in Figure 6.24.

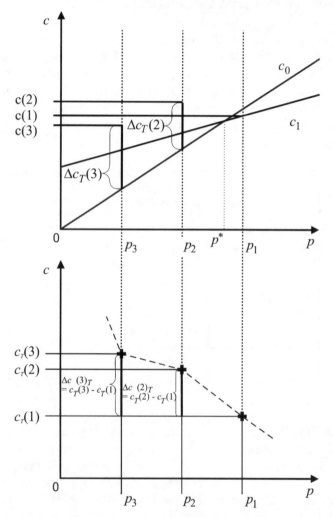

Figure 6.24. Development costs for the achieved reduction of the defect rate

The starting point for the considerations in the example shown in Figure 6.24 are the technological costs $c_T(1)$ of the initial solution for the technology T for the average defect rate p_1. The quality strategy resulting from this defect rate has a quality process directly after the technology T ($p^* < p_1$). To achieve a significantly lower defect rate p_2 for the considered process, new equipment must be obtained for the technological process T. This investment leads to an increase of the technological costs per unit by $\Delta c_T(2)$. The associated quality strategy for the defect rate p_2 results from the basic model and means in this case the noninspection after the technology T (see Figure 6.24). If the change costs for the improved technological process $\Delta c_T(2)$ are now added to the quality costs for defect rate p_2, this produces the cost value $c(2)$. In the example shown in Figure 6.24, this value is higher than the cost $c(1)$ of the initial solution. This means that despite a significant reduction of the defect rate, the complete costs per unit for this technology with the new solution are higher than those for the old solution. In practice, it is often the case that another, smaller investment can further positively affect the properties of the technological process in the direction of an additional reduction of the defect rate. This case is represented with the defect rate p_3 in the example. In this case, the cost component $\Delta c_T(3)$ for the improvement of the technological process compared with the initial solution must be added to the quality costs. The graphs (Figure 6.24) show that the resulting costs $c(3)$ are lower than the costs $c(1)$ of the initial solution. The optimisation goal of the reduction of the defect rate and the unit costs for this technological process has been achieved.

This concept of optimisation of the process costs can also be used for the sample test. The individual change costs of the technological costs must be added to the cost value of the quality costs (graph c_2 in Figure 6.22) for the associated defect rate.

The example shown in Figure 6.24 certainly occurs in practice, e.g., for automatic component placement. Because an automatic placement machine operates at (or even outside) the limits of its specification, it has a relatively high defect rate for the placement of Finepitch components. An investment is made in a new automatic placement machine that has a higher placement accuracy (high investment → higher technological costs $c_T(2)$). The operation shows that the placement of chip components of the size 0201 still has a high defect rate. The manufacturer of the new automatic placement machine recommends the installation of a picture-processing system with a higher resolution for the inspection and positioning of the components (additional, but lower investment → higher technological costs $c_T(3)$). This solution can be used for the placement of the complete component spectrum for a smaller process defect rate. Whether the optimisation goal of the reduction of the defect rate and costs has been achieved depends on the actual costs.

6.4 Dynamic Programming to Solve Complex Process Chains

If a process chain is to be analysed and optimised, in the normal case an existing solution consisting of technological processes and quality processes is used as the basis. Consequently, quality data for the considered product from previously manufactured lots are available for the analysis. The behaviour of the technological processes with regard to their defect rates is particularly interesting here.

How can such an existing system now be analysed and optimised? The dynamic programming, already discussed in Section 4.3, provides a concept here. It is "a general procedure for determining an optimum control [...] for various (*e.g.*, physical) systems" ([19]: 5). The dynamic programming is based on research conducted by the American economist Richard Bellman between the end of the 1940s and the mid-1950s ([3, 4]).

In dynamic programming, as for distribution problems, a differentiation is made between discrete- and continuous-type tasks. In the first case, finite or countable infinitely many mutually separate steps are present. In the continuous problem case, no mutually separate steps are present. Furthermore, a differentiation can be made between deterministic and stochastic problems.

For deterministic problems, the state is uniquely defined after passing through some number of decision steps from the initial state and the decisions made in the steps. For stochastic problems, the state resulting after making the decision is determined only by the quantities of the probability distributions that describe this state ([17]: 150). Consequently, the actual task of optimisation of manufacturing chains from individual technological processes is a discrete deterministic problem.

"The nature of the dynamic programming involves replacing the determination of the extreme values [...] of a function for several variables by the stepwise determination of the extreme values of a function of a variable or a small number of variables" ([17]: 150).

For the optimisation of the quality strategy for the manufacturing of product, this means the determination of the minimum of the quality costs. Because, in principle, every technological process produces defects, a basic model in accordance with Section 6.2.3 can also be specified for each process. The defect rates are then: p_1, p_2, ... , p_n, ... , p_N. The quality costs that result from the immediate inspection are then: Q_{11}, Q_{22}, ... , Q_{nn}, ... , Q_{NN}. If the process chain contains just a final inspection, the quality costs reduce to: Q_{1N}, Q_{2N}, ... , Q_{nN}, ... , Q_{NN}. This makes it possible to calculate the separation defect rates $p_1^*(1,N)$, $p_2^*(2,N)$, ..., $p_n^*(n,N)$, ..., $p_N^*(N,N)$. If the average defect rates of the processes, the test costs and the rejection costs are known for all processes in the considered chain, the addition of the associated minimum values of the quality costs produces the overall minimum.

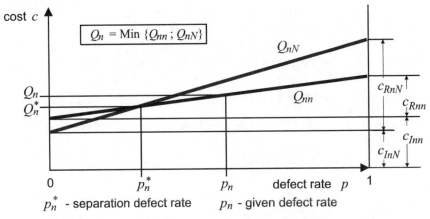

Figure 6.25. Minimum quality costs for a given defect rate p_n

The minimum of the *quality costs* Q_n from the two "test immediately" and "test later" strategies are calculated for each individual technological process:

$$Q_n = Min\{Q_{nn}; Q_{nN}\}$$

(6.31)

Figure 6.25 shows an example of this relationship. For a given defect rate $p_n > p_n^*$, the cost minimum in this example is achieved by performing of the quality process Q_{nn} (directly after the technology that caused the defect).

The minimum *complete quality costs* Q is determined by superposition:

$$Q = \sum_{n=1}^{N} Q_n = \sum_{n=1}^{N} Min\{Q_{nn}; Q_{nN}\}$$

(6.32)

Now, however, as already shown in Section 6.2.4, a quality process is not necessarily performed immediately after the technological process that caused the defect or after the last technological process, but rather also after any process step in between. For example,

$$Q_{mn} = c_{Imn} + p_m \cdot c_{Rmn}$$

(6.33)

means the following:

The technological process T_m causes defects with the defect rate p_m. Immediately after the later process T_n ($n > m$), the quality process Q_{mn} takes place with the test costs c_{Imn} and the rejection costs c_{Rmn}. The prerequisite here is, in general, that the quality process Q_{nn} is also performed here. This representation allows an overall model form that permits the use of dynamic programming. Figure 6.26 shows such a model for three technological processes in a line with a subsequent quality process.

A manufacturing line contains the processes: solder paste printing (T_1), automatic component placement (T_2) and reflow soldering (T_3). After T_3, all products are subjected to the quality process Q_3, the test department. The technologies T_1 to T_3 are characterised by their defect rates p_1, p_2 and p_3. When, however, for cost reasons, is it now desirable to add another quality process Q after one (or both) of the processes T_1 or T_2? The graph in Figure 6.26 helps to make this decision.

Three technological processes with possible quality process arrangements:

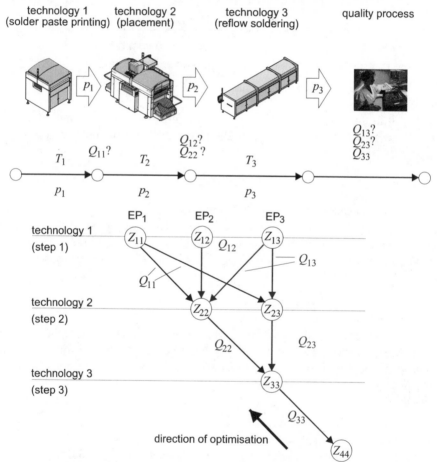

EP$_x$ - entry point depending on the chosen quality strategy

Figure 6.26. Basic model with three processes and possible optimisation paths

The nodes (circle), designated with Z_{xy}, specify states and are used to number a matrix. For each state, the cost function S_{mn} (Bellman function) recursively calculated counter to the process flow direction or counter to the time direction ([19]: 13) always represents the associated minimum of the state costs until the physical end of the process chain (namely, the starting point of the optimisation).

Consequently, the cost functions for the example in Figure 6.26 are prepared starting at the last state, namely the state Z_{44}. The minimum of the state costs S_{1n} is then found for all n. This is then used to produce the optimum complete costs and the most cost-effective path through the graph. This path represents the optimum test strategy for the considered product and the considered process chain.

Bellman uses the following formulation for determining the optimum strategy: "An optimum strategy has the property that, independent of the entry state in the first step and the control made in this step, the controls of the following steps represent an optimum strategy for the state that results from the decision of the first step" ([17]: 151).

For the example in Figure 6.26, this produces the following mathematical description:

$S_{44} = 0$

$S_{33} = Q_{33}$

$S_{22} = Q_{22} + S_{33}$
$S_{23} = Q_{23} + S_{33}$

$S_{11} = Q_{11} + \text{Min} \{S_{22} ; S_{23}\}$
$S_{12} = Q_{12} + S_{22}$
$S_{13} = Q_{13} + \text{Min} \{S_{22} ; S_{23}\}$

The associated quality costs Q_{xy} are calculated for given defect rates using Equation (6.33). The shown procedure corresponds to dynamic programming. For the overall optimisation of the selected example, this produces the following equation:

$$Q = \text{Min}\left\{ [Q_{12} + Q_{22}] ; [\text{Min}\{Q_{11} ; Q_{13}\} + \text{Min}\{Q_{22} ; Q_{23}\}] + Q_{33} \right\} \qquad (6.34)$$

Thus, for the given average defect rates p_1 and p_2, the associated minimum must be determined from the quality costs for the "immediate" or the "later" inspection from the two sums and in accordance with the above equation:

$$Q = \text{Min}\{S_{11} ; S_{12} ; S_{13}\} \qquad (6.35)$$

The Bellman recursion equation can be derived from this example:

$$S_{n;n} = Q_{n;n} + \text{Min}\left\{ S_{n+1;n+1}; S_{n+1;n+2}; \ldots; S_{n+1;N} \right\} \tag{6.36}$$

with $n = 1; 2; \ldots; N-1$
and

$$S_{n;n+k} = Q_{n;n+k} + \text{Min}\left\{ S_{n+1;n+1}; S_{n+1;n+2}; \ldots; S_{n+1;n+k} \right\}$$

with $k = 1; 2; \ldots; N-n$.
The minimum complete costs using the appropriate Equation (6.35) give:

$$Q = \text{Min}\left\{ S_{11}; S_{12}; \ldots; S_{1N} \right\} \tag{6.37}$$

An example with four processes shown in Figure 6.27 then gives the following system of equations:

$S_{55} = 0$

$S_{44} = Q_{44}$

$S_{33} = Q_{33} + S_{44}$
$S_{34} = Q_{34} + S_{44}$

$S_{22} = Q_{22} + \text{Min}\{S_{33} \,; S_{34}\}$
$S_{23} = Q_{23} + S_{33}$
$S_{24} = Q_{24} + \text{Min}\{S_{33} \,; S_{34}\}$

$S_{11} = Q_{11} + \text{Min}\{S_{22} \,; S_{23} \,; S_{24}\}$
$S_{12} = Q_{12} + S_{22}$
$S_{13} = Q_{13} + \text{Min}\{S_{22} \,; S_{23}\}$
$S_{14} = Q_{14} + \text{Min}\{S_{22} \,; S_{23} \,; S_{24}\}$

with the minimum
$Q = \text{Min}\left\{ S_{11} \,; S_{12} \,; S_{13} \,; S_{14} \right\}$

The example shows the number of possible paths when the number of processes increases with the subsequent increase of the size of the system of equations. Thus, the number of totals increases from step to step by the associated level number (*e.g.*, step 3 – add 3), where, however, the last state in each chain (in Figure 6.27: Z_{55}) is not counted as a step. The number of totals Z_S used to determine the minimum for the optimum test strategy is calculated as follows:

$$Z_S = 1 + \sum_{k=1}^{n} k \quad \text{für} \quad n > 0 \tag{6.38}$$

Because process chains with 14 or more technological processes certainly occur in real electronics manufacturing processing, in accordance with Equation (6.38), the minimum from 106 totals must be found for *one* product with 14 technological steps. In a typical manufacturing plant, more than 100 different products are produced during the course of a year.

This shows that although the computing effort for such an optimisation is very high, it can certainly be mastered using modern computers. The associated algorithm is implemented in a quality-planning program tailored to each manufacturing process.

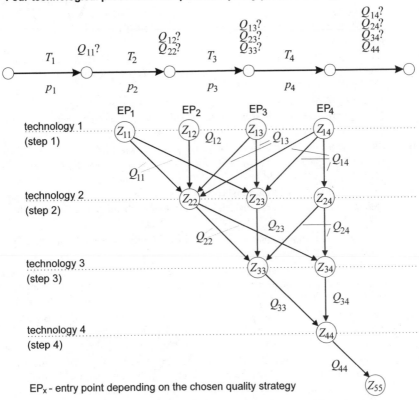

Figure 6.27. Dynamic programming for four processes

References

[1]   BARAD, M.: *Using break-even quality level for selecting acceptance sampling plans given a prior distribution.* International Journal of Production Research, Vol. 24, No. 1, 1986.

[2]   BARAD, M.: *A break-even quality level approach to location of inspection stations in a multi-stage production process.* International Journal of Production Research, Vol. 28, No. 1, 1990.

[3]   BELLMAN, R.: *Dynamic Programming.* Princeton: University Press, 1957.

[4]   BELLMAN, R.: *The Theory of Dynamic Programming.* In: Bulletin of the American Mathematical Society, No. 6, p. 503–516, November 1954.

[5]   BRONSTEIN, I. N.; SEMENDJAJEW, K. A.: *Taschenbuch der Mathematik.* Joint Edition – Moskow: Verlag Nauka; Leipzig: BSB B. G. Teubner Verlagsgesellschaft, 1979.

[6]   DEUTSCHES INSTITUT FÜR NORMUNG: *DIN ISO 2859-1; Annahmestichprobenprüfung.* Berlin; Vienna; Zurich: Beuth Verlag GmbH, 1993.

[7]   HERRMANN, H. P.: *Ausbeuteberechnung und –optimierung von integrierten Schaltungen.* Dissertation Karlsruhe, Universität Fridericiana (Technische Hochschule), 1980.

[8]   IPC – THE INSTITUTE FOR INTERCONNECTING AND PACKAGING ELECTRONIC CIRCUITS: *IPC-A-610 Revision C; Acceptability of Electronic Assemblies.* Northbrook, Illinois: IPC, 2000.

[9]   IPC – THE INSTITUTE FOR INTERCONNECTING AND PACKAGING ELECTRONIC CIRCUITS: *IPC-7912; Calculationof DPMO and Manufacturing Indices of Printed Board Assemblies.* Northbrook, Illinois: IPC, 2000.

[10]   IPC – THE INSTITUTE FOR INTERCONNECTING AND PACKAGING ELECTRONIC CIRCUITS: *IPC-CM-770 Revision D; Guidelines for Printed Board Component Mounting.* Northbrook, Illinois: IPC, 1996.

[11]   KLEIN WASSINK, R. J.; VERGULD, M. M. F.: *Manufacturing Techniques for Surface Mounted Assemblies.* Port Erin (Isle of Man): Electrochemical Publications Ltd., 1979.

[12]   LINß, G.: *Qualitätsmanagement für Ingenieure.* Munich; Vienna: Fachbuchverlag Leipzig im Carl Hanser Verlag, 2002.

[13]   LINß, G.: *Trainingsbuch Qualitätsmanagement – Trainingsfragen – Praxisbeispiele – Multimediale Visualisierung.* Munich; Vienna: Fachbuchverlag Leipzig im Carl Hanser Verlag, 2003.

[14]   OPPERMANN, M.: *Modellierung und Optimierung des Qualitätsverhaltens von Fertigungsprozessen in der Elektronik: Elektronik-Technologie in Forschung und Praxis.* 1. Auflage – Templin: Verlag Dr. Markus A. Detert, 2002.

[15]   OPPERMANN, M., SAUER, W., WOHLRABE, H.: *Optimization of Quality Costs.* In: Robotics and Computer Integrated Manufacturing – An international journal of manufacturing and product and process development, Volume 19, No. 1–2, p. 135–140, Elsevier Science Ltd., February–April 2003.

[16]   SCHEEL, W. (HRSG.): *Baugruppentechnologie der Elektronik.* 1st Edition - Berlin: Verlag Technik; Saulgau: Leuze Verlag, 1997

[17]   STANEK, W. ET AL.: *Rechnergestützte Fertigungssteuerung. Mathematische Methoden.* 1st Edition - Berlin: VEB Verlag Technik, 1988.

[18]   STORM, R.: *Wahrscheinlichkeitsrechnung, mathematische Statistik und statistische Qualitätskontrolle :mit 120 Beispielen.* 11th, revised edition, Munich; Vienna: Fachbuchverlag Leipzig im Carl Hanser Verlag, 2001.

[19]   WENZEL, J. S.: *Elemente der Dynamischen Optimierung.* Leipzig: B. G. Teubner Verlagsgesellschaft (= Reihe Kleine naturwissenschaftliche Bibliothek), 1966.

[20]   WOHLRABE, H.: *Maschinen und Prozessfähigkeit von Bestückausrüstungen der SMT: Elektronik-Technologie in Forschung und Praxis.* 1st Edition – Templin: Verlag Dr. Markus A. Detert, 2000.

[21]   WOLTER, K.-J.: *Elektroniktechnologie* lecture script. Technische Universität Dresden, Fakultät Elektrotechnik und Informationstechnik, Institut für Elektronik-Technologie, 2000 edition.

7 Design of Experiments and Regression Analyses

7.1 General Goals

Chapter 5 introduced the methods used to ensure the quality, in particular in a running manufacturing process. The prerequisite here is that the process to be controlled is capable of producing quality. It is assumed that the quality capability of the process is present. Thus, the process monitoring is used to show that nonpermitted deviations have occurred in the process that lead or can lead to defects. It must be emphasised that all these indications of nonpermitted deviations may only be evaluated as *notification* of such a deviation. The cause of this deviation and the possible means and methods for the correction of the indicated defect cannot normally be determined from this notification, but rather provide only information about the type of defect. If, for example, a mean chart indicates a control-limit violation, in all probability (considering the defect probabilities) the mean or the operating point of the process has moved. This means knowledge must be available as to how to explicitly change this operating point so that it again agrees with the setpoint of the process.

Another important task involves initially giving a process-quality capability so that techniques such as quality-control charts may be used.

Many means and methods are available to achieve these tasks, of which the FMEA (failure mode and effect analysis), QFD (quality function deployment) and TQM (total quality management) should be mentioned. However, the following discussion concentrates on the use of statistical methods. The following terms are used here:

A *response variable* is a quality characteristic to be explicitly influenced (*e.g.*, minimised, maximised, attain a specific target value, minimise the standard deviation). Response variables are generally designated in equations with Y or the individual components with y_i.

A *factor* designates the quantities that should be changed in order to achieve the required change of the response variable. Factors are generally designated as X or x_i.

In addition to the factors, there are other quantities that have an effect on the target quantity. These include quantities that cannot be measured, quantities that cannot be explicitly influenced and unknown quantities. The quantities are grouped as *noise variable*, designated as S.

The relationships between the quantities should be determined as accurately as possible (see Figure 7.1).

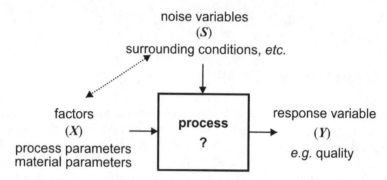

Figure 7.1. General relationship between factors, response variables and noise variables

$$Y = f(X) + g(S) \tag{7.1}$$

Thus, the central tasks can be formulated as:

- Function f: *determine* the dependency between the response variable and the factors!
- Function g: *minimise* the dependency between the response variables and the noise variables!

7.2 Regression Analysis

7.2.1 Problem Specification

A factor is explicitly changed in a technological process. With each actual value, a product is manufactured and the values for the required quality characteristic measured. What needs to be found is a relationship between the change of the factor and the change of the quality characteristic. Or, in other words, can an explicit change of the factor produce a required change of the quality characteristic?

Example:
For a dispenser, a compressed air blast through a dispenser needle (a special hollow needle) is used to place a paste medium (*e.g.*, solder paste or SMT adhesive) on a substrate. The principal quality characteristic is the applied volume that must be variable for each task. It is obvious that the duration of the compressed air blast has a large effect on this volume, one would assume a relatively clear linear relationship. The task is now to prove this relationship and to determine the associated coefficients of the corresponding linear equation. This equation can now be used to control the dispenser so that an explicit choice of the compressed air blast duration applies the solder paste with a specified volume on the substrate.

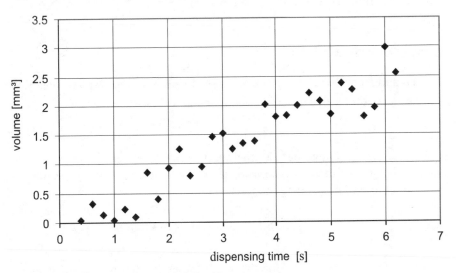

Figure 7.2. Test results

Figure 7.2 shows the results of such a test. The dispensing time t was systematically increased 29 times from 0.4 s in 0.2 s increments and the applied volume V measured. As expected, the figure shows that the volume also increases with increasing dispensing time. However, the figure also shows that in the test not every increase of the dispensing time resulted in an increase of the volume, but rather sometimes an unexpected reduction occurred.

The problem situation is now: does a linear relationship exist between the dispensing time and the volume, and how good is this relationship?

In this case, the volume is the response variable and the dispensing time is the factor. The noise variable includes, for example, the temperature that has a direct effect on the viscosity of the medium to be dispensed and thus also on the applied volume. A fluctuating pressure also plays a role.

7.2.2 Simple Linear Regression

For the dispenser test, there is now the assumption that a linear relationship exists between the dispensing time t and the volume V that can be described with the following equation:

$$V = V_o + bt \quad \text{resp. generally} \quad y = a + bx \qquad (7.2)$$

Such an equation is designated as a model. The task is to determine the coefficients a and b so that the model has the best possible agreement with the performed test. This linear equation is the simplest conceivable model here. In general, however, other mathematical relationships can also be used as a model.

Least Squares Method

If the a and b coefficients are known, an estimate $\hat{y}_i$ of the response variable can be calculated for each setting value of size x_i.

$$\hat{y}_i = a + b x_i \qquad (7.3)$$

For the least squares method, the coefficients are determined so that the sum of squares of differences between the estimated values from Equation (7.3) and the actual observations y_i has a minimum.

$$\sum_{i=1}^{n} (y_i - \hat{y}_i)^2 = \sum_{i=1}^{n} (y_i - (a + b x_i))^2 \rightarrow \text{minimum}$$

The estimates $\hat{a}$ and $\hat{b}$ for the sought coefficients are obtained using partial differentiation.

$$\hat{b} = \frac{Q_{XY}}{Q_{XX}} = \frac{s_{xy}}{s_x^2} \qquad (7.4)$$

$$\hat{a} = \bar{y} - b \cdot \bar{x}$$

$$Q_{XY} = \sum_{i=1}^{n} (x_i - \bar{x})(y_i - \bar{y}) \qquad Q_{XX} = \sum_{i=1}^{n} (x_i - \bar{x})^2$$

The quantities Q_{XY} and Q_{XX} are auxiliary quantities often used in this form (or a slightly modified form) in the following discussion. The size s_{xy} is also designated as (empirical) covariance. The dispenser example yields the following results:

$$\hat{b} = 0.4313 \text{ mm}^3/\text{s}$$

$$\hat{a} = -0.064 \text{ mm}^3$$

This means that for a lengthening of the dispensing time by 1 s, a volume increase by 0.4313 mm^3 is expected. The axis section a (negative here) does *not* mean that for a dispensing time of 0 the dispenser absorbs 0.064 mm^3 but rather that a certain minimum time (here 0.15 s) of the compressed air blast must be present to allow the application. This can be explained by the friction effects, *etc*. This also shows that such calculated results must always be questioned critically. Furthermore, the results must always be processed only within the experimental range (here 0.4 s – 6.2 s). This includes, in particular, an *interpolation* of the values. *Extrapolations* are not generally permitted because the results of such calculations can be incorrect but cannot always be immediately recognised as such (negative volumes) in the shown example.

Goodness of Fit and Correlation Coefficient

Figure 7.2 shows that a linear relationship is certainly apparent. However, this raises the question as to how good is this relationship. A quality that measures this relationships is the *correlation coefficient* r_{xy}. It is defined as:

$$r_{xy} = \frac{S_{xy}}{S_x S_y} = \frac{\sum_{i=1}^{n}(x_i - \bar{x})(y_i - \bar{y})}{\sqrt{\sum_{i=1}^{n}(x_i - \bar{x})^2 \sum_{i=1}^{n}(y_i - \bar{y})^2}} = \frac{Q_{XY}}{\sqrt{Q_{XX}Q_{YY}}} \qquad (7.5)$$

Using the definition, the correlation coefficient can assume values in the interval $-1 \le r_{xy} \le 1$. If the coefficient is exactly –1 or 1, then a purely linear relationship exists. The points in Figure 7.2 must then all lie on a straight line. This means in the example that the applied volumes can *only* be determined by the dispensing time. The counterpart is a correlation coefficient of zero. This means that no relationship exists between the two considered quantities. In this case, the graphical representation would show only a "diffuse" cloud. The sign of the correlation coefficient is identical with the sign for b, the gradient of the straight lines. In practice, such extreme cases hardly ever occur. The dispenser example produces a value of

$$r_{xy} = 0.929$$

This represents quite a good relationship. Another measure for the quality of the relationship is the goodness of fit B (or r-squared)

$$B_{xy} = r_{xy}^2 \qquad\qquad (7.6)$$

The dispenser example has as a result $B_{xy} = 86.3\%$. This means 86.3% of the changes of the target quantity volumes can be described by the change of the dispensing time with the remaining 13.7% resulting from factors not considered here and by the effects of noise variables.

Graphical Evaluation of the Test Results – Residues

The calculated linear equation can be used to compare the observed test results y_i with the model $\hat{y}_i$. The deviations $(y_i - \hat{y}_i)$ are also designated as residues. After each regression, these residues should be represented graphically in order to determine any features that, for example, make themselves apparent by an inadequate goodness of fit. However, even if the goodness of fit has values that are actually good, the subsequent analysis can show further reserves for additional improvements.

The residues can be shown as four types: the representation depending on the factor, depending on the order of the experiments, depending on the calculated response variable, and a test for the normal distribution of the residues. The representations of the dependencies should not exhibit any systematic effects, but rather the residues must be dispersed purely randomly in all representations.

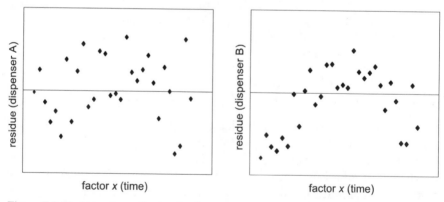

Figure 7.3. Residues depending on the factor x; on the left-hand side the example dispenser on the right-hand side the graph of the residues when not a linear but rather a quadratic relationship exists

Figure 7.3 shows on the right-hand side the dispersion of the residues. This indicates that the chosen model concept of a linear relationship is either not present or only partially present. A quadratic relationship can be presumed.

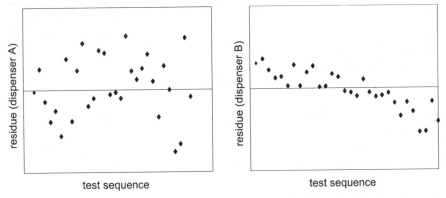

Figure 7.4. Residues depending on the test sequence; a unique trend is apparent on the right-hand side

Figure 7.4 shows a unique trend of the residues on the right-hand side. This can be caused, for example, by test conditions that change during the course of the tests.

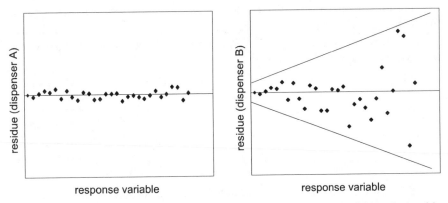

Figure 7.5. Representation of the residues depending on the response variable calculated in the model

Another analysis tested the prerequisite that the standard deviation of the results is constant. The setpoint of the response variable calculated by the model is used as a sort criterion for the x-axis. Although the left-hand side of the figure does not show any change of the standard deviation depending on the response variable, such a change is readily apparent on the right-hand side. If such dependencies are apparent, this problem can normally be solved using the Box–Cox transformation [3].

The normal distribution of the residues can be tested using the probability network.

All these investigations are used to test the "rigour" of the associated model. Each determined peculiarity can impair the reusability of the model. Whether the peculiarities have severe effects must be decided depending on the actual technical background. In each case, such peculiarities have effects on the subsequent statistical evaluations, because the calculations of the confidence intervals assume the validity of the model.

Confidence Intervals of the Regression Parameters

The calculated parameters of the linear equation (7.4) are estimates and thus random. Confidence intervals provide information about the quality of the estimated parameters. The gradient b is the more important parameter here. The confidence intervals are calculated using the residual variance s_R^2. It specifies the average squared deviations of the observed values of the model and is thus also indicates the dispersion of the residues.

$$s_R^2 = \frac{1 - B_{xy}}{n - 2} Q_{YY} \tag{7.7}$$

The number of degrees of freedom is two less than the test size because two estimated quantities are contained in the calculation. The confidence interval for the gradient b of the regression lines is specified by the following equation:

$$g_{l,u} = \hat{b} \mp \frac{s_R}{\sqrt{Q_{XX}}} \cdot t_{n-2;1-\alpha/2} \tag{7.8}$$

For a type I error for $\alpha = 1\%$, this yields for the dispenser example:

$$g_l = 0.3405 \quad ; \quad g_u = 0.5194$$

The confidence interval does not contain the value zero, namely it can be assumed that the value of the gradient differs significantly (as part of the selected error probability for $\alpha = 1\%$) from zero, thus an increase results. As a consequence, the analysed dispenser example contains a proven linear component.

For the evaluation of the significance of this increase, it is, however, usual to analyse the quality of this agreement. To do this, the type I error is varied.

$$\alpha = 5\% \quad ; \; g_l = 0.3636 \quad ; \quad g_u = 0.4963$$
$$\alpha = 0.1\% ; \; g_l = 0.3110 \quad ; \quad g_u = 0.5489$$

If the confidence interval for $\alpha = 0.1\%$ does not contain the zero value, this is considered to be a highly significant increase and has the general identification or evaluation (***). In the described example, such a highly significant increase is present. If this is not the case, but the interval for $\alpha = 1\%$ does not contain the zero value, this is called a significant increase (**). For $\alpha = 5\%$, an indifferent increase (*) is present, otherwise there is said to be no indication of the presence of a linear dependency (–).

A confidence interval can also be determined for the axis section a:

$$g_{l,u} = \hat{a} \mp s_R \sqrt{\frac{1}{n} + \frac{\bar{x}^2}{Q_{XX}}} \, t_{n-2;1-\alpha/2} \tag{7.9}$$

$$g_l = -0.3671 \quad ; \quad g_u = 0.2684 \qquad (\alpha = 1\%)$$

The knowledge of these confidence intervals also allows a confidence range to be specified for the regression lines.

$$g_{l,u}(x) = \hat{a} + \hat{b}x \mp s_R \sqrt{\frac{1}{n} + \frac{(x - \bar{x})^2}{(n-1)s_x^2}} \, t_{n-2;1-\alpha/2} \tag{7.10}$$

This equation describes the confidence range as a hyperbola shown dotted in Figure 7.6.

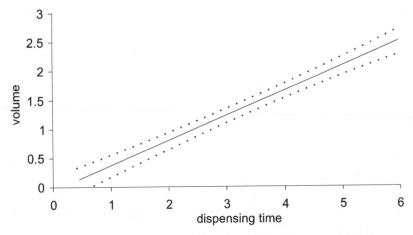

Figure 7.6. Confidence range (dotted; $\alpha = 1\%$) of the regression lines

Quasilinear Regression

A result of the analysis of the residues in Section 7.2.2 can be that the assumed linear relationship is not (or only partially) confirmed. Nonlinear relationships, such as quadratic or exponential dependencies, are also possible. In addition to these analysis results, a technical consideration can also lead to the result that a linear relationship either does not (or, at best, only partially) describe the investigated behaviour.

For example, the volume in the dispenser can also be varied by using needles with different internal diameters. The relationship between diameter d and the applied volume V is a quadratic dependency here, because the volume is initially linearly affected by the effective area of the needle aperture. The following relationship can now be prepared:

$$V = V_0 + c_d \, d^2$$

The transformation $y = V$ and $x = d^2$, and an adaptation of the coefficients ($a = V_0$; $b = c_d$), produces the linear relationship

$$y = a + b \, x$$

This allows this quadratic relationship to be treated as a simple linear regression.

Dispenser needles exist with different lengths l, such as for specific repair purposes. The increased friction for a longer needle means that the applied volume sinks with increasing needle length. An inverse relationship is expected.

$$V = a + \frac{c}{l} \quad \Rightarrow \quad y = V; x = \frac{1}{l} \quad \Rightarrow \quad y = a + cx$$

Exponential relationships can also be handled in this way:

$$z = a \, e^{bt} \Rightarrow \ln z = \ln a + b t \Rightarrow y = \ln z; x = t \Rightarrow y = \ln a + bx$$

7.2.3 Dual Linear Regression

Previously, it was assumed that only one factor acts on the response variable. In practice, normally several quantities exist. This is shown in the following discussion using two factors. The dispenser example will be extended with the needle length l factor. The following relationship can be used as the expected model:

$$V(t,l) = V_0 + c_t \, t + \frac{c_l}{l}$$

Based on the discussions in Section 7.2.2, this is described in the following discussion by:

$$y(x_1, x_2) = a + b_1 x_1 + b_2 x_2 \quad ; \quad x_1 = t \quad ; \quad x_2 = \frac{1}{l} \tag{7.11}$$

The method of least squares of differences is also used here:

$$\sum_{i=1}^{n} \left(y_i - \left(a + b_1 x_{1i} + b_2 x_{2i} \right) \right)^2 \Rightarrow \text{minimum}$$

this implies:

$$\hat{b}_1 = \frac{Q_{x_2 x_2} \cdot Q_{x_1 y} - Q_{x_1 x_2} \cdot Q_{x_2 y}}{Q_{x_1 x_1} \cdot Q_{x_2 x_2} - Q_{x_1 x_2}^2} \tag{7.12}$$

$$\hat{b}_2 = \frac{Q_{x_1 x_1} \cdot Q_{x_2 y} - Q_{x_1 x_2} \cdot Q_{x_2 y}}{Q_{x_1 x_1} \cdot Q_{x_2 x_2} - Q_{x_1 x_2}^2}$$

$$\hat{a} = \overline{y} - \hat{b}_1 \overline{x}_1 - \hat{b}_2 \overline{x}_2$$

For the coefficients of the regression model, confidence intervals can be determined analogously to the discussion for the single regression [5].

7.2.4 Restrictions for Performing Regressions

When regressions are performed, a differentiation is made between planned tests and unplanned tests. For planned tests, the factors are explicitly set, whereas for unplanned tests, only those factors that result during the test are determined. In general, planned tests are preferable.

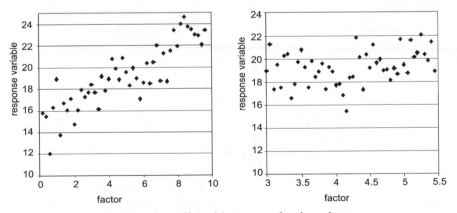

Figure 7.7. Effect of the test area for planned tests

The left-hand side of Figure 7.7 shows that the planned variation of the factor should cover a specific test area. If the factor is only changed in a small area, as the right-hand part of the figure shows, it is quite possible that the residual scattering "covers" any existing effect. However, the test area should not be chosen too large, because it is quite possible that quality jumps occur that then can lead to wrong conclusions.

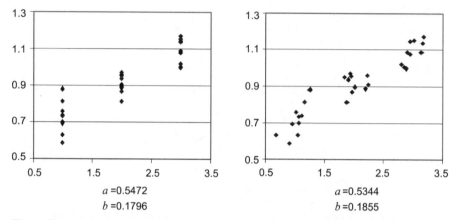

$a = 0.5472$ $a = 0.5344$

$b = 0.1796$ $b = 0.1855$

Figure 7.8. Comparison of a regression test (on the left-hand side using the values to be set for the factor; on the right-hand side using the real measured values of the factor)

If a planned variation of the factors is performed, a test should at least be made whether the planned setting has actually also been realised. It is certainly possible that a setting quantity does not realise *the* value that should be set.

Figure 7.8 shows such an example. It is apparent that the values of the line parameters are different, which, in a specific case, can lead to different conclusions.

The following principle applies: what can be measured (obviously with an acceptable effort), should be measured. It is much more difficult to search afterwards for specific data suddenly required for the evaluation of tests.

7.3 Variance Analysis

A mean comparison was performed in Section 5.6.3. It is often necessary to compare more than two processes, machines, *etc.* If, for example, a manufacturing line has several equivalent dispensers, this raises the question whether the applied volume for the same parameter settings is the same for each machine. The machines can now be compared pairwise. However, because of the large number of comparisons involved, this is very expensive. The variance analysis is a tool with which this comparison can be performed independently of the number of investigated objects.

Table 7.1. Test results of the dispenser comparison for the applied volumes

Test	Dispenser 1	Dispenser 2	Dispenser 3	Dispenser 4
1	0.689	0.600	0.812	0.854
2	0.839	0.594	0.742	0.909
3	0.830	0.745	0.818	0.819
4	0.682	0.570	0.795	0.747
5	0.588	0.692	0.690	0.895
Mean $\bar{y}_i$	0.725	0.640	0.771	0.845
Standard deviation s_i	0.107	0.075	0.055	0.065

For the simple balanced variance analysis, a factor A is considered at various levels (number k). For the "comparison of dispensers" example, the levels here mean the compared machines. The variance analysis is balanced when the number of tests is identical for each machine. Although this is not essential, it is desirable.

Table 7.1 shows the results from the investigation of the dispensers. It is now assumed that all dispensers follow a normal distribution and there is no difference in the standard deviations between the machines. The overall variance s^2_{sum} and the variance of the means $s^2_{\bar{y}}$ gives:

$$s^2_{sum} = \frac{1}{k}\sum_{i=1}^{k} s_i^2 \qquad (= 0.00606) \qquad (7.13)$$

$$s^2_{\bar{y}} = \frac{1}{k-1}\sum_{i=1}^{k}\left(\bar{y}_i - \bar{\bar{y}}\right)^2 \qquad (= 0.00733) \qquad (7.14)$$

If the means of the individual machines are identical, the test quantity

$$F_{test} = \frac{n \cdot s^2_{\bar{y}}}{s^2_{sum}} \qquad (= 6.052)$$

may only randomly differ from 1. Thus, it is a deviation comparison (for a further example, see Section 5.7.6) whose critical values can be determined using the F-distribution.

For the described case, the test quantity is compared with the following quantiles ($n = 5$ tests, $k = 4$, Table 7.1):

$$F_{f_1;f_2;1-\alpha} = F_{k-1;k(n-1);1-\alpha}$$

$F_{3;16;0.95} = 3.239$ $\Rightarrow$ evaluation *

$F_{3;16;0.99} = 5.292$; $F_{test} = 6.052$ $\Rightarrow$ evaluation **

$F_{3;16;0.999} = 9.006$ $\Rightarrow$ evaluation ***

In the described case, the means of the dispensers differ with the evaluation (**: significant difference). This evaluation is analogous to the evaluation of the gradients of the regression lines (see Section 7.2.2).

This very simple calculation can also be performed differently. Although the subsequent evaluation appears somewhat more difficult, it has, however, a systematic for which the multiple variance analysis is used. This systematic is also used for the statistical test designing. It is a so-called variance analysis (actually decomposition of the sums of squares of differences).

Table 7.2. Variance analysis table for the simple variance analysis

Deviation cause	Sum of the squares	Degrees of freedom	Variance	Test quantity	p-value
Ffactor A	Q_A	$f_A = k - 1$	$s_A^2 = \dfrac{Q_A}{f_A}$	$F_{test} = \dfrac{s_A^2}{s_R^2}$	
Total error	Q_R	$f_R = k(n-1)$	$s_R^2 = \dfrac{Q_R}{f_R}$		
Total	$Q = Q_R + Q_A$	$f = f_A + f_R$			

The sum of the squares of differences Q_A between the levels (between the dispensers in the example) within the levels Q_R and the complete sum Q is formed.

$$Q_A = \sum_{i=1}^{k} n(\bar{y}_i - \bar{\bar{y}})^2 \quad ; \quad Q = \sum_{i=1}^{k}\sum_{j=1}^{n}(y_{ij} - \bar{\bar{y}})^2 \quad ; \quad Q_R = Q - Q_A$$

The variance analysis table is then prepared. Table 7.2 and Table 7.3 show the general table for the simple variance analysis and for the considered dispenser example, respectively.

Table 7.3. Variance analysis table for the dispenser example

Deviation cause	Sum of the squares	Degrees of freedom	Variance	Test quantity	p-value
Factor A	0.1100	3	0.0367	6.052	0.0059
Total error	0.0969	16	0.0061		
Total	0.2069	19			

The p-value in the last column specifies for which value the type I error α "tips" between significant and not significant. The smaller this value, the more significant is the difference. The following relationships exist

$$p\text{-value} \geq 0.05 \qquad \text{evaluation} -$$

0.05   > p-value ≥ 0.01 evaluation *
0.01   > p-value ≥ 0.001 evaluation **
0.001 > p-value   evaluation ***

This evaluation is identical with the analysis of the gradient of the regression lines in Section 7.2.2.

Another use of the simple variance analysis is the process type B discussed in Section 5.7.2. Thus, it is possible to determine the numerical values of the internal and external standard deviation.

For the multiple variance analysis, not only one factor is considered, but rather several factors can be analysed. The basis is the variance analysis, which, however, requires significantly more time. Another evaluation is performed analogously in an appropriate variance analysis table.

7.4 Design of Experiments

The statistical test planning, also designated as DoE (design of experiments), permits the analysis of technical processes for the minimisation, maximisation or specific setting of response variables. Its planned approach allows a lot to be achieved for relatively many factors with a minimum of test effort. Its successful use, however, requires certain knowledge, because even a few errors can produce unsatisfactory results.

7.4.1 Complete Factorial Test Designing

Two Factors on Two Levels

The following test design varies two factors, where each factor is set to two different steps or levels and each possible combination is performed. It is usual in the test design to indicate the cause variables as factors and label them with letters. Instead of the actual values for the settings of the levels, in general + and − are used for the upper and the lower setting, respectively. If a value that lies between the levels is used, it is usually labelled with 0. The assignment into 1, 2 and 3 is also sometimes found.

The dispenser example is considered again. In this case, the length of the compressed air blasts is varied (factor A: lowest level 1 s $\Rightarrow$ "−"; upper level 3 s $\Rightarrow$ "+") and the needle diameter (factor B: lowest level 0.3 mm $\Rightarrow$ "−"; upper level 0.5 mm $\Rightarrow$ "+"). Because the needle diameter has a quadratic relationship to the volume (see Section 7.2.2), the needle cross-sectional area is used for the following calculations.

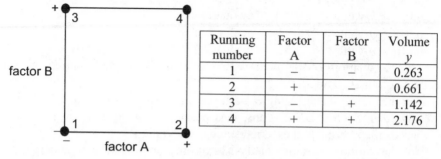

Figure 7.9. Test design for the dispenser investigation with the measured results

Thus, the test design consists of four combinations, because two factors at two levels are combined. This means four tests must be performed. Such a design is also designated as a 2^2 design.

When factor A is considered,

$$\overline{A}_- = \frac{y_1 + y_3}{2} \quad ; \quad \overline{A}_+ = \frac{y_2 + y_4}{2}$$

are the associated means of the factor A at the lower and upper level, respectively. The difference between the two means is designated as effect A and is a measure of the effect of the dispensing time on the volume.

$$\text{effect } A = \overline{A}_+ - \overline{A}_- = \frac{y_2 + y_4}{2} - \frac{y_1 + y_3}{2} \qquad (A = 0.716) \qquad (7.15)$$

Effect B is calculated similarly:

$$\text{effect } B = \overline{B}_+ - \overline{B}_- = \frac{y_3 + y_4}{2} - \frac{y_1 + y_2}{2} \qquad (B = 1.197)$$

The effects that depend only on one factor are also designated as main effects.

The difference $y_4 - y_3$ is the effect of A when B is at the upper level. The difference $y_2 - y_1$ is the effect of A when B is at the lower level. The half-difference between the two is a measure of the degree to which the dispensing time A depends on the actual value of the needle cross-sectional area B. This is called the interaction AB.

$$\text{interaction } AB = \frac{y_4 - y_3}{2} - \frac{y_2 - y_1}{2} \qquad (= 0.6359) \qquad (7.16)$$

The left part of Figure 7.10 shows the action of the effects A and B. It shows that the needle area causes a larger change of the volume than a change of time. This statement, however, applies only to the selected levels!

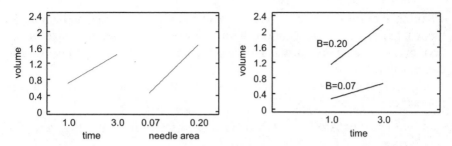

Figure 7.10. Graphical representation of the effects (left) and the interactions (right) for the dispenser test

In general, however, it is possible to say the greater the gradient of the straight lines in the *common* representation, the greater is the action of the corresponding effect. This is, however, not an evaluation in the statistically significance sense. The right-hand part shows a representation of the interaction AB. The change over the time is considered here, where the change of the two factors is made using two straight lines. If the two straight lines are parallel, there is no interaction. Interactions can have various causes. In the dispenser example, there is a multiplicative relationship between time and needle area on the volume.

The test results can also be used to derive a model of the considered process:

$$V = -0.271 + 0.0401 \cdot A + 1.52 \cdot B + 3.97 \cdot A \cdot B$$

This equation is comparable with the linear regression models (see Section 7.2.2). Analyses such as the significance of the coefficients are, however, not yet possible here, because four tests have been performed here and in this model four coefficients have been calculated. This model can also be represented as a graph, see Figure 7.11. An interaction exists if the surface is curved in this representation.

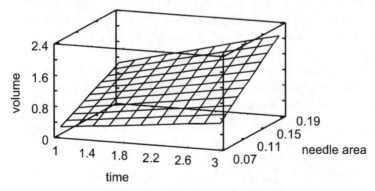

Figure 7.11. Illustration of the model

Another example shows a very simple illustration of the interaction. In this example, we consider a door with its associated frame. The size of the door and the frame are varied. The target quantity is the air-tightness of this combination.

Using Equations (7.15) and (7.16) to calculate the effects and the interactions, this produces:

$$\text{effect A} = 0 \quad ; \quad \text{effect B} = 0 \quad ; \quad \text{interaction AB} = 1$$

Table 7.4. Test design 2^2 for the door-frame combination and their results

Line number	Door size (A)	Frame size (B)	Air-tightness
1	Small	Small	Air-tight (= 0)
2	Large	Small	Not air-tight (= 1)
3	Small	Large	Not air-tight (= 1)
4	Large	Large	Air-tight (= 0)

This means that this "process" is described just by its interaction. Although it is apparent for the considered example that the air-tightness of the combination is given only when the door and frame have the same size. In practice, however, there are sufficient processes where hidden interactions exist, and such interactions often tend to be neglected.

k Factors at Two Levels

The considerations in the previous section can be transferred from two factors to any number of k. There are

$$m = 2^k \tag{7.17}$$

combinations here to which a corresponding number of tests are assigned.

For $k=3$, Figure 7.12 shows such a design for the dispenser example using the third factor l (inverse length of the needle; factor C); this design is often designated as a dice design.

Table 7.5 is needed to calculate the effects. The sign for the interactions results from the multiplication of the corresponding sign of the factors. The table shows that there are three main effects, three double interactions and one triple interaction.

$$\text{effect} = \frac{2}{m} \sum_{i=1}^{m} \left(\text{sign} \cdot y_i \right) \tag{7.18}$$

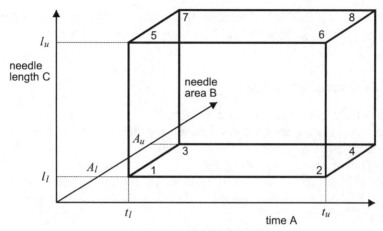

Figure 7.12. Illustration of a 2^3 test design

Table 7.5. Sign of a 2^3 test design

Line no.	A	B	C	AB	AC	BC	ABC
1	−	−	−	+	+	+	−
2	+	−	−	−	−	+	+
3	−	+	−	−	+	−	+
4	+	+	−	+	−	−	−
5	−	−	+	+	−	−	+
6	+	−	+	−	+	−	−
7	−	+	+	−	−	+	−
8	+	+	+	+	+	+	+

The designs can be extended with any number of factors. The interpretation, the graphical representations of the results and the calculation of models are analogous to the test design with two factors; only the associated effort increases.

7.4.2 Reduced Test Designs

For complete test designs, the number of required tests increases when the number of considered factors grows. Furthermore, a rapidly increasing number of interactions occurs, which often (but not always) have little effect and so scarcely play any role for the investigated process. Furthermore, test designs for which only one test is performed for each factor combination do not provide any information about the test scattering, unless effects and interactions are eliminated. The most reliable path to obtain information about the test scattering is the repetition of tests. This, however, raises the question whether all possible factor combinations need to be proved using tests.

The Fractional 2^{4-1} Design

For fractional test designs, additional factors take the place of existing higher-order interactions but that can be neglected. For the transition from a complete 2^3 design (Table 7.5) to a fractional 2^{4-1} design, the additional factor D is used for the existing interaction ABC. The sign column for ABC from the 2^3 design is used unchanged for D. Table 7.6 shows the new test design. Rather than 16 tests for the 2^4 design, only 8 tests are required.

Table 7.6. Sign of a 2^{4-1} test design

Line no.	A	B	C	AB	AC	BC	D
1	−	−	−	+	+	+	−
2	+	−	−	−	−	+	+
3	−	+	−	−	+	−	+
4	+	+	−	+	−	−	−
5	−	−	+	+	−	−	+
6	+	−	+	−	+	−	−
7	−	+	+	−	−	+	−
8	+	+	+	+	+	+	+

The advantage of this effort saving has, however, also disadvantages. For the 16 tests of the complete 2^4 designs, 16 coefficients (4 main effects, 6 double, 4 triple and one quadruple interactions, and the mean) can be calculated. This is obviously not possible for the 8 tests of the fractional test designs available here, only 8 effects or interactions can be calculated. This does not mean that specific interactions merely disappear, but rather they are contained in hidden form in the results. Thus, in the added factor D, the associated main effect D is obviously mixed with the triple interaction ABC. A separation is not possible. This behaviour is also called *confounding* or *aliases*. The following confoundings occur for the 2^{4-1} design:

  A + BCD
  B + ACD
  C + ABD
  D + ABC
  AB + CD
  AC + BD
  AD + BC
  ABCD + mean

The mean would receive a column analogue to the factors A–D for the complete notation of the test design. Because the corresponding sign column, however, contains only + signs, this column is generally not shown. The mean is also designated as *identity* with letter I in the test designing. The construction of the 2^{4-1} design in which the sign column of the triple interaction ABC is transferred directly from the 2^3 design to the sign column for D leads to the term *generator*. The generator is the product from the additionally introduced factor and the interaction it replaces; thus, in the described case, $D \cdot ABC = ABCD$. Consequently, the generator itself always has the same sign column as I (all +).

The reduced test design has a confounding (see above) in each column. The associated effect or the associated interaction is multiplied by the generator. Because the multiplication of any effect with itself gives I (*e.g.*, $A \cdot A = I$) and the multiplication of an effect with I again gives the effect (*e.g.*, $A \cdot I = A$), such factors can be eliminated in the previously obtained result.

Example

- Confounding for A:
  $A \cdot generator = A \cdot ABCD = AA \cdot BCD = I \cdot BCD = BCD$
- Confounding for AB:
  $AB \cdot generator = AB \cdot ABCD = AA \cdot BB \cdot CD = I \cdot I \cdot CD = CD$

The further confoundings are calculated similarly. The considerations of the 2^{4-1} design can be generalised so that 2^{k-p} designs can be prepared. They have p generators and so multiple confoundings. The approach is similar.

Confoundings and Resolution

The list of aliases or confoundings provides an overview of the interactions and main effects that cannot be separated. What are the consequences?

Obviously, for example, the factor A is confounded with the interaction BCD. For the evaluation of the test, the associated coefficient is normally "added" to the factor A. This neglects the interaction BCD. An important rule for the preparation of a test with reduced designs is to test whether interactions that then disappear in the confounding really can be neglected. This can only be done by performing an analysis, including any existing preliminary information, experience, *etc.*, of the investigated process.

The example of the air-tight door from Section 7.4.1 is changed slightly to illustrate this behaviour. A third cause variable, the colour of the door, is added. For cost reasons, the number of tests is retained, namely, a 2^{3-1} design is constructed. Table 7.7 shows this modified design.

Table 7.7. 2^{3-1} test design for the door-frame-colour combination and their results

Running number	Door size (A)	Frame size (B)	Colour (C=AB)	Air-tightness
1	Small	Small	Black	Aair-tight (=0)
2	Large	Small	White	Not air-tight (=1)
3	Small	Large	White	Not air-tight (=1)
4	Large	Large	Black	Air-tight (=0)

The evaluation of this design provides the result: to achieve maximum air-tightness, the door must be painted black. The door and frame quantities do not play any role. Such a result is obviously nonsensical. The cause is here, the effect that plays the main role, namely the reduction of the design, has assigned the interaction between door size and frame size to a new factor (colour of the door) so that this nonsensical result from the viewpoint of the test planning is formally correct. This inappropriate example serves only to show that interactions do not automatically play a small role but rather their possible importance should be tested in advance.

To evaluate reduced designs for the risks that result with regard to the confoundings, there is an evaluation designated as resolution. Table 7.8 shows the usual resolutions with their evaluation. Table 7.9 lists the actual resolution for a selected number for reduced designs.

Table 7.8. Resolutions of test designs

Resolution	Confounding	Evaluation
III	Factor with double interaction (2FWW)	Critical
IV	Factor with 3FWW 2FWW with 2FWW	Less critical
V	Factor with 4FWW 2FWW with 3FWW	Not critical
VI	Factor with 5FWW 2FWW with 4FWW 3FWW with 3FWW	Not critical

Table 7.9. Resolution of some reduced test designs

2^{3-1}	2^{4-1}	2^{5-2}	2^{6-3}	2^{5-1}	2^{6-2}	2^{7-3}	2^{6-1}	2^{7-2}	2^{8-1}
III	IV	III	III	V	IV	IV	VI	IV	VIII

7.4.3 Test Designs with Repetitions

Some disadvantages of the previously considered test designs have already been mentioned. The available results have already been completely "exhausted" for the calculation of the coefficients of the models. An estimate of a residual scattering is not possible. Consequently, goodness of fits and regression coefficients also cannot be calculated. There is, in particular for complete or only little-reduced test designs above a factor number of $k \geq 5$, a larger number of effects and interactions whose numerical values are very small (near zero), whereas only a few assume numerically larger values. This leads to the conclusion that a large part of the factors does not have any effect on the target quantity and the calculated values only randomly differ from zero. Such effects and interactions can now be eliminated, although a true criterion for the elimination is not present.

The elimination makes it possible to perform a variance analysis with the remaining factors and draw conclusions about the confidence level of the determined model. This procedure is called *pooling*. It is very subjective and can easily lead to wrong conclusions. The better method is to perform repetitions of the individual test points. The disadvantage is, however, the increased test cost. If, however, for example, preliminary tests have shown that a significant test scattering is to be expected, repetitions should be planned and the design for the compensation of the fractional effort reduced. There are no special considerations for the preparation, execution and evaluation of such designs. A multiple-variance analysis is used to calculate the significance of the effects and interactions.

7.4.4 Test Designs for Nonlinear Relationships

Two-level designs can be used to determine linear or quasilinear relationships. If, however, there is the expectation that nonlinear components, just because of the technical consideration of the analysed process, must exist, alternative designs should be considered. Section 7.4.6 contains further comments about this topic. One possibility is the transformation of the response variables to obtain a linearisation. This functions, however, only when, for example, a pure quadratic dependency without linear components exists. Test designs can also be modified so that nonlinear relationships can be determined. One possibility are the so-called surface response designs. The basic principle involves adding central (or zero) and star points to the previously considered complete or fractional test designs, which sometimes are also designated as *dice designs*. Figure 7.13 shows such a design for two factors.

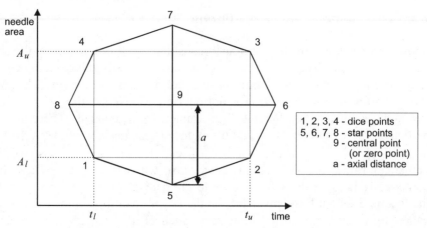

Figure 7.13. General representation of a surface response design

Such designs can be used to determine quadratic models in the following form:

$$y = a + b_1 x_1 + b_2 x_2 + b_{12} x_1^2 + b_{22} x_2^2 \tag{7.19}$$

An important quantity for such test designs is the axial distance a. The choice of this distance affects an important property of test designs, the *orthogonality*. If a design is orthogonal, the coefficients in the model (7.19) are mutually independent. If this property is not present, the coefficients change when nonsignificant factors are eliminated as part of an evaluation. Orthogonal designs also guarantee the smallest-possible confidence intervals for the coefficients.

A surface response design is orthogonal when the axial distance is determined as follows:

$$a = \sqrt{0.5\, c_{st} \sqrt{n_{sum}\, c_w\, N_w} - c_w\, N_w} \tag{7.20}$$

$$n_{sum} = c_w N_w + c_{st} N_{st} + c_o N_o \tag{7.21}$$

$N_w = 2^k$ dice tests (c_w repetitions)

$N_{st} = 2k$ star point tests (c_{st} repetitions)

$N_o = 1$ zero point tests (c_o repetitions)

k number of factors

A disadvantage of these designs is that 5 different levels (2 for the dice points, 1 for the central point and 2 for the star points) are necessary per factor. This is sometimes not technically possible. In such cases the axial distance $a = 1$ can be used for all or for the affected factors. This is then called a face-centred design.

A comprehensive example of the use of such a design is discussed in Section 7.4.7.

Further designs for determining nonlinear relationships are 3^k or 3^{k-p} designs (see [4]). In principle, they are comparable with the two-level designs. However, they are very complex, for example, a 3^4 design requires 81 tests. Other possible designs are the Box–Behnken designs [1] and Draper–Lin designs [2].

7.4.5 Special Response Variables

A prerequisite for the evaluation of test designs is that the response variable has a normal distribution. If this is not the case, there is the danger that all statistical analyses, such as the evaluation of the significance of the factors, are incorrect and thus, at least, inaccurate models are determined. However, it is not always possible to make the distribution test for a response variable. Normally the number of tests does not suffice, because process settings cannot be changed for a distribution test. This means only the process knowledge and experience can be used to assume whether a normal distribution exists. However, there are response variables that prevent the assumption of a normal distribution. These include good-bad or attributive quality characteristics. Although Chapter 4 mentioned that measurable results should be used only when they can be determined in some appropriate manner, there are, however, restrictions here. For example, in the module production, results such as "the number of defective modules in a lot" or "the number of soldering defects on a module" cannot be replaced by other quantities.

When we consider the defect rate as a response variable in a lot with

$$p_i = \frac{x_i}{n}$$

this quantity has a binomial distribution. One way to approximate this distribution is the so-called angular transformation.

$$y_i = \arcsin \sqrt{p_i} \tag{7.22}$$

Rather than the original data from the test, the transformed results are used in the test evaluation.

Another possible response variable is the number of defects per unit. This response variable has a Poisson distribution. The so-called square root transformation:

$$y_i = \sqrt{\text{number of defects}} \tag{7.23}$$

7.4.6 Practical Recommendations for Designing Tests

To design a successful test, several restrictions, in particular, with the preparation, must be observed. This preparation is also designated as brainstorming. This initially involves the selection of the factors, the response variables and the type of test design to be used. The selection of the factors initially always involves the listing of a large number of variables. Anything and everything that could affect the response variables should be listed. This list must obviously be reduced, where experience must be carefully evaluated as to whether the removed quantities really play a subordinate role. Also, considerations concerning the form of the determined model are very important.

Many practical processes have factors that mutually affect each other, namely, changing factor A automatically changes factor B. Such factors must not be used in this type of test planning.

Another important discussion point is the setting of the levels. For some factors, there are empirical setting values that permit optimum results. There is the danger that the levels are chosen too restrictive. The test design then has the same result as for a regression, there is a high probability that significant effects are not determined as significant (see Figure 7.7). The conclusion could be that the considered factor does not have any effect on the target quantity and so this quantity can be neglected for manufacturing monitoring. This can certainly lead to errors. Furthermore, the execution of a statistical test design should obtain additional process knowledge and not just find an optimum. This additional knowledge is obtained, in particular, when the process is also "run" in areas that have previously never been tested. A level setting that is too broad, however, should also be avoided. Negative effects here can be the increased occurrence of nonlinear proportions (see Figure 7.14) and the occurrence of quality jumps (see Figure 7.15).

In addition to the consideration of factors, there are many additional factors that then act as noise variables on the test design. These quantities must be noted. The test execution must ensure that these quantities remain constant. Any changes should be documented, as should any known noise variables, such as environmental conditions.

A major problem is the size of the complete test. From the viewpoint of the statistical analyses, the test size should be as large as possible. This often cannot be realised because a test consumes limited resources (in particular, time and materials). The mean comparison discussed in Section 5.6.3 provides an orientation for the necessary test size. In general, a test planning is nothing other than a multiple mean comparison between the levels of the individual factors. Thus, if it is specified which mean differences d will be recognised with defined error probabilities and preliminary tests have provided information about the process standard deviation σ, the total test size can be estimated using:

$$n = 2 \cdot \left(u_{1-\alpha} + u_{1-\beta}\right)^2 \cdot \left(\frac{\sigma}{d}\right)^2 \qquad (7.24)$$

where α is the probability that a nonexistent difference is incorrectly recognised as being an existing difference and β is the probability that an existent difference will not be recognised. The values 0.1% and 1% are normally used in the test planning for α and β, respectively. This simplifies Equation (7.24) to:

$$n \approx 60 \cdot \left(\frac{\sigma}{d}\right)^2 \qquad (7.25)$$

This is obviously only an orientation that must be adapted to the actual conditions. For the planning of material resources, one should be aware that, at least some, test points have factor settings that represent the rejection, corrective work, *etc.*, for the product produced in the test. This, however, is normal and should not necessarily be considered as being a disadvantage of the test planning because these test points are used to obtain new process knowledge.

The specification of the response variables is also important. It should be oriented on the question: what should the test achieve? What conclusions can be expected from the determined response variables? *Several* response variables are quite usual. The measurability (or also the testability) of the response variables should also be discussed. A test is not useful if once it has been performed the response variables cannot be measured with the required accuracy or the resources for the complete measurement are not available. If several response variables are envisaged, the associated weightings must also be defined, because it is quite possible that the evaluation of specific response variables produces conflicting results.

If a two-level plan is used, good results are obtained only when the actual relationships between factors and response variables are nearly linear. If this is not the case, incorrect models are produced and the conclusions about changes to the process, *etc.*, can have the opposite effect. Figure 7.14 demonstrates this behaviour. To protect against the presence of strong nonlinearity in the test area, a zero point (possibly with test repetition) should be provided. This zero point is at the centre of the "cube". Although the test results for this point cannot determine any nonlinear model parts, as, for example, the surface response designs permit, but strong deviations of the observations from the calculated model in the zero point indicate the presence of larger nonlinearities.

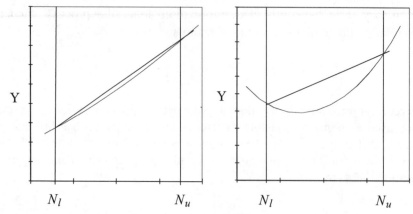

Figure 7.14. Different possible relationships between factors and response variables (bent curves: true but unknown relationship; straight line: relationship determined by the test planning)

Another aspect to be observed for the preparation of a test planning is the possible presence of quality jumps within the test area. Figure 7.15 illustrates this behaviour. If such effects are to be expected, then a test planning yields very faulty results. Even a zero point cannot help here.

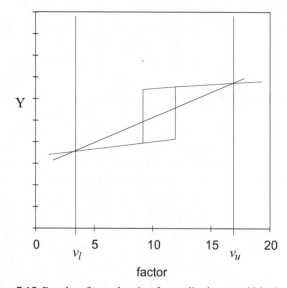

Figure 7.15. Results of test planning for quality jumps within the test area

Two important terms in the test planning are block formation and the randomisation. *Block formation* is the grouping of the tests. A typical block formation for designs with repetition is a single complete test section (*e.g.*, a 2^3 design). The blocks should be processed successively so that the random scattering in a block will be minimised. A consistent block formation can detect regularities during the test execution. Within a block, the tests should be performed in a random sequence. This is called *randomisation*. This is a measure to minimise the effects of trends during the test execution. However, it must be analysed as to whether block formation and randomisation permit a rational test execution. If the effort for the conversion of the process between the individual tests is too large (*e.g.*, the time required for heating or cooling a plant), it can be omitted.

7.4.7 Investigation of a Wave Soldering Bath

A manufacturing plant for electronic modules has the task of optimising the setting parameters of a wave soldering bath for a complex module to minimise the solder bridges and wetting defects response variables. The restriction is also given that approximately 100 modules are available for this test.

In a brainstorming with the operators of the wave soldering bath, the cause variables initially to be considered were defined, as shown in Table 7.10.

Table 7.10. Summary of the factors

Process section	Considered factors	Ignored factors	Noise variables
Fluxer	Flux quantity	Fluxer speed; spraying head speed	Flux properties
Preheating section	Heating zone settings (12 items)	Preheating speed	
Soldering bath	Height of the chip wave; surge plate advance	Soldering pot height; angle; height main soldering wave height; transport speed; soldering bath temperature	Soldering bath residual oxygen; soldering bath composition (additives)
Material effects			Wettability of the circuit board and of the components
Environmental conditions		Air humidity; ambient temperature	

The operators of the wave soldering bath, in particular, have additional requirements for factors that should be varied because they have known empirical effects on the response variables. For example, the soldering pot height could be suggested for use as a factor.

Table 7.11. Test design for the wave soldering process (selection)

Test point type	Line no.	Ssurge plate	Chip wave	Tempe-rature	Flux quantity	Position in the frame
Zero	1	5.15	1600	110	4.4	Left
Zero	2	5.15	1600	110	4.4	Left
Corner	3	4.4	1500	100	4	Left
Corner	4	5.9	1500	100	4	Left
Corner	5	4.4	1700	100	4	Left
Corner	6	5.9	1700	100	4	Left
Corner	7	4.4	1500	120	4	Left
Corner	8	5.9	1500	120	4	Left
Corner	9	4.4	1700	120	4	Left
Corner	10	5.9	1700	120	4	Left
Corner	11	4.4	1500	100	4.8	Left
Corner	12	5.9	1500	100	4.8	Left
Corner	13	4.4	1700	100	4.8	Left
Corner	14	5.9	1700	100	4.8	Left
Corner	15	4.4	1500	120	4.8	Left
Corner	16	5.9	1500	120	4.8	Left
Corner	17	4.4	1700	120	4.8	Left
Corner	18	5.9	1700	120	4.8	Left
Star	19	4.04	1600	110	4.4	Left
Star	20	6.26	1600	110	4.4	Left
Star	21	5.15	1452	110	4.4	Left
Star	22	5.15	1748	110	4.4	Left
Star	23	5.15	1600	100	4.4	Left
Star	24	5.15	1600	120	4.4	Left
Star	25	5.15	1600	110	3.81	Left
Star	26	5.15	1600	110	4.99	Left
Zero etc.	27	5.15	1600	110	4.4	Right
Star	104	5.15	1600	110	4.99	Right

Evaluation

The following abbreviations are used for the subsequent evaluations:

A   main effect - surge plate
B   main effect - chip wave
C   main effect - temperature
D   main effect - flux quantity
E   position (newly introduced levels: -1 = left; $+1$ = right)
AB   interaction between the surge plate and the chip wave, *etc.*
CD   interaction between the temperature and the flux quantity
AA   quadratic effect - surge plate.

However, because there is a dependency on other factors (an increase of the soldering pot height always increases the heights of both waves), this wish cannot be accepted. It must be re-emphasised that the existence of dependencies can cause incorrect test results. The knowledge of additional preliminary information about possible nonlinearities and the consideration of the available resources means that a 2^{4-1} surface response design with three repetitions is used that produces a total of 104 modules. The level settings are defined in agreement with the plant operators. Table 7.11 shows a section from this design.

The position in the soldering frame was also recorded as an additional factor but not mentioned in the brainstorming. For the setting of the levels for the temperature of the preheating section, the star points cannot be realised as envisaged (line numbers 23 and 24). These settings are face-centred. The minimisation of solder bridges and wetting defects are specified as target quantities. The test is performed and the bridges and wetting defects counted.

Table 7.12. Variance analysis for the bridges (s.s.d. = sum of the squared differences)

Factor	s.s.d.	Degrees of freedom	Variance	Test quantity	p-value
A surge plate	0.9468	1	0.9468	3.85	0.0532
B chip wave	0.0718	1	0.0718	0.29	0.5905
C temperature	0.1761	1	0.1761	0.72	0.4001
D flux	0.0344	1	0.0344	0.14	0.7094
E position	9.0059	1	9.0059	36.59	0.0000
AA	0.1471	1	0.1471	0.6	0.4416
AB	0.7493	1	0.7493	3.04	0.0847
AC	0.3790	1	0.3790	1.54	0.2181
AD	1.1914	1	1.1914	4.84	0.0305
AE	0.0444	1	0.0444	0.18	0.6721
BB	0.0019	1	0.0019	0.01	0.9298
BC	0.0251	1	0.0251	0.1	0.7502
BD	1.0996	1	1.0996	4.47	0.0375
BE	0.2236	1	0.2236	0.91	0.3432
CC	0.4137	1	0.4137	1.68	0.1983
CD	0.0134	1	0.0134	0.05	0.8163
CE	0.4294	1	0.4294	1.74	0.1901
DD	0.1907	1	0.1907	0.77	0.3812
DE	0.0344	1	0.0344	0.14	0.7094
Total error	20.6734	84	0.2461		

Evaluation of solder bridges

The solder bridges specify the number of defects per module. The results of the response variable are converted using the transformation (7.23).

A variance analysis is initially performed to determine the significance of the factors. Table 7.12 shows this analysis. A large part of the effects has a p-value significantly higher than 0.05. This means that these effects are not significant. These initial analyses also show that the position of the module plays a large role that was not envisaged prior to the test and was also added to the design for the evaluation as an additional factor.

Another possibility for the clear representation of the initial first results is a Pareto diagram that shows this significance in standardised form. All effects with values less than 2.0 are not significant (for a 95% confidence level). This means that these effects can be successively eliminated. Because the actual design is not orthogonal, this elimination can also cause changes to those effects that have not been eliminated. Figure 7.17 shows the Pareto diagram after the elimination of all nonsignificant effects.

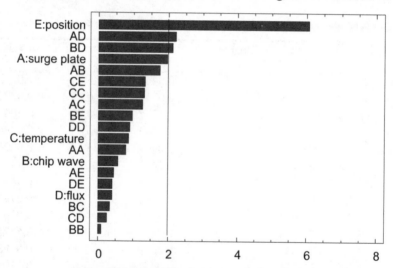

Figure 7.16. Pareto diagram of all effects for the bridges

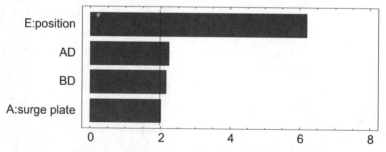

Figure 7.17. Pareto diagram of the significant effects

The position of the module thus has the principal effect on the solder bridges. Two interactions and the position of the surge plates are also added. The consequences of the main effects can be shown as follows:

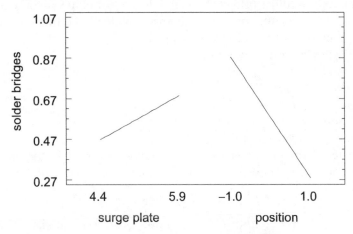

Figure 7.18. Representation of the significant main effects for the solder bridges

This diagram is easy to interpret: a low surge plate produces few bridges and a position $= +1$ (corresponds to right) also produces few bridges. It should be noted for the interpretation of the position that statements about intermediate values (*e.g.*, position $= 0$) may not be made because no test values are available for them!

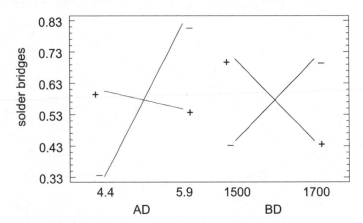

Figure 7.19. Interaction plot for the significant interactions of the bridges

The interaction plot shown in Figure 7.19 can be interpreted as follows:
AD interaction:
If the surge plate is set low, this produces a small number of bridges for a small flux quantity. If the surge plate is set high, then few bridges can be

achieved with a large flux quantity. If the flux quantity is large, the setting of the surge plate plays scarcel almost no role.

BD interaction:

The number of solder bridges is small when a low chip wave is combined with a small flux quantity or a high chip wave with a large flux quantity.

An optimum result with regard to the bridges can consequently be achieved using the following setting:

surge plate   4.4
chip wave   1500
temperature   (no shown effect)
flux quantity 4
position   right.

The determined dependency of the number of bridges on the cause variables can be specified using the following regression equation:

$$\sqrt{\text{bridges}} = -33.54 + 2.145 \cdot \text{surge plate} + 0.0144 \cdot \text{chip wave}$$
$$+ 7.58 \cdot \text{flux} - 0.30 \cdot \text{position} - 0.455 \cdot \text{surge plate} \cdot \text{flux}$$
$$- 0.00328 \cdot \text{chip wave} \cdot \text{flux}$$

The following restrictions must be observed when this equation is used:

- This equation is an interpolation, namely it is true only in the range of the performed tests for the associated cause variables; extrapolations are not permitted.
- Only values of −1 or +1 may be used for the position of the module.

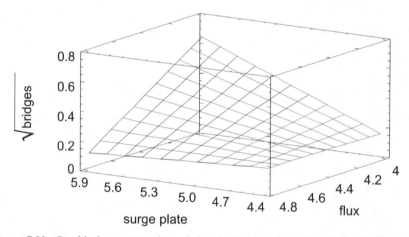

Figure 7.20. Graphical representation of the determined dependencies fixed chip wave settings = 1700; position right

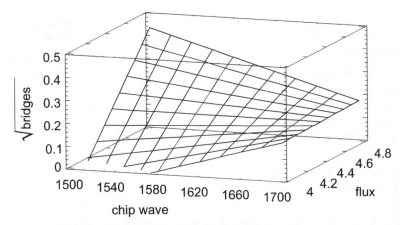

Figure 7.21. Another graphical representation of the determined dependencies fixed surge plate settings = 4.4; position right

Figure 7.20 shows a graphical representation of the model. To make the result surface easier to see, the x-axis (surge plate) and the y-axis (flux) are shown decreasing. Because only a maximum of two cause variables can be shown, the remaining quantities must be fixed.

Figure 7.21 shows that a high surge plate combined with a large flux quantity produces a low number of bridges. If the surge plate is low, this produces only a few additional bridges, but this does not depend on the flux quantity. Figure 7.22 shows the limits of such models. The left-hand lower corner contains values that correspond to a negative number of bridges, which although not possible in practice, can occur in the calculation.

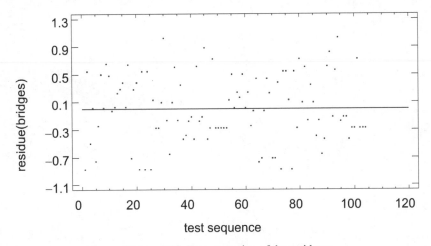

Figure 7.22. Representation of the residues

Optimum here is a low chip wave, but this, however, still produces good "bridge values" for a high chip wave (for a low surge plate) and this relatively independent of the flux quantity. This independence is a different conclusion from that for the interpretation of the interaction plot (see above). The cause of this difference is the fixed setting of the surge plate (here at the lower level). This conclusion is also true for the left position (but with a larger number of bridges). The analysis of the residues in Figure 7.22 shows there are no peculiarities for the sequence analysis.

Evaluation of the wetting
Because this is similar, no details are described here.

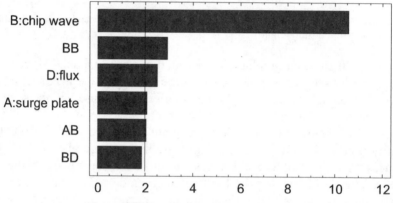

Figure 7.23. Pareto chart of the significant effects

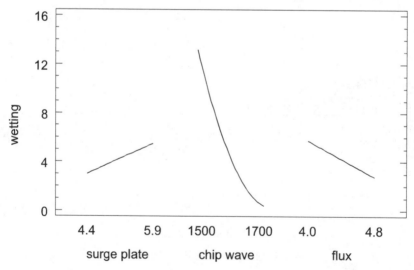

Figure 7.24. Representation of the main effects for the wetting

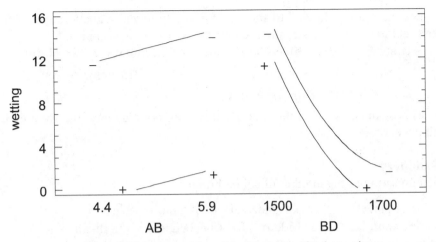

Figure 7.25. Representation of the interactions for the wetting

The figure shows that the height of the chip wave (both as the main effect and as a quadratic effect) plays the largest role for the wetting.

An optimum result for the wetting thus can be attained with the following setting:

surge plate   4.4; (for an arbitrarily large flux quantity)
chip wave   1700 (and possibly higher)
temperature   (without any shown effect)
flux quantity   4.8 (for an arbitrarily high chip wave)
position: (without any shown effect)

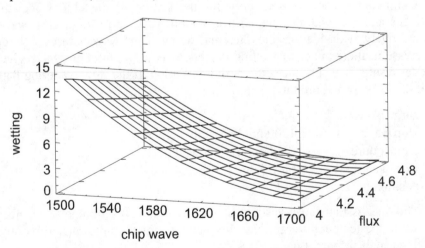

Figure 7.26. Graphical representation of the determined dependencies (fixed settings: surge plate = 4.4; position left)

For the "extrapolation" of the wetting, the following equation can again be specified (similar restrictions as for the use for bridges apply):

$$\text{wetting} = 835.02 + 30.99 \cdot \text{surge plate} - 0.922 \cdot \text{chip wave} - 53.74 \cdot \text{flux}$$
$$- 0.0183 \cdot \text{surge plate} \cdot \text{chip wave} + 0.000255 \cdot \text{chip wave}^2$$
$$+ 0.031 \cdot \text{chip wave} \cdot \text{flux}$$

This again shows that the flux quantity does not play any role for a high chip wave.

Summary

Minimisation of the number of solder bridges:

- surge plate   4.4 (independent of the flux quantity)
- chip wave   1500 or 1700 (independent of the flux quantity)
- temperature   (without any shown effect)
- flux quantity   4 (no effect for a low surge plate/high chip wave)
- position   right

Optimum wetting setting

- surge plate   4.4 (for an arbitrarily flux quantity)
- chip wave   1700 (and possibly higher)
- temperature   (without any shown effect)
- flux quantity   4.8 (for an arbitrarily high chip wave)
- position   (without any shown effect)

Contradictions are present only for the setting of the chip wave, the bridges and the wetting demand a low chip wave and a high chip wave, respectively. Considering the fact that a low chip wave causes a large increase in the number of wetting defects, overall, a high chip wave gives better results. This produces the optimum setting of the flow soldering bath for the investigated product:

- surge plate   4.4
- chip wave   1700 (and possibly higher)
- temperature   (without any shown effect)
- flux quantity   (any value between 4 and 4.8)
- position   right

This means that the investigated plant is relatively immune to changes made to the temperature and the flux quantity for the investigated module and the analysed target parameters.

References

[1]   BOX, G.E.P./DRAPER, N.R.: *Empirical Model-Building and Response Surfaces.* John Wiley, New York 1987

[2]   DRAPER, N.R./LIN, D.K.J. *"Small response surface designs"* in: Technometrics 32 (1990), p. 187–194

[3]   KLEPPMANN, W.: *Taschenbuch Versuchsplanung*; Hanser Verlag 1999

[4]   SCHEFFLER, E.: *Statistische Versuchsplanung und Auswertung*; DVG; 3rd edition, Stuttgart 1997

[5]   STORM, R.:*Wahrscheinlichkeitsrechnung, mathematische Statistik und statistische Qualitätskontrolle.* 11th revised edition, Hanser Verlag, Munich/Vienna 2001

8 Reliability of Products and Processes

8.1 Reliability Characteristics and Models

Reliability is the quality (also designated as usability) of a unit (product) under given use conditions during or after a specified time. The reliability of a unit is primarily described using reliability characteristics, namely quality characteristics that affect the reliability. A differentiation is made between qualitative reliability characteristics (temperature stability, the function of a component, the function of a solder point, *etc.*) and quantitative reliability characteristics (*e.g.* the resistance value of a solder point, switching times).

8.2 Terms and Reliability Characteristics

The lifetime of a unit is the time from the start of use until the time of failure. Time with regard to the reliability is not just an operational time (*e.g.*, hours), but the usage cycles (*e.g.*, number of load changes, number of times a switch is operated, *etc.*) is often also used. These cycles are generally designated by N. To avoid confusion with the universal set, t will be used for the time measure in the following discussion; t must be replaced by N for the corresponding cyclical uses cases.

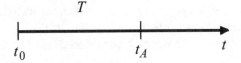

Figure 8.1. Lifetime T, t_0 usage begins, t_A time of failure

Important for all reliability considerations is that at the start of use ($t = 0$) the corresponding unit was fault free with regard to reliability. A *failure* occurs when a nonpermitted deviation of at least one characteristic has occurred. It must be carefully defined what is nonpermitted. For example, a solder point at an electronic component fails, at the latest, when it is interrupted by a breakage (electrical resistance infinitely large). In specific cases, however, a finite increase of the resistance of the solder point suffices for the associated module to fail. A differentiation is made between two basic types of the failures:

- Sudden failure (solder point interrupted, resistance layer burnt through, *etc.*)
- Degradation failure (a defined end caused by a continuous change process such as wear, fatigue, *etc.*)

By specifying a defined limit value in its description, a degradation failure can be evaluated like a sudden failure. Furthermore, the considered units are differentiated into those that can be repaired and those that cannot be repaired.

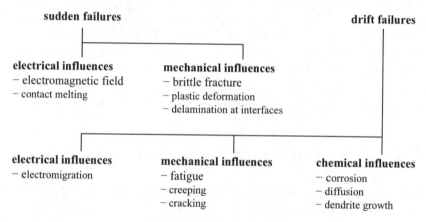

Figure 8.2. Failure mechanisms

The time of the failure of a unit is a (continuous) random value T. Consequently, lifetime distributions are constructed to describe the failure behaviour. In analogy to the probability density $f(x)$, the corresponding function in the reliability theory is called the failure density $f(t)$. This density, however, is scarcely used in practical reliability work. The distribution function $F(x)$ corresponds to the *failure probability* $F(t)$. It specifies the probability with which a considered unit has failed by time t.

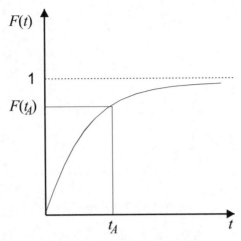

Figure 8.3. Failure probability $F(t) = P(T \leq t)$

The complement of the failure probability $F(t)$ is the survival probability $R(t)$.

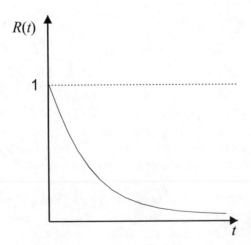

Figure 8.4. Survival probability $R(t)$

Another characteristic value is the *failure rate* $\lambda(t)$. It is generally the probability of the failure of a unit in the interval between t and $t+dt$ (dt differentially small) under the prerequisite that the unit was not yet failed at time t.

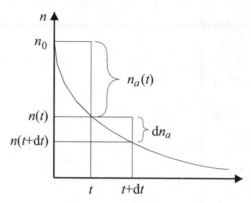

Figure 8.5. Failure rate $\lambda(t)$

dn_a	number of failures in dt
n_0	number of considered units
$n(t) = n_0 - n_a(t)$	number of functioning units at time t
$n_a(t)$	number of failed units up to time t

These quantities have the following relationships:

$$F(t) = \int_{x=0}^{t} f(x)\,dx$$

$$R(t) = 1 - F(t) \qquad F'(t) = f(t) = -R'(t)$$

$$\lambda(t) = \frac{f(t)}{R(t)} = -\frac{R'(t)}{R(t)}$$

These allow the reliability characteristics to be derived. The most important characteristic is the characteristic lifetime T_c (sometimes also designated as T or N_f for the use of cycles). It is the time by when a unit has failed a the probability of 63.21%. Derivations from the average lifetime are:

MTTF	Mean time to failure; average lifetime for units that cannot be repaired
MTBF	Mean time between failure; average duration between two failure times for units that can be repaired
MTTR	Mean time to repair; average duration for the repair of a unit

8.3 Lifetime Distributions

8.3.1 Exponential Distribution

The exponential distribution describes failures *only* caused by chance. The failure density and the failure probability of this distribution have already been discussed in Section 5.4.5.

$$f(t) = \lambda \, e^{-\lambda t} = \frac{1}{T_c} e^{-t/T_c} \tag{8.1}$$

$$F(t) = 1 - e^{-\lambda t} = 1 - e^{-t/T_c} \text{ with } \lambda = \frac{1}{T_c} \tag{8.2}$$

The survival probability $R(t)$ can be calculated from the failure probability.

$$R(t) = 1 - F(t) = e^{-\lambda t} = e^{-t/T_c} \tag{8.3}$$

The average lifetime μ_T (expected value of the lifetime) is calculated as:

$$E\{T\} = \mu_T = \frac{1}{\lambda} = T_c \tag{8.4}$$

The parameter λ is the failure rate, independent of the time and constant *only* for the exponential distribution. The characteristic lifetime T_c is (only) identical to the expected value of the lifetime μ_T for the exponential distribution.

When time $t = \mu_T$ is considered, then:

$$R(t = T_c) = e^{-\mu_T/T_c} = e^{-1} = 36.79\%$$

This means that 63.21% of the considered units have failed after the expiration of the characteristic lifetime T_c (and here also after the average lifetime).

8.3.2 Weibull Distribution

The Weibull distribution is the most important distribution for reliability investigations. It is named after the Swede Weibull, who first showed the wide use of this probability distribution ([1, 3, 4]). Initially the two-parameter Weibull distribution is considered with b form parameters (failure gradient, sometimes also β) of the Weibull distribution.

$$\text{Failure density} \quad f(t) = \frac{b}{T_c} \left(\frac{t}{T_c} \right)^{b-1} \cdot e^{-\left(\frac{t}{T_c} \right)^b} \tag{8.5}$$

$$\text{Failure probability} \quad F(t) = 1 - R(t) = 1 - e^{-\left(\frac{t}{T_c} \right)^b} \tag{8.6}$$

$$\text{Failure rate} \quad \lambda(t) = \frac{b}{T_c} \left(\frac{t}{T_c} \right)^{b-1} \tag{8.7}$$

$$\text{Expected value of the lifetime} \quad \mu_T = \Gamma\left(\frac{1}{b} + 1 \right) \cdot T_c \tag{8.8}$$

$$\text{Dispersion} \quad \sigma_T^2 = \left(\Gamma\left(1 + \frac{2}{b} \right) - \Gamma\left(1 + \frac{1}{b} \right)^2 \right) \cdot T_c^2 \tag{8.9}$$

The exponential distribution is a special case of the Weibull distribution for $b = 1$. The definition of the gamma distribution $\Gamma(x)$ is described in Section 1.3.2.

The characteristic curves of failure densities, failure probability and failure rate are shown in Figure 8.6. It is easy to see that at the time of the characteristic lifetime, the failure probability is independent of the form parameter b.

Some products have a so-called *minimum life duration* t_0 (or N_0), namely a product never fails before t_0 (incubation time). Only afterwards does the use of the products start to cause failures to occur. Such a behaviour can be described by the three-parameter Weibull distribution. The failure density of this distribution is defined in Equation (8.10).

$$f(t) = \begin{cases} 0 & \text{for} \quad t < t_0 \\[2ex] \frac{b}{T_c - t_0} \left(\frac{t - t_0}{T_c - t_0} \right)^{b-1} \cdot e^{-\left(\frac{t - t_0}{T_c - t_0} \right)^b} & \text{for} \quad t \geq t_0 \end{cases} \tag{8.10}$$

The figure shows that only a parallel offset of the time axis occurs. This also illustrates the failure probability in Figure 8.6, where for the form factor $b = 4$ the corresponding curve for a minimum lifetime duration of $t_0 = 2000\,\text{h}$ is represented. This ensures that a three-parameter Weibull distribution can be transferred to a two-parameter Weibull distribution.

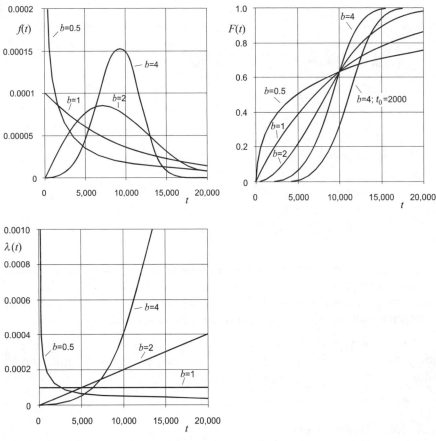

Figure 8.6. Curve of the failure density $f(t)$ (above left), the failure probability $F(t)$ (above right) and the failure rate $\lambda(t)$ for a characteristic lifetime T_c of 10,000 h and for different form parameters of the Weibull distribution depending on time

8.3.3 Use of Lifetime Distributions

Three main use areas must be differentiated for the two-parameter Weibull distribution:

1. $0 < b < 1$: The failure rate decreases monotonously. This allows so-called early failures to be described. Early failures are generally caused by hidden manufacturing faults. A typical example are cold solder points that provide contact during the inspection but that break completely after a short time.

2. $b = 1$: The failure rate is constant. The failure occurs by chance. An example is a sudden (but unforeseen) loading that causes failure. This special case of the Weibull distribution is described by the exponential distribution.

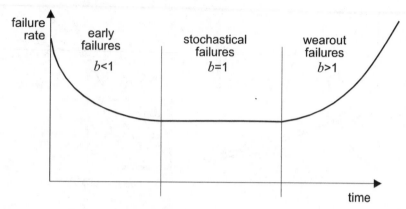

Figure 8.7. Possible curve of the failure rate for complex units

3. $b > 1$: The failure rate increases monotonously. This describes late failures primarily caused by wear or ageing effects. This group includes, above all, degradation failures. For solder points, fissures that continuously increase are the main cause of such failures.

For complex products (electronic devices, cars, computers), a mixing of the failure mechanisms occurs. The so-called "bath-tub life curve" illustrates the general curve for the failure rate.

8.4 Reliability Tests

8.4.1 Graphical Evaluations of Reliability Data

For the evaluation of the determined reliability data, it is important to determine which lifetime law (lifetime distribution) governs these data. This raises the question as to whether a Weibull distribution (two or three-parameter) is capable of describing the origin of these reliability data sufficiently accurately. In practice, graphical evaluation methods, in particular, have established themselves, of which the lifetime network is the most important.

Lifetime Network

The use of the lifetime network is very common because it allows the very easy evaluation of data for reliability tests and the lifetime law (parameters of the Weibull distribution) can be determined graphically. Analogous to the procedure in the probability network of the normal distribution, the network is constructed so that the representation of the survival probability $R(t)$ produces a straight line.

$$R(t) = e^{-\left(\frac{t}{T_c}\right)^b}$$

$$-\ln R = \left(\frac{t}{T_c}\right)^b$$

$$\ln(-\ln R) = b \cdot (\ln t - \ln T_c)$$

$$\ln \ln \frac{1}{R} = b \cdot 2.3026 (\lg t - \lg T_c) \tag{8.11}$$

Equation (8.11) is a linear equation so that the desired network must have a logarithmic scale on the abscissa and the ordinate a ln (1/R) scale. Figure 8.8 shows the general appearance of the lifetime network with an evaluation example.

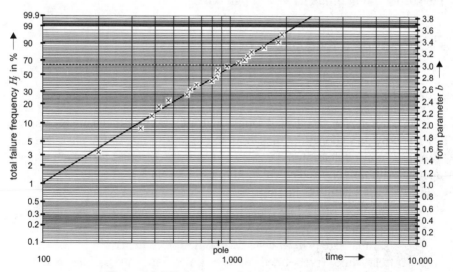

Figure 8.8. Lifetime network (with an actual example)

Uses of the Lifetime Network

For the use of the lifetime network, proceed as follows:

The failure times t_i (lifetime values, unit in cycles or time) are sorted according to size. For samples $n > 30$, a classification can also be made (recommended number of classes $k = 10$–20, preferably with constant class width; although not mandatory). The class frequencies n_i must be determined for a classification. The corrected frequency sums H_j (as percentage) must be calculated afterwards:

$$H_j = 100 \cdot \left(\frac{j - 0.3}{n + 0.4} \right) \quad \text{(for individual values)} \tag{8.12}$$

$$H_j = 100 \cdot \frac{\sum_{i=1}^{j} n_i}{n} \quad \text{(for classed values)} \tag{8.13}$$

If the entered points can be *easily* approximated by a straight line, the parameters of the lifetime law can be *estimated* directly from the network.

The intersection point of the entered straight lines with the line $H_j = 63.2\%$ (Figure 8.8) supplies an estimation of the characteristic lifetime T_c on the abscissa. A parallel shift of the entered straight line through the pole of the network (on the abscissa of the network) allows the form factor b to be read from the right-hand edge of the network at the intersection point of the moved straight line.

8.4.2 Computed Evaluations of the Results

Calculations with an Exponential Distribution

If it is known for the tested units that the failure mechanism can be described by an exponential distribution, the failure rate or the characteristic lifetime T_c can be determined easily. There are 4 different test plans, shown together with the required calculation rules in Table 8.1.

The specific quantities in Table 8.1 are:

n_{pr} number of testing positions (or the simultaneously used test specimens)

n sample size ($=n_{pr}$ if tested without replacement)

t_{pr} test time (predefined or the time until the failure count is reached)

t_i failure times of the individual units

i failure unit count (predefined or failed after the specified test time)

E   test plan with replacement, namely each failed unit will be replaced

O   test plan without replacement

Table 8.1. Test plans for calculating the reliability characteristic values

Plan (above designation, below example)	Estimate of the parameters	Confidence limits
$[n_{pr}\text{–}E\text{–}t_{pr}]$ $n_{pr}=5;\ n=10;\ i=5$	$\hat{T}_c=\dfrac{n_{pr}t_{pr}}{i}$ $\hat{\lambda}=\dfrac{1}{\hat{T}_c}$	$\lambda_l=\dfrac{\hat{\lambda}}{2i}\chi^2_{2i;l}$ $\lambda_u=\dfrac{\hat{\lambda}}{2i}\chi^2_{2(i+1);u}$
$[n_{pr}\text{–}E\text{–}i]$ $n_{pr}=5;\ n=10;\ i=6$	$\hat{T}_c=\dfrac{n_{pr}t_{pr}}{i}$ $\hat{\lambda}=\dfrac{1}{\hat{T}_c}$	$\lambda_l=\dfrac{\hat{\lambda}}{2i}\chi^2_{2i;l}$ $\lambda_u=\dfrac{\hat{\lambda}}{2i}\chi^2_{2i;u}$
$[n_{pr}\text{–}O\text{–}t_{pr}]$ $n_{pr}=5;\ n=5;\ i=3$	$\hat{T}_c=\dfrac{\left(n_{pr}-i\right)t_{pr}+\sum t_i}{i}$ $\hat{\lambda}=\dfrac{1}{\hat{T}_c}$	$\lambda_l=\dfrac{\hat{\lambda}}{2i}\chi^2_{2i;l}$ $\lambda_u=\dfrac{\hat{\lambda}}{2i}\chi^2_{2(i+1);u}$
$[n_{pr}\text{–}O\text{–}i]$ $n_{pr}=5;\ n=5;\ i=4$	$\hat{T}_c=\dfrac{\left(n_{pr}-i\right)t_{pr}+\sum t_i}{i}$ $\hat{\lambda}=\dfrac{1}{\hat{T}_c}$	$\lambda_l=\dfrac{\hat{\lambda}}{2i}\chi^2_{2i;l}$ $\lambda_u=\dfrac{\hat{\lambda}}{2i}\chi^2_{2i;u}$

—○ failure   ━● end of test

The actual plan to be used depends on the current conditions. This includes the number of available testing positions and the number of units to be tested (normally an economic problem because the used units can no longer be sold after the test). In each case, it does not matter for the result (estimate of the failure rate and its accuracy) whether n_1 units are tested for t_1 time or n_2 units are tested for t_2 time if the product is identical for the unit count and test time. This is also called a constant number of component-hours (this statement is true only for an exponential distribution!).

The specification of the end of the reliability test is critical. There are two possibilities: the specification of a defined test time t_{pr} (failure-time terminated plan) or the specification of a defined failure count i (failure-count terminated plan). The failure-time terminated plan has the advantage that the relative conclusion accuracy (confidence limits) can be specified before the test. The disadvantage is the unknown test duration. The reverse is true for the failure-time terminated plan. Plans with replacement should be used if it is easy to replace a failed unit with a new unit.

The mistake probabilities with l or u have been used for calculating the confidence limits (last column of Table 8.1). $l = \alpha / 2$ (or $l = \alpha$ for a one-sided estimate) or $u = 1 - \alpha / 2$ (or $u = 1 - \alpha$) should be used for the lower confidence limit or the upper confidence limit, respectively. The reciprocal values of the confidence limits for the failure rate can be used to obtain the confidence limits for the characteristic lifetime, where the lower confidence limit for the failure rate λ_l is used to calculate the upper confidence limit of the characteristic lifetime.

Table 8.2. Results

$n_{pr} =$ / $i =$	Estimated average lifetime (in h)	Failure rate (in h⁻¹)	Confidence limits for the failure rate (in h⁻¹) upper/lower/single-side-upper	Confidence limits; characteristic lifetime lower/upper/single-side-lower
50			$\lambda_l=0.107\times10^{-3}$	$T_l=1,285.4$ h
5	3,000	0.33×10^{-3}	$\lambda_u=0.777\times10^{-3}$	$T_u=9,311.6$ h
			$\lambda_u=0.701\times10^{-3}$	$T_l=1,427.5$ h
50			0.361×10^{-3}	815.5 h
10	1,500	0.67×10^{-3}	1.226×10^{-3}	3,133.7 h
			1.131×10^{-3}	884.4 h
100			0.0536×10^{-3}	2,570.7 h
5	6,000	0.167×10^{-3}	0.389×10^{-3}	18,623.1 h
			0.350×10^{-3}	2,855.0 h
100			0.159×10^{-3}	1,631.0 h
10	3,000	0.33×10^{-3}	0.613×10^{-3}	6,267.5 h
			$0.565v10^{-3}$	1,768.9 h

In general, one-sided estimates should be preferred because a reliability test normally must make a statement such as: "The failure rate is less than or the characteristic lifetime is greater than ..."

Example:
50 (100) ($= n_{pr}$) printed-circuit boards are tested for $t_{pr} = 300$ h. Each failed module will be replaced. When the test time has completed, $i = 5(10)$ modules have failed. A mistake probability of 5% is specified.

The consideration of the confidence limits shows that they are relatively wide apart. This is the largest problem for reliability tests. Precise conclusions require a large number of tested units or a long test time. Economic considerations mean, however, that limits are set here. The results also show that more accurate conclusions are possible when many units have failed during the test, where a high failure count normally is associated with poor reliability characteristic values. In particular for long-life units (components, solder points, *etc.*), it is difficult to obtain exact reliability data. In this case, the units must be tested under stress (accelerated reliability test) and the obtained results transformed to real conditions. When the equations from Table 8.1 are used, note, however, that good results can be expected only when the failures are random, namely they obey an exponential distribution. In particular, the first failures must be analysed for the possibility of early failure behaviour and the last failures for late failure behaviour (wear).

Calculations with the Weibull Distribution

If the lifetime network is not used for the evaluation, the characteristic parameters of the Weibull distribution can also be approximated using the following equations:

$$\hat{b} = \frac{0.577}{s_{\lg(t)}} \quad ; \quad \hat{T}_c = 10^{\left(\lg(t) + \frac{0.2507}{\hat{b}}\right)} \tag{8.14}$$

$$s_{\lg(t)}^2 = \frac{1}{n-1}\left[\sum_{i=1}^{n}\left(\lg t_i\right)^2 - \frac{1}{n}\left(\sum_{i=1}^{n}\lg t_i\right)^2\right] \quad ; \quad \lg(t) = \frac{1}{n}\sum_{i=1}^{n}\lg t_i$$

The confidence intervals (for $n > 50$) are:

$$T_{l,u} = \hat{T}_c \mp u_{1-\alpha/2}\frac{1.052}{\sqrt{n}}\frac{\hat{T}_c}{\hat{b}} \tag{8.15}$$

$$b_{l,u} = \hat{b} \mp u_{1-\alpha/2}\frac{0.78}{\sqrt{n}}\hat{b} \tag{8.16}$$

Note, however, that the use of these calculation rules alone does not allow a more accurate analysis of the failure behaviour because effects such as the existence of a minimum life duration or the occurrence of mixed distributions cannot be detected.

8.5 Markov Chains

The Markov chain is a mathematical model used to describe the behaviour of technological processes on the basis of states and state transitions, and their probabilities. The technological process is considered depending on the time, where, however, only at equidistant points in time. The following model describes a process with two states.

8.5.1 Technological Process with Two States

The two states are designated with x_1 and x_2:
x_1 – the technological process is operational
x_2 – the technological process has failed.
The term event (see Section 5.1) can also be used for the states:
$x_1 = f$ (operational)
$x_2 = a$ (failed)

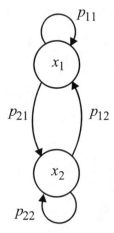

Figure 8.9. State graph for 2 states

These two events are complementary. A probability can also be assigned to the events.

A state graph represents the states as nodes and the state transitions as arrows (see Figure 8.9).

Obviously arrows must also be provided for the retaining of a state. The following consideration makes this clear:

A *transition probability* is assigned to each arrow in Figure 8.9.

p_{11} – probability for retaining the state x_1

p_{21} – probability for the transition of the state x_1 to the state x_2

p_{12} – probability for the transition of the state x_2 to the state x_1

p_{22} – probability for retaining the state x_2

These probabilities can be represented as a square matrix:

$$\mathbf{P} = \begin{pmatrix} p_{11} & p_{12} \\ p_{21} & p_{22} \end{pmatrix} \tag{8.17}$$

P is called the matrix of the transition probabilities, shortened to: *transition matrix*.

It is easy to see that the column sum in this matrix always must have the value 1. A Markov chain can, however, only be fully appreciated when the time behaviour is considered.

The times should have the distance (the time interval) Δt, *i.e.*

$$t_k = k \cdot \Delta t \tag{8.18}$$

with $k = 0, 1, 2, \ldots$

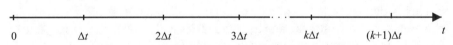

Figure 8.10. Times for the Markov chain

The transition probabilities are now easy to interpret. This is explained using the example for p_{21}: p_{21} is the conditional probability that the technological process at time $t_{k+1} = (k + 1)\Delta t$ can assume the state x_2 under the condition that in the time $t_k = k\Delta t$ (*i.e.* an earlier time) it was in the state x_1. In other words: p_{21} is the failure probability based on the time interval Δt.

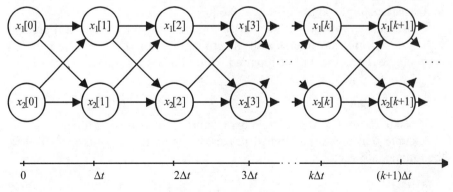

Figure 8.11. Markov chain

If the previous relationships are brought into a single figure, this gives the optical picture of a chain. Such a Markov chain is called homogeneous when the transition probabilities do not change over time, *i.e.* when the transition matrix P is constant or independent of k. The mathematical modelling still requires the introduction of state probabilities.

$p_1(k \cdot \Delta t) = p_1[k]$ probability that the technological process in the time $k \cdot \Delta t$ assumes the state x_1.

$p_2(k \cdot \Delta t) = p_2[k]$ probability that the technological process in the time $k \cdot \Delta t$ assumes the state x_2.

By combining both probabilities together to produce a vector

$$\mathbf{p}(k \cdot \Delta t) = \mathbf{p}[k] = \begin{pmatrix} p_1(k \cdot \Delta t) \\ p_2(k \cdot \Delta t) \end{pmatrix} = \begin{pmatrix} p_1[k] \\ p_2[k] \end{pmatrix} \tag{8.19}$$

the Markov chain equation can be formulated, where the short notation for the times are now used.

$$\mathbf{p}[k+1] = \mathbf{P} \cdot \mathbf{p}[k] \tag{8.20}$$

This equation can be used to calculate a technological process with regard to its chronological behaviour provided the transition matrix P and the initial state vector $p[0]$ are given.

This is demonstrated with an example. A technological process with the two states x_1 = operational and x_2 = failed is analysed in the separation of $\Delta t = 1$ h. The following values are calculated for this time interval

$p_{21} = \lambda$ failure rate

$p_{12} = \mu$ renewal rate

This produces the transition matrix

$$\mathbf{P} = \begin{pmatrix} 1-\lambda & \mu \\ \lambda & 1-\mu \end{pmatrix} \tag{8.21}$$

numerical example: $\lambda = 0.2$ h^{-1} $\mu = 0.6$ h^{-1}

Thus, with a probability of $\lambda = 20\%$, an operational technological process will fail within an hour; with a probability of $\mu = 60\%$, a failed technological process will be repaired within an hour.

$$\mathbf{P} = \begin{pmatrix} 0.8 & 0.6 \\ 0.2 & 0.4 \end{pmatrix}$$

These probabilities should not change during the course of time.

Assume that the technological process is operational at time $t = 0$, then the initial state vector is

$$\mathbf{p}[0] = \begin{pmatrix} 1 \\ 0 \end{pmatrix}$$

The probability that the technological process at time $t = 0$ has the state x_1 is 1.

Equation (8.20) is now used to calculate

$$\mathbf{p}[1] = \mathbf{P} \cdot \mathbf{p}[0] = \begin{pmatrix} 0.8 & 0.6 \\ 0.2 & 0.4 \end{pmatrix} \cdot \begin{pmatrix} 1 \\ 0 \end{pmatrix} = \begin{pmatrix} 0.8 \\ 0.2 \end{pmatrix}$$

$$\mathbf{p}[2] = \mathbf{P} \cdot \mathbf{p}[1] = \begin{pmatrix} 0.8 & 0.6 \\ 0.2 & 0.4 \end{pmatrix} \cdot \begin{pmatrix} 0.8 \\ 0.2 \end{pmatrix} = \begin{pmatrix} 0.76 \\ 0.24 \end{pmatrix}$$

$$\mathbf{p}[3] = \mathbf{P} \cdot \mathbf{p}[2] = \begin{pmatrix} 0.8 & 0.6 \\ 0.2 & 0.4 \end{pmatrix} \cdot \begin{pmatrix} 0.76 \\ 0.24 \end{pmatrix} = \begin{pmatrix} 0.752 \\ 0.248 \end{pmatrix}$$

etc. (see Figure 8.12).

Thus, Equation (8.20) allows the so-called *transient response* to be easily calculated.

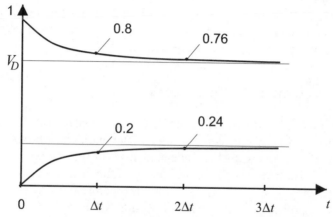

Figure 8.12. Markov chain of a technological process for $\lambda = 0.2$ and $\mu = 0.6$

For large times, *i.e.* for large values of k, the state vector apparently no longer changes and the solution independent of k

$$\lim_{k \to \infty} \mathbf{p}[k] = \mathbf{p}_{st} \tag{8.22}$$

is also called the stationary solution p_{st}. Using Equation (8.20), Equation (8.22) produces the calculation equation for the stationary solution

$$\mathbf{p}_{st} = \mathbf{P} \cdot \mathbf{p}_{st} \tag{8.23}$$

with

$$\sum_{n=1}^{N} p_{nst} = 1$$

In the example, using Equation (8.21) yields

$$\begin{pmatrix} p_{1st} \\ p_{2st} \end{pmatrix} = \begin{pmatrix} 1-\lambda & \mu \\ \lambda & 1-\mu \end{pmatrix} \cdot \begin{pmatrix} p_{1st} \\ p_{2st} \end{pmatrix} \qquad (8.24)$$

$$p_{1st} + p_{2st} = 1$$

$$p_{1st} = \frac{\mu}{\lambda + \mu} = V_D$$

$$p_{2st} = \frac{\lambda}{\lambda + \mu}$$

With numbers: $p_{1st} = 0.75$; $p_{2st} = 0.25$

p_{1st} is also called the *stationary availability* or *asymptotic availability*, often also shortened to availability ($p_{1st} = V_D$). In the numerical example, the availability of the technological process is $V_D = 0.75\%$.

8.5.2 Two Technological Processes

The previous considerations can also be easily transferred to a manufacturing system that consists of two technological processes. Each technological process can fail independently of the other process and similarly can also be renewed independently of the other process.

The following systematic is used for the operational process events (f) and the failed process (a) introduced in Section 8.5.1:

- Technological process 1 is operational $\triangleq$ event f_1
- Technological process 1 is failed $\triangleq$ event a_1
- Technological process 2 is operational $\triangleq$ event f_2
- Technological process 2 is failed $\triangleq$ event a_2.

The manufacturing system has 4 states:

$$x_1 \triangleq f_1 \cap f_2 \text{ (both processes operational)} \qquad (8.25)$$

$$x_2 \triangleq a_1 \cap f_2 \text{ (process 1 failed, process 2 operational)}$$

$$x_3 \triangleq f_1 \cap a_2 \text{ (process 1 operational, process 2 failed)}$$

$$x_4 \triangleq a_1 \cap a_2 \text{ (both processes failed)}$$

($\cap$ means "and" for the events and a multiplication of the probabilities.)

Using the two transition matrixes of the individual technological processes

$$\mathbf{P}_1 = \begin{pmatrix} 1-\lambda_1 & \mu_1 \\ \lambda_1 & 1-\mu_1 \end{pmatrix} \text{ and } \mathbf{P}_2 = \begin{pmatrix} 1-\lambda_2 & \mu_2 \\ \lambda_2 & 1-\mu_2 \end{pmatrix} \tag{8.26}$$

the transition matrix of the manufacturing system that consists of 16 transition probabilities produces

$$\mathbf{P} = \begin{pmatrix} \begin{pmatrix} 1-\lambda_1 & \mu_1 \\ \lambda_1 & 1-\mu_1 \end{pmatrix} \cdot (1-\lambda_2) & \begin{pmatrix} 1-\lambda_1 & \mu_1 \\ \lambda_1 & 1-\mu_1 \end{pmatrix} \cdot \mu_2 \\ \\ \begin{pmatrix} 1-\lambda_1 & \mu_1 \\ \lambda_1 & 1-\mu_1 \end{pmatrix} \cdot \lambda_2 & \begin{pmatrix} 1-\lambda_1 & \mu_1 \\ \lambda_1 & 1-\mu_1 \end{pmatrix} \cdot (1-\mu_2) \end{pmatrix} \tag{8.27}$$

For the stationary solution, this produces

$$p_{1st} = \frac{\mu_1}{\lambda_1 + \mu_1} \cdot \frac{\mu_2}{\lambda_2 + \mu_2} \quad ; \quad p_{2st} = \frac{\lambda_1}{\lambda_1 + \mu_1} \cdot \frac{\mu_2}{\lambda_2 + \mu_2} \tag{8.28}$$

$$p_{3st} = \frac{\mu_1}{\lambda_1 + \mu_1} \cdot \frac{\lambda_2}{\lambda_2 + \mu_2} \quad ; \quad p_{4st} = \frac{\lambda_1}{\lambda_1 + \mu_1} \cdot \frac{\lambda_2}{\lambda_2 + \mu_2}$$

Both the matrix $\mathbf{P}$ and the stationary solution $\mathbf{p}_{st}$ with regard to the state vector $\mathbf{x}$ have a systematic structure that can be used as a basis for manufacturing systems with more than two technological processes.

Thus, manufacturing systems with an independent state behaviour of the individual technological processes that each can only assume two states are relatively easy to model using Markov chains. For example, the product rule for the availability applies here.

$$V_{Dsum} = V_{D1} \cdot V_{D2} \cdot V_{D3} \cdot ... \tag{8.29}$$

8.5.3 Technological Processes with More than Two States

A technological process with two states is obviously the simplest possible case. For practical purposes, often more than two states must be added to the Markov chain. Sometimes, not all state transitions are possible, *i.e.* the transition matrix has a zero at the appropriate places.

Otherwise Equations (8.20) and (8.23) as specified apply.

8.5.4 Markov Chains with Infinitely Many States

Sometimes it is mathematically easier to assume infinitely many states for a process, although this obviously violates the practical conditions.

The example of a buffer storage demonstrates this [2].

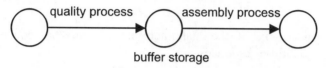

quality process   assembly process

buffer storage

Figure 8.13. Markov chain for buffer storage (example)

Example: At the end of a test and repair process (quality process), electronic modules are subjected to an additional assembly process. The test and repair process is stochastic with regard to time; the assembly process is deterministic with duration Δt. A buffer storage is needed between quality process and assembly process. There is an upper limit for the number of modules contained in the buffer storage. The maximum number of modules that the quality process supplies to the buffer storage is 2. At each time $t_k = k \cdot \Delta t$ a module is taken from the buffer storage (except for when the buffer storage is empty). The state x_n of the buffer storage corresponds to the number of modules in the buffer storage ($n = 0, 1, 2, \ldots$). The modelling of the buffer storage as a Markov chain is made possible by the following details: the probabilities q_m, with which exactly m modules leave the quality process in a time interval Δt and are supplied to the buffer storage is analysed. m should be maximum 2 here, *i.e.* the following values are required:

Number of supplied modules m	0	1	2
Probability q_m	q_0	q_1	q_2

Numerical example:

m	0	1	2
q_m	0.4	0.4	0.2

This allows the transition matrix to be calculated

$$
\mathbf{P} = \begin{pmatrix}
0.8 & 0.4 & 0 & 0 & \cdots \\
0.2 & 0.4 & 0.4 & 0 & \\
0 & 0.2 & 0.4 & 0.4 & \\
0 & 0 & 0.2 & 0.4 & \\
0 & 0 & 0 & 0.2 & \cdots \\
\vdots & & \vdots & & \ddots
\end{pmatrix}
$$

with the stationary solution

$$\mathbf{p}_{st} = \begin{pmatrix} \frac{1}{2} \\ \frac{1}{4} \\ \frac{1}{8} \\ \frac{1}{16} \\ \vdots \end{pmatrix}$$

The average number of modules contained in the buffer storage is $\bar{n} = 1$, the average number modules supplied to the buffer storage is $\bar{m} = 0.8$.

8.6 Multidimensional Process and Test Characteristics

8.6.1 Two-dimensional Process Characteristic

A technological process T is characterised by two process characteristic values X and Y with the two-dimensional process characteristic $f(x,y)$. (Using the example of the production of chip resistors, sometimes not only the resistance value but also the temperature coefficients are of interest.) In this section, X and Y are generally used to avoid increasing the number of designations unnecessarily. As for a one-dimensional process characteristic (Section 1.4), the methods from probability theory are used for further analysis. X and Y are random values. $f(x,y)$ is thus a two-dimensional probability density with

$$\iint f(x,y)\,dx\,dy = 1 \qquad\qquad (8.30)$$

If not specified otherwise, the integration limits are always $-\infty$ and $+\infty$. The marginal density

$$f_R(x) = \int f(x,y)\,dy \qquad\qquad (8.31)$$

is the usual (*one-dimensional*) *process characteristic* in the previous sense and so can be considered as if the process characteristic value Y has not been investigated. The marginal density means that a relationship to the size Y also cannot be determined.

The $f_R(x)$ function can be used with the usual equations to calculate the expected value μ_x and the dispersion σ_x^2; the specification of tolerances also allows the yield or the defect rate to be determined.

The same is also the case for the second marginal density with regard to the process characteristic value Y.

$$f_R(y) = \int f(x,y)\,dx \qquad (8.32)$$

The reducing of the two-dimensional process characteristic $f(x,y)$ to two one-dimensional densities $f_R(x)$ and $f_R(y)$, however, does not yet reflect the complete complexity of a two-dimensional process characteristic value that is best written as a vector,

$$\begin{pmatrix} X \\ Y \end{pmatrix}$$

Namely, the analysis of the relationships between X and Y is missing. This can be most simply described using the covariance σ_{xy}.

$$\sigma_{xy} = \iint (x - \mu_x)\cdot(y - \mu_y)\cdot f(x,y)\,dx\,dy \qquad (8.33)$$
$$= \iint x\cdot y\cdot f(x,y)\,dx\,dy - \mu_x \cdot \mu_y$$

The measure for this relationship "normalised" to the standard deviation σ_x and σ_y is the correlation coefficient

$$\rho = \frac{\sigma_{xy}}{\sigma_x \cdot \sigma_y} \qquad (8.34)$$

which is a dimensionless value between -1 and $+1$. Where $\rho = 0$ means no relationship ($\sigma_{xy} = 0$) exists between X and Y. In this case

$$f(x,y) = f_R(x)\cdot f_R(y) \qquad (8.35)$$

If the process characteristic is a *two-dimensional normal distribution*, then it is completely described by the two expected values and deviations, and the correlation coefficient.

$$f(x,y) = \frac{1}{2\pi\sigma_x\sigma_y\cdot\sqrt{1-\rho^2}}\cdot\ldots \qquad (8.36)$$

$$\ldots\cdot e^{-\frac{1}{2(1-\rho^2)}\left[\left(\frac{x-\mu_x}{\sigma_x}\right)^2 - 2\rho\cdot\left(\frac{x-\mu_x}{\sigma_x}\right)\left(\frac{y-\mu_y}{\sigma_y}\right) + \left(\frac{y-\mu_y}{\sigma_y}\right)^2\right]}$$

Many technological processes have a normal distribution so that the two-dimensional normal distribution can be used both for forming the theory and also for the empirical calculations.

8.6.2 Process Equation

In practice, we would like to have an equation that describes the relationship between X and Y. This equation is obtained using the so-called *regression*

$$\mu_y(x) = \frac{\int y \cdot f(x,y)\,dy}{\int f(x,y)\,dy} \tag{8.37}$$

This means: the expected value for Y is defined as the process equation depending on x. This equation is a straight line for the two-dimensional normal distribution.

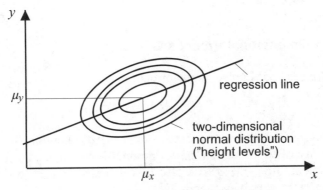

Figure 8.14. Two-dimensional normal distribution

$$\mu_y(x) = \mu_y + \beta \cdot (x - \mu_x) \tag{8.38}$$

with the gradient β

$$\beta = \frac{\sigma_{xy}}{\sigma_x^2} = \rho \cdot \frac{\sigma_y}{\sigma_x} \tag{8.39}$$

This regression line can also be specified empirically as an estimate using experiments:

$$y = \bar{y} + b \cdot (x - \bar{x}) = a + b \cdot x \tag{8.40}$$

with the gradient

$$b = \frac{S_{xy}}{S_x^2} = r \cdot \frac{S_y}{S_x} \qquad (8.41)$$

$$a = \bar{y} - b \cdot \bar{x}$$

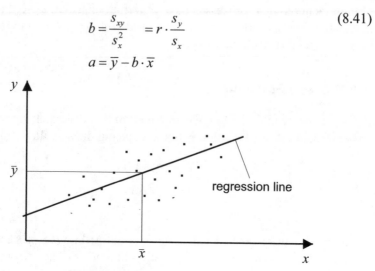

Figure 8.15. Empirical regression line

8.6.3 Two-dimensional Process Operator

The two-dimensional Laplace transformation

$$\iint f(x,y) \cdot e^{-px} \cdot e^{-qy} \, dx \, dy = F(p,q) \qquad (8.42)$$

can be used to produce the two-dimensional $F(p,q)$ operator. Similar to the procedure described in Section 2.3 for product-flow graphs, the 1st- and 2nd-degree moments from $F(p,q)$ can also be calculated here by differentiation and by setting the variables to zero, *e.g.*,

$$\mu_x = -\frac{\partial F(p,q)}{\partial p}\bigg|_{\substack{p=0 \\ q=0}} \qquad (8.43)$$

Because of the similarity to the described procedure, no further equations need to be described here.

8.6.4 Chain Structure of Two Technological Processes

For two technological processes in a chain it gives in the operator range

$$H(p,q) = F(p,q) \cdot G(p,q) \qquad (8.44)$$

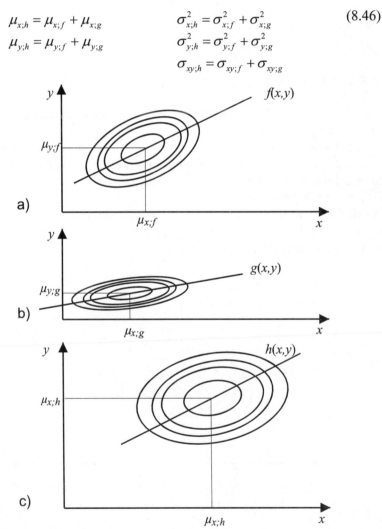

Figure 8.16. Chain process (two-dimensional)

The two-dimensional convolution is also true in the original range.

$$\mu(x,y) = f(x,y) * g(x,y) \tag{8.45}$$

It is easy to show that the 1st- and 2nd-degree moments can be added, namely

$$\mu_{x;h} = \mu_{x;f} + \mu_{x;g} \qquad\qquad \sigma^2_{x;h} = \sigma^2_{x;f} + \sigma^2_{x;g} \tag{8.46}$$

$$\mu_{y;h} = \mu_{y;f} + \mu_{y;g} \qquad\qquad \sigma^2_{y;h} = \sigma^2_{y;f} + \sigma^2_{y;g}$$

$$\sigma_{xy;h} = \sigma_{xy;f} + \sigma_{xy;g}$$

Figure 8.17. Process equations for two normally distributed processes in a chain structure (**a**) process 1, (**b**) process 2, (**c**) complete process

If both processes have a normal distribution, the complete process also has a normal distribution and gives for the process equation:

$$\mu_{y;h}(x) = \mu_{y;h} + \beta_h \cdot (x - \mu_{x;h}) \tag{8.47}$$

with

$$\mu_{x;h} = \mu_{x;f} + \mu_{x;g} \quad ; \quad \mu_{y;h} = \mu_{y;f} + \mu_{y;g} \tag{8.48}$$

$$\beta_h = \frac{\sigma_{xy;f} + \sigma_{xy;g}}{\sigma_{x;f}^2 + \sigma_{x;g}^2} \tag{8.49}$$

8.6.5 Measured Process Characteristic

If $f(x,y)$ and $g(x,y)$ are interpreted as a process characteristic and a measurement characteristic, respectively, then $h(x,y)$ is the measured process characteristic. This allows all characteristic values from Section 1.4 also to be specified for the two-dimensional case, in particular: acceptance and rejection rates, test blurring type 1 and 2, and acceptance and rejection average outgoing quantity.

Important here is obviously the two-dimensional tolerance range. If both tolerance ranges are mutually independent, both together form a rectangle. Otherwise complicated calculations result.

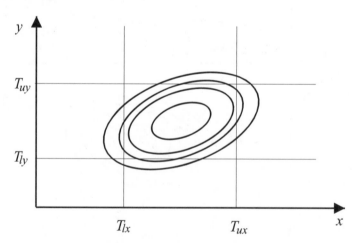

Figure 8.18. Two-dimensional tolerance range

8.6.6 Measured Process Equation

The technological process with the true (theoretical) process characteristic $f(x,y)$ has the true (theoretical) process equation $\mu_{y;f}(x)$. The measurement process with the measurement characteristic $g(x,y)$ and the measurement equation $\mu_{y;g}(x)$ produces the measured process characteristic $h(x,y)$ with the measured process equation $\mu_{y;h}(x)$ as

$$\mu_{y;h}(x) = \frac{\left[\mu_{y;f}(x) \cdot f_R(x)\right] * g_R(x) + \left[\mu_{y;g}(x) \cdot g_R(x)\right] * f_R(x)}{f_R(x) * g_R(x)} \quad (8.50)$$

where $f_R(x)$ and $g_R(x)$ are the marginal densities (see Equation (8.32)). The proof for Equation (8.50) can easily be provided over the operator range. If the technological process and the measurement process have a normal distribution, Equation (8.50) yields the linear Equation (8.52). Because the quantities x and y represent errors of dimension for the measurement process (see Section 1.5), this gives for the noncorrelated measurement

$$\sigma_{xy;g} = 0$$

and for the unbiased measurement

$$\mu_{x;g} = 0 \quad \text{and} \quad \mu_{y;g} = 0.$$

This gives from Equation (8.52)

$$\mu_{y;h} = \mu_{y;f} + \frac{\sigma_{xy;f}}{\sigma_{x;f}^2 + \sigma_{x;g}^2}\left(x - \mu_{x;f}\right) \quad (8.51)$$

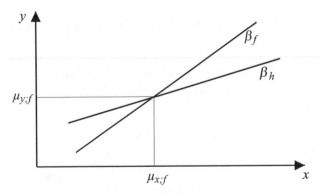

Figure 8.19. True and measured process equation

A stochastic measurement error $\sigma_{x;g}$ also always leads to the reduction of the size of the measured gradient β_h compared with the theoretical (true) gradient β_f, i.e.,

$$\beta_h = \beta_f \cdot \eta \tag{8.52}$$

with

$$\eta = \frac{1}{1 + \left(\dfrac{\sigma_{x;g}}{\sigma_{x;f}}\right)^2} < 1$$

8.7 Reliability of Technological Processes

The reliability methods, characteristics and models are also very powerful tools for the analysis of technological processes. The starting point for the modelling is the process characteristic $f(x)$ from which the tolerance limits can be used to calculate the defect rate p.

For some time, the defect rate has been regarded as the *procedure reliability*. It is, however, more exact to consider the defect rate p as a *characteristic value of the process reliability*. After all, a technological procedure (*e.g.*, soldering) does not have a defect rate as such, but rather always only together with the electronic product and the fault criteria. In the following discussion, in addition to the defect rate, other characteristic values are also described, such as those that result from the drift of the technological processes, but also the reliability characteristic values of the inspection processes also are very important in process technology.

8.7.1 Process Defect Rate for Several Quality Characteristics

In practice, the quality of electronic products generally depends on several characteristics x. The operating point x_A of the technological process is then also multidimensional and should be included in the process characteristic, *i.e.*

$$f(x) = f(x \,|\, x_A) \tag{8.53}$$

In other words: the form of the process characteristic depends on the set operating point.

Each quality characteristic has a tolerance range, where it is also possible that for a given quality characteristic, the tolerance range can depend on the value of another quantity. Consequently, a tolerance range G_T should normally be assumed.

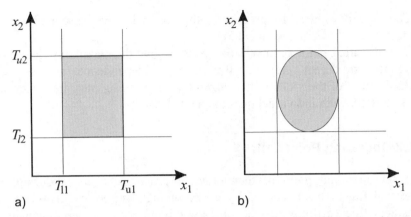

Figure 8.20. Independent (**a**) and dependent (**b**) two-dimensional tolerance range

Figure 8.20 shows this case of a tolerance range for two quality characteristics X_1 and X_2. The rectangle means that the tolerance ranges are mutually independent. In contrast, the elliptical tolerance range is a typical case of the tolerance limits for a quality characteristic that depend on the value of a different characteristic.

The effects of the dependency of the quality characteristics amongst themselves have already been described in Section 8.6, for example, the covariance. This must now be generalised and transferred to all characteristics. This section, however, primarily considers the calculation or the determination of the defect rate p as the characteristic value of the process reliability.

If the process characteristic $f(x \mid x_A)$ and the tolerance range G_T are known, the yield y can then be calculated initially

$$y = \underbrace{\int \cdots \int f\left(\mathbf{x} \mid \mathbf{x}_A\right) d\mathbf{x}}_{G_T} \tag{8.54}$$

This gives the defect rate $p = 1 - y$. If all quality characteristics are mutually independent (and also the tolerance ranges), the multiple integral Equation (8.54) can be separated into single integrals that each represent their own yield y_n with regard to the quality characteristic x_n and must be multiplied together.

$$y_{sum} = y_1 \cdot y_2 \cdots y_n \cdots y_N \tag{8.55}$$

with

$$y_n = \int_{T_{ln}}^{T_{un}} f\left(x_n \mid x_{An}\right) dx_n \tag{8.56}$$

The overall yield is the product of the yields for the associated quality characteristic (given independence).

The operating point x_A must now be set to give the maximum yield y (or the minimum defect rate). For the case of independence (and thus the multiplication of the yields), this shows that this setting must be performed separately for each individual quality characteristic.

8.7.2 Operating Point Drift

The set operating point of technological processes changes during the course of time. This is caused by wear, ageing, fatigue, *etc.* This section considers this "drifting" of the operating point using a one-dimensional case.

A further restriction is that the chronological operating point change should be linear, which is also certainly the case for many processes and for a large time interval after making the setting. The linear operating point drift can be represented by three different coefficients:

- drift time constant τ_D
- drift rate λ_D
- drift constant b_D.

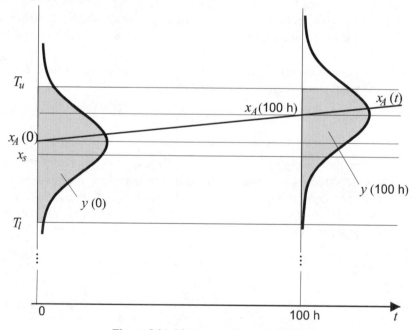

Figure 8.21. Linear operating point drift

Drift Time Constant τ_D

The linear change of the operating point is described by the following equation (linear drift equation)

$$x_A(t) = x_A(0) \cdot \left(1 + \frac{t}{\tau_D}\right) \tag{8.57}$$

The drift time constant has hours (h) as dimension. The value of the drift time constants is in general very high, because it means that after a time duration $t = \Delta t = \tau_D$, the operating point would be twice as large as at the beginning. An example is shown to demonstrate this behaviour.

The manufacturing of ohmic resistors using the thick-film procedure with the setpoint $x_s = 1{,}000\ \Omega$ and a relative tolerance range of $\pm 5\%$ ($T_u = 950\ \Omega$, $T_o = 1{,}050\ \Omega$) is realised by a normally distributed process characteristic with $\mu = 1{,}010\ \Omega$ and $\sigma = 25\ \Omega$. The manufacturing process has a linear operating point drift with $\tau_D = 5{,}000$ h.

Thus, $x_A(0) = 1{,}010\ \Omega$ and the initial defect rate is $p(0)$

$$p(0) = \Phi(\tau + \varepsilon) + \Phi(\tau - \varepsilon) + 1 = 6.3\%$$

$(\tau = 2.0; \quad \varepsilon = 0.4)$

The effect of the drift is to be calculated after $\Delta t = 100$ h. The operating point then has the value

$$x_A(100\,\mathrm{h}) = x_A(0) \cdot \left(1 + \frac{100}{5{,}000}\right) = 1.02 \cdot x_A(0) = 1{,}030.2\ \Omega$$

Thus, the yield after 100 h reduces to the value
$y(100\ \mathrm{h}) = 78.6\ \%$, namely $p(100\ \mathrm{h}) = 21.4\%$
$(\tau = 2.0; \quad \varepsilon = 1.2)$.

Drift Rate λ_D

The drift rate is defined by

$$\lambda_D = \frac{1}{\tau_D} \tag{8.58}$$

The drift equation is

$$x_A(t) = x_A(0) \cdot (1 + \lambda_D \cdot t) \tag{8.59}$$

where λ_D specifies the value by which the operating point changes per hour relative to the set operating point.

For the numerical example, this gives
$$\lambda_D = 2 \times 10^{-4}\,\mathrm{h}^{-1}$$

In other words: each hour the operating point increases by 0.2‰, *i.e.* the operating point is 2% larger after 100 h. This makes the drift rate λ_D a quantity that is easy to interpret.

It is also possible to consider a time dependency of the drift rate $\lambda_D(t)$ in Equation (8.59) for the case that a change of the drift value is observed after some time.

Drift Constant b_D

The drift constant b_D is defined as the gradient of the drift equation, *i.e.* it specifies the absolute change of the operating point per hour

$$b_D = \frac{dx_A(t)}{dt} = x_A(0) \cdot \lambda_D \qquad (8.60)$$

b_D thus has the quality characteristic value per hour as dimension. The drift equation is

$$x_A(t) = x_A(0) + b_D \cdot t \qquad (8.61)$$

In the described case

$$b_D = 1{,}010\,\Omega \cdot 2 \times 10^{-4}\,\mathrm{h}^{-1} = 0.202\,\Omega \cdot \mathrm{h}^{-1}$$

In other words, this manufacturing process for ohmic resistors is characterised by the operating point increasing linearly by 202 mΩ per hour.

Characteristic values for the drift of the standard deviation σ also exist, because the standard deviation can change over time, in general it increases. The procedure is similar to the previously described method.

8.7.3 Momentary Yield

Because the operating point can be considered as being time dependent, the process characteristic that results from Equation (8.61)

$$y(t) = \underbrace{\int \cdots \int f(\mathbf{x} \mid \mathbf{x}_A(t))\, d\mathbf{x}}_{G_T} \qquad (8.62)$$

$$= \underbrace{\int \cdots \int f(\mathbf{x}, \mathbf{x}_A(0), t)\, d\mathbf{x}}_{G_T}$$

is also time dependent. For the one-dimensional case, the momentary yield is demonstrated using a normally distributed process characteristic with a linear operating point drift.

Equation (1.66) together with Equations (1.62), (1.52) and (1.50) gives

$$y = \Phi(\tau + \varepsilon) + \Phi(\tau - \varepsilon) - 1$$

$$\tau = \frac{\Delta T}{\sigma} \quad ; \quad \varepsilon = \frac{\Delta \mu}{\sigma}$$

Because $\Delta\mu$ is the deviation of the operating point x_A from setpoint x_s, it follows here that this deviation is time dependent, *i.e.*

$$\Delta\mu(t) = x_A(t) - x_s$$

and with Equation (8.61) this gives

$$\Delta\mu(t) = x_A(0) - x_s + b_D \cdot \tau \qquad (8.63)$$

This eccentricity (Section 1.4.3) is also time dependent.

$$\varepsilon(t) = \varepsilon(0) + \beta_D \cdot t \qquad (8.64)$$

with the specific drift constant β_D

$$\beta_D = \frac{b_D}{\sigma}$$

The momentary yield $y(t)$ is then

$$y(t) = \Phi(\tau + \varepsilon(0) + \beta_D \cdot t) + \Phi(\tau - \varepsilon(0) - \beta_D \cdot t) - 1 \qquad (8.65)$$

with

$$\tau = \frac{\Delta t}{\sigma} = \frac{T_o - T_u}{2\sigma}$$

$$\varepsilon(0) = \frac{x_A(0) - x_s}{\sigma}$$

$$\beta_D = \frac{b_D}{\sigma}$$

For the resistor example

$(x_s = 1{,}000\ \Omega,\ x_A(0) = 1{,}010\ \Omega,\ \Delta T = 50\ \Omega,\ \sigma = 25\ \Omega,\ b_D = 202\ \text{m}\Omega\cdot\text{h}^{-1})$, this yields $\tau = 2$, $\varepsilon(0) = 0.4$ and $\beta_D = 8.08 \times 10^{-3}$.

For $t = \Delta t = 100$ h,
this yields $y\ (100\ \text{h}) = \Phi\ (3.28) + \Phi\ (0.792) - 1 = 78.6\%$
and thus $p(100\ \text{h}) = 21.4\%$.

For comparison, the values at the start of the manufacturing process:

$y\ (0) = \Phi\ (2.4) + \Phi\ (1.6) - 1 = 93.7\%,\ p(0) = 6.3\%$.

If no setting error was present, *i.e.* $x_A(0)$ was set to $1{,}000\ \Omega$, this would produce the optimum values, maximum yield $y^* = 2 \cdot \Phi\ (2.0) - 1 = 95.45\%$ and minimum defect rate $p^* = 4.55\%$.

8.7.4 Test Blurring

The test blurring type 1 (α) and type 2 (β) have already been introduced in Sections 1.5 and 6.3. Both characteristic values indicate the *reliability of the inspection processes* and must be analysed in electronics process technology.

The quantities α and β can often assume high values for many inspection processes (*e.g.*, the X-ray inspection of solder points) and for the measuring inspection (*e.g.*, the quality characteristic values for electronic components and mechanical parts).

The following discussion shows how α and β can be calculated for the measurement. This requires that three quantities are known:

- process characteristic $f(x)$
- tolerance limits T_l and T_u
- measurement characteristic $g(x')$

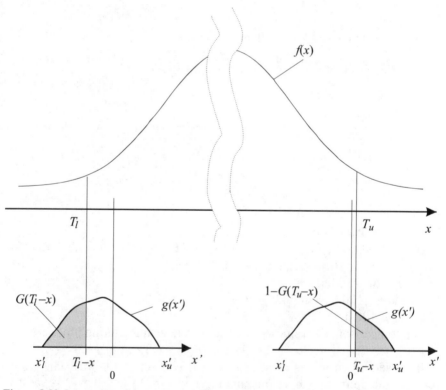

Figure 8.22. Interaction of the process and the measurement characteristic for the calculation of α for measurement near the lower or upper tolerance limit

Calculation of the test blurring type 1 (α)
For products whose quality parameter x lies in the two ranges

$$T_l \le x \le T_l - x_l' \qquad \left(x_l' < 0 \right)$$

and

$$T_u - x_u' \le x \le T_u$$

an associated probability can be specified with which it will be rejected (using the measurement characteristic) although they are quality-conform, because the previously mentioned values lie within the tolerance range (see Figure 8.22). These ranges are called the *indecision ranges*.

If the value from the process characteristic $f(x)$ is used and the integration made over the previously mentioned ranges, this gives α when this integral applies to the yield, *i.e.*

$$\alpha = \frac{1}{y} \left(\int_{T_l}^{T_l - x_l'} f(x) \cdot G(T_l - x)\, dx + \int_{T_u - x_u'}^{T_u} f(x) \cdot \left[1 - G(T_u - x) \right] dx \right) \qquad (8.66)$$

with

$$y = \int_{T_l}^{T_u} f(x)\, dx \;\; = 1 - p \qquad (8.67)$$

and

$$G(T_l - x) = \int_{x_l}^{T_l - x} g(x')\, dx' \qquad (8.68)$$

$$G(T_u - x) = \int_{x_u}^{T_u - x} g(x')\, dx'$$

Calculation of the test blurring type 2 (β)
The test blurring type 2 (β) is calculated in the same way. The products lie within the two ranges (indecision ranges)

$$T_l - x_u' \le x \le T_l$$

and

$$T_u \le x \le T_u - x_l' \qquad \left(x_l' < 0 \right)$$

i.e. outside the tolerance range!
The probability for an acceptance is given because of the measurement characteristic. This produces

$$\beta = \frac{1}{p}\left(\int_{T_l - x_u'}^{T_l} f(x) \cdot \left[1 - G(T_l - x)\right] dx + \int_{T_u}^{T_u - x_l'} f(x) \cdot G(T_u - x)\, dx \right) \qquad (8.69)$$

Two very simple examples are used to demonstrate these relationships, where the measurement characteristic should be unbiased and symmetric.

Example 1:   Process characteristic with uniform distribution, measurement characteristic with uniform distribution.

This is the simplest of all imaginable possibilities.

The yield is calculated as

$$y = \frac{T_u - T_l}{x_u - x_l}$$

and produces

$$\alpha = \frac{1}{2} \frac{\tau}{T_u - T_l}$$

$$\beta = \frac{1}{2} \frac{\tau}{(x_u - T_u) + (T_l - x_l)}$$

These results are very easy to interpret:

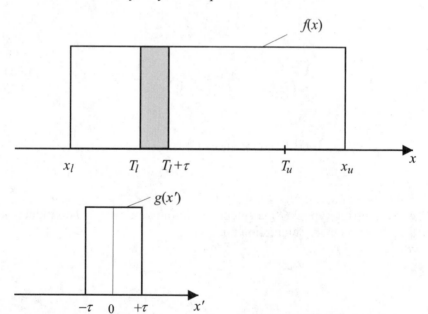

Figure 8.23. Process characteristic with uniform distribution; measurement characteristic with uniform distribution

First interpretation:

A product with the probability p_τ (corresponds to the area with grey background in Figure 8.23) lies within the $T_l \leq x \leq T_l + \tau$ range

$$p_\tau = \frac{\tau}{x_u - x_l}$$

It can now be shown that such a product (because of the measurement characteristic) will be rejected with ¼ p_τ probability even though it lies within the tolerance range.

The symmetry means that corresponding conclusions can be made for all ranges with width τ above and below the tolerance limits.

Second interpretation:

The generalisation of this conclusion leads to the following statement:

An arbitrary process characteristic in the vicinity of the lower tolerance limit has an (approximately) constant form with height $f(T_u)$. The uniform distributed measurement characteristic produces an indecision range with width τ. Then

$$p_\tau = \tau \cdot f(T_l)$$

is the probability that the true value of the quality characteristic value of a product lies within this range.

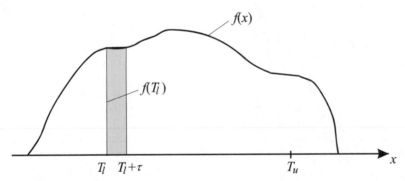

Figure 8.24. Process characteristic with approximately constant behaviour at the tolerance limits

The measurement characteristic means that such a product will be rejected with the probability

$$\alpha_l' = \frac{1}{4} p_\tau$$

even though it lies within the tolerance range.

Because a similar behaviour is also present at the upper tolerance limit, the following equations can be generally used for an arbitrary process characteristic $f(x)$ with a uniformly distributed measurement characteristic

$$\alpha = \frac{1}{4} \cdot \tau \cdot \frac{f(T_l) + f(T_u)}{\int_{T_l}^{T_u} f(x)\,dx}$$

$$\beta = \frac{1}{4} \cdot \tau \cdot \frac{f(T_l) + f(T_u)}{1 - \int_{T_l}^{T_u} f(x)\,dx}$$

Prerequisites here are the conditions

$$f(T_l + \tau) = f(T_l - \tau) = f(T_l)$$
$$f(T_u + \tau) = f(T_u - \tau) = f(T_u)$$

which may, however, only be satisfied approximately.

Because a constant behaviour cannot necessarily be assumed at the tolerance limits, a constant *gradient* should be assumed. This is illustrated with the next example.

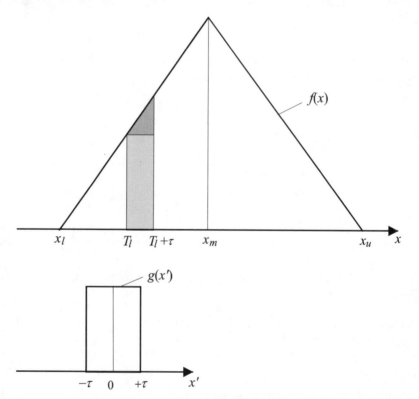

Figure 8.25. Process characteristic with a triangular distribution; measurement characteristic with uniform distribution

Example 2:   Process characteristic with a triangular distribution, measurement characteristic with uniform distribution. This case is shown in Figure 8.25.

Indecision ranges with width τ at the two tolerance limits also occur here.

Again, initially only the lower tolerance limit T_l is considered. The area shown with grey background in Figure 8.25 is the probability p_τ that the value of the quality characteristic value of a product lies within the range between T_l and $T_l + \tau$. The area is the sum of the rectangle area and the triangle area, *i.e.*

$$p_\tau = f(T_l) \cdot \tau + \frac{1}{2} f'(T_l) \cdot \tau^2$$

in the shown case is

$$f(T_l) = 4 \cdot \frac{T_l - x_l}{(x_u - x_l)^2} = \frac{T_l - x_l}{(\Delta T)^2} \quad ; \quad f'(T_l) = \frac{4}{(x_u - x_l)^2} = \frac{1}{(\Delta T)^2}$$

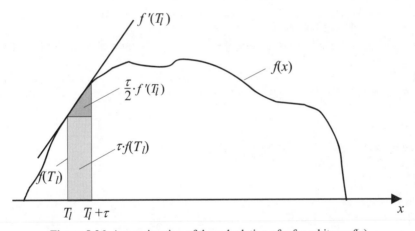

Figure 8.26. Approximation of the calculation of α for arbitrary $f(x)$

The uniform distribution of the measurement characteristic means 1/4 and 1/6 of the products will now be rejected in accordance with the rectangle area and the triangle area, respectively, although they actually lie within the tolerance range. Thus, α is calculated as

$$\alpha = \frac{\tau \cdot \left[f(T_l) + f(T_u) \right] + \frac{\tau^2}{3} \left[f'(T_l) + | f'(T_u) | \right]}{4 \cdot \int_{T_l}^{T_u} f(x) \, dx}$$

β is also calculated as

$$\beta = \frac{\tau \cdot \left[f(T_l) + f(T_u) \right] - \frac{\tau^2}{3} \left[f'(T_l) + | f'(T_u) | \right]}{4 \cdot \left(1 - \int\limits_{T_l}^{T_u} f(x) \mathrm{d}x \right)}$$

The associated conditions are

$$f'(T_l - \tau) = f'(T_l + \tau) = f'(T_l)$$

and

$$f'(T_u - \tau) = f'(T_u + \tau) = f'(T_u)$$

References

[1]   BIROLINI, A.: *Quality and Reliability of Technical Systems, Theory - Practice - Management*, Springer-Verlag, Berlin Heidelberg 1994

[2]   DREYER, H.; SAUER, W.: *Prozessanalyse, Elementare Stochastische Methoden*, VEB Verlag Technik, Berlin 1982

[3]   HÄRTLER, G.: *Statistische Methoden für die Zuverlässigkeitsanalyse*, VEB Verlag Technik, Berlin 1983

[4]   REINSCHKE, K.: *Zuverlässigkeit von Systemen, Band 1: Systeme mit endlich vielen Zuständen*, VEB Verlag Technik, Berlin 1973

9 Assembly Accuracy Theory

9.1 Introduction

Positioning processes play a decisive role in every production process in which parts need to be inserted in some way. This also applies to the surface-mounted technology (SMT) for electronic modules.

The goal for the assembly of electronic or microtechnical modules is always the production of high-quality mechanical and electrical (in future also optical, *etc.*) connections between a board and one or more components. This forms the basis for a reliable module, which means the quality stays within defined limits over the product lifetime. The typical process sequence for the assembly of an electronic module with components only on one side in a pure surface-mounted technology involves the three process steps shown in Figure 9.1 [4, 11, 13].

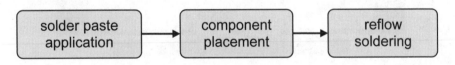

Figure 9.1. Simplified representation of the process sequence for a one-sided SMT

Nowadays, mask printing is typically used for the *solder paste application*. Alternative procedures include dispensing or the still-experimental procedure of the solder paste printing using techniques related to the ink-jet principle. Highly flexible automatic machines are normally used in series production for the component *placement*. Machine concepts based on the so-called pick-and-place principle and the collect-and-place principle prevail here. The subsequent *reflow soldering* then establishes the electrical and mechanical connection between the board substrate and the components. The detailed discussion of the technological procedure is not part of this book; only references to the appropriate literature are provided here [9].

The solder paste application and the positioning subprocesses associated with the assembling process steps, in particular, have a decisive effect on the achieved quality. They are discussed in more detail in the following sections.

9.2 Positioning Tasks for the Solder Paste Application

9.2.1 Machine Accuracy (Positioning Accuracy)

For the solder paste application using mask printing, the mask must be positioned to match with the board substrate to be printed. The mask has openings for those positions where the solder paste depot has to be placed on the board, the so-called pads. A squeegee then squeezes the solder paste contained on the mask through the openings that were positioned onto the board substrate (Figure 9.2, for more detailed discussions, see [9]).

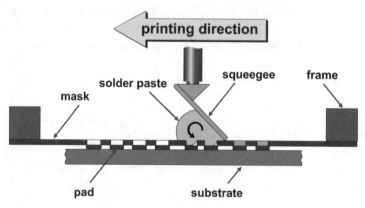

Figure 9.2. Principle of the solder paste printing

A relative positioning action is required to bring the mask into aligment with the board substrate. In real solder paste printers, this is done by the substrate located on a work piece board being recognised with a suitable camera system using defined structures on the surface (setting marks, fiducials).

This permits the determination of the position of the substrate in the coordinate system of the machine. This is used with a X–Y positioning system to bring the substrate and the mask into alignment with each other after which a movement in the Z direction is realised. Simplified, it can be considered as initially being the positioning of a surface A on an (equally sized) surface B (Figure 9.3). Technical positioning actions obviously have a remaining inaccuracy because of the involved mechanical components.

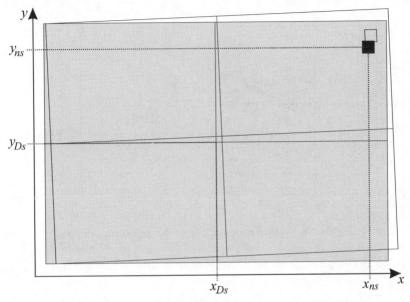

Figure 9.3. Positioning of the mask relative to the substrate

The fact that both the substrate and the mask are rigid bodies means the resulting deviation between both components differs depending on the location because normally not only a positioning error in the X– and Y– directions, but also a rotation, are present.

Real X-Y positioning systems can have very different constructive forms in such technological equipment. To correct an angular offset between the substrate and the mask, normally two drives are realised in one of the two axes driven counter-rotating for a rotary motion. This has the consequence that the rotation is not made at a fixed point, but rather can take place at quite different points in the coordinate system of the machine. Indeed, it is also possible that the rotation point for a specific angle correction lies far outside the actual machine.

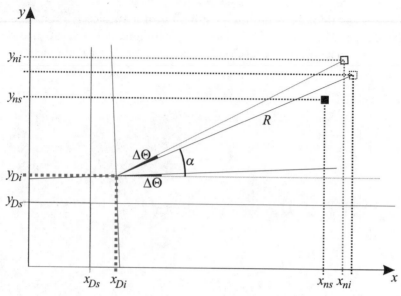

Figure 9.4. Detailed representation of the positioning deviation

The following relations for the positioning error at an arbitrary point (x_n, y_n) for Figure 9.4 result.

$$\Delta x_n = x_{ni} - x_{ns} \tag{9.1}$$
$$= x_{Di} + R \cdot \cos(\alpha + \Delta\Theta) - (x_{Ds} + R \cdot \cos\alpha)$$
$$= (x_{Di} - x_{Ds}) + R(\cos\alpha \cdot \cos\Delta\Theta - \sin\alpha \cdot \sin\Delta\Theta) - R \cdot \cos\alpha$$
$$= (x_{Di} - x_{Ds}) + (x_{ns} - x_{Ds}) \cdot (\cos\Delta\Theta - 1) - (y_{ns} - y_{Ds}) \cdot \sin\Delta\Theta$$

where

- x_{Ds}, y_{Ds} set position of the centre of rotation
- x_{Di}, y_{Di} actual position of the centre of rotation
- $\Delta\Theta$ angular misalignment
- x_{ns}, y_{ns} set position of an arbitrary point of the surface
- x_{ni}, y_{ni} actual position of this point

The following designations are used for the specific position deviations:

$\Delta x_D = x_{Di} - x_{Ds}$; $\Delta y_D = y_{Di} - y_{Ds}$ position deviation of the centre of rotation

$\Delta x_n = x_{ni} - x_{ns}$; $\Delta y_n = y_{ni} - y_{ns}$ position deviation any point and

$x_{nr} = x_{ns} - x_{Ds}$; $y_{nr} = y_{ns} - y_{Ds}$ position of the point relative to the centre of rotation

Substituting the values in Equation (9.1) gives:

$$\Delta x_n = \Delta x_D + x_{nr} \cdot (\cos \Delta \Theta - 1) - y_{nr} \cdot \sin \Delta \Theta \qquad (9.2)$$

Similarly:

$$\Delta y_n = \Delta y_D + y_{nr} \cdot (\cos \Delta \Theta - 1) + x_{nr} \cdot \sin \Delta \Theta \qquad (9.3)$$

The position deviation of an arbitrary point thus consists of the position deviation of the centre of rotation and a component caused by the angular misalignment, which, depending on the position of the point relative to the centre of rotation and the direction of the rotation, can increase or decrease the offset. The following example shows the general sizes the values for Δx_n and Δy_n can assume.

Example: A printed-circuit board in the Double European Standard size (233 mm × 160 mm) is positioned in a screen-printing automatic machine relative to the mask. The centre of rotation should lie exactly at the middle of the board. The position deviation of the upper right corner point, which is at the relative distance of $x_{nr} = 80$ mm and $y_{nr} = 116.5$ mm from the centre of rotation, is shown dependent on the angular misalignment in Table 9.1. The distance of the considered point from the centre of rotation means even relatively small angles have a large effect on the resulting position deviation in the X or Y direction.

Table 9.1. Resulting position deviation in μm for different angles

Δx_D and Δy_D in μm	Angular misalignment $\Delta \Theta$ in °							
	0.01		0.02		0.03		0.04	
	Δx_n	Δy_n	Δx_n	Δy_n	Δx_n	Δy_n	Δx_n	Δy_n
10	−10.33	23.96	−30.67	37.92	−51.01	51.87	−71.35	65.82
20	−0.33	33.96	−20.67	47.92	−41.01	61.87	−61.35	75.82
30	9.67	43.96	−10.67	57.92	−31.01	71.87	−51.35	85.82
40	19.67	53.96	−0.67	67.92	−21.01	81.87	−41.35	95.82
50	29.67	63.96	9.33	77.92	−11.01	91.87	−31.35	105.82

Ultimately, for the consideration of the effects of positioning deviations on the result of an assembly process, it does not matter whether these deviations are caused by purely translatory and/or rotary components. For an arbitrary point in the assembly area, the result is always a lateral deviation in the X and Y directions.

Actually, the previously discussed position deviation should be designated as the machine-dependent position deviation, namely a result of the machine accuracy.

> The *machine accuracy* determines the positional deviation between two ideal parts to be positioned caused by the positioning system of the technological equipment subject to tolerances.

It takes account of the deviations caused by the function units of the solder paste printer used for the positioning action, and assumes an ideal substrate and an ideal mask. Both are, however, components that are manufactured with real technological procedures. This means both have tolerances and also affect the final positioning event, the required alignment between the mask and the substrate.

9.2.2 Assembly Accuracy

If, as described initially, the substrate is positioned to the mask, the previously introduced characteristic variables can be interpreted as follows:

- $[x_{ns}, y_{ns}]$ is the position of a point on the *ideal* mask.
- $[x_{ni}, y_{ni}]$ is the position of the associated point on the *ideal* substrate after the positioning using the position deviation.

Considering the tolerances for the substrate and the mask, this gives an additional deviation for the real points as shown in Figure 9.5.

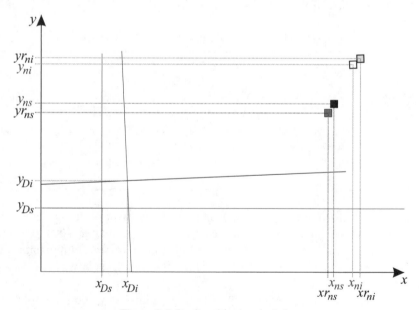

Figure 9.5. Real positioning deviation

With

$$\Delta xr_{ns} = xr_{ns} - x_{ns} \quad ; \quad \Delta yr_{ns} = yr_{ns} - y_{ns} \quad \text{deviation of the real point mask}$$

$$\Delta xr_{ni} = xr_{ni} - x_{ni} \quad ; \quad \Delta yr_{ni} = yr_{ni} - y_{ni} \quad \text{deviation of the real point substrate}$$

this gives the real positioning deviations

$$\Delta xr = \Delta x_n + \Delta xr_{ni} - \Delta xr_{ns} \tag{9.4}$$

and

$$\Delta yr = \Delta y_n + \Delta yr_{ni} - \Delta yr_{ns} \tag{9.5}$$

The deviations of the real points on the mask and the substrate from their associated ideal position are signed. Thus, in the actual case, they can further increase or reduce the positioning deviation caused by the machine accuracy.

The *assembly accuracy* designates the positioning that can be achieved by the machine accuracy together with the position deviations caused by the tolerances of the two assembly partners (Figure 9.6).

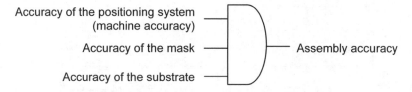

Figure 9.6. Factors that affect the assembly accuracy

The term *accuracy of the positioning system* is defined as being the sum of all factors that influence the machine accuracy. In the specific case of the solder paste printing, these can include:

- the accuracy of the vision system that determines the position of the substrate
- the accuracy of the vision system that measures the position of the mask
- the accuracy of the mechanical components of the positioning system.

The accuracy of the mask and the accuracy of the substrate are both "created" by their production.

All the mentioned accuracies, or more correctly, the associated deviations from the nominal sizes, are, like every other technological quantity, statistical parameters. Namely, the result of the positioning action of a substrate to a mask in the solder paste printer, even when all influencing variables are known, can only be determined with a probability. Each repetition of such a positioning action produces a different result (dispersion), where the random deviations of the substrate and the mask for these repetitions, can, in effect, be considered as being "frozen". If, however, a complete set of substrates is processed, the random deviations of the individual substrates must also be added.

9.3 Positioning Actions for Component Placing

Ignoring the actual construction principle of an automatic placement machine in the electronics surface-mounted technology, the theoretical considerations from Section 9.2 for the positioning accuracy for the solder paste printing can also be used for the placement action.

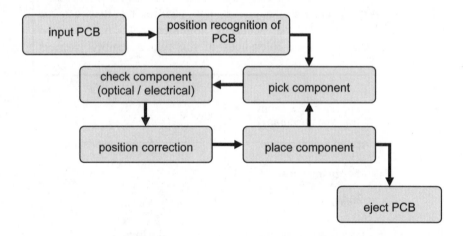

Figure 9.7. Procedure sequence for the automatic assembly in the SMT

We again initially assume an ideal component and an ideal substrate, namely, the dimensions of both joining partners do not have any deviations from the setpoints from the CAD system or from the component datasheets. The placement is made using the procedure shown in Figure 9.7.

As for the solder paste printing, the position of the printed-circuit board is determined using the measurement of defined structures, the *fiducials*. With this basis, the placement positions for the individual components can be corrected appropriately. Each individual component then passes through a cycle in which the position of the component at the placement tool (mechanical gripper, vacuum pipette, *etc.*) is measured and the component then transported to the placement position using the obtained data.

Thus, an arbitrary component is positioned with a dispersion in the X, Y directions (Δx_S and Δy_S) and an angle deviation $\Delta\Theta$. The index S indicates that initially the deviations are based on the centre of rotation or the area centre of gravity of the component. The angle deviation is always based on this point and thus shown without an index. The further consideration can be taken from Section 9.2. The positioning of the component in relation to its nominal position also corresponds to the relative positioning of two surfaces to each other, here, in a "local" coordinate system, orthogonal to the nominal position of the component. For the assignment in the global coordinate system of the automatic placement machine, only a transformation between the nominal position of the component and the machine coordinate system needs to be considered.

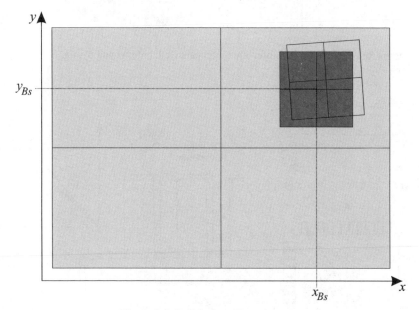

Figure 9.8. Positioning of a component

This means that the conclusions made in Section 9.2.1 and Section 9.2.2 also apply. The positioning system of the automatic placement machine always places a component with a positioning error $\{\Delta x_S, \Delta y_S, \Delta\Theta\}$ relative to its nominal position. The placement position normally refers to the area centre of gravity of the component. Thus, the angular positioning error can only be decomposed into its lateral components and the deviations Δx and Δy when the specific position deviation of the individual component lead (respectively, ball) is examined.

This will become clear when the following special case is considered: a component is to be placed with the positioning deviations $\Delta x_S = 0$, $\Delta y_S = 0$ and $\Delta\Theta > 0$ (Figure 9.9, right-hand part). If the centre of rotation for the angle deviation lies exactly in the area centre of gravity of the component, no additional lateral displacement of the area centre of gravity results from this rotation. Every other point of the component will, however, be moved from its nominal position because of the rotation.

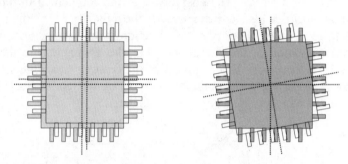

Figure 9.9. Positioning deviation of a component (left: only Δx and Δy, right: only $\Delta\Theta$)

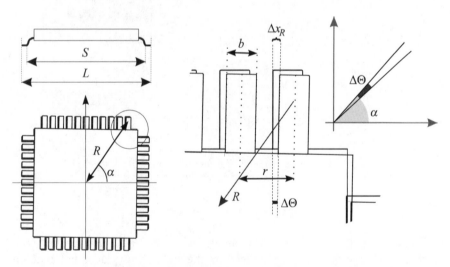

Figure 9.10. Rotation of a component using the example QFP

The effect that a rotation of the component by $\Delta\Theta$ can have on the translatoral offset at the peripheral of the component is now illustrated. The following applies:

$$\Delta x_R = R \cdot \cos(\alpha - \Delta\Theta) - R \cdot \cos\alpha \qquad (9.6)$$
$$= R \cdot (\cos\alpha \cdot \cos\Delta\Theta - \cos\alpha + \sin\alpha \cdot \sin\Delta\Theta)$$
$$= R \cdot (\cos\alpha \cdot (\cos\Delta\Theta - 1) + \sin\alpha \cdot \sin\Delta\Theta)$$

with

- S inside separation of the pins (actually the pin foot zone)
- L the corresponding outside separation (also the overall outer dimension)
- R separation of the area centre of gravity pin/component
- α angle of the outermost pin
- b width of the pin
- r separation pin to pin (component pitch)
- $\Delta\Theta$ angular positioning deviation
- Δx_R translatoral offset at the pin caused by angular misalignment.

For small $\Delta\Theta$, $\cos\Delta\Theta = 1$ and $\sin\Delta\Theta = \Delta\Theta$ (in radians) can be set. Equation (9.6) then reduces to:

$$\Delta x_R = R \cdot \Delta\Theta \cdot \sin\alpha \qquad (9.7)$$
$$= \Delta\Theta \cdot (L + S)/4$$

The translatoral offset resulting from the rotation is easy to calculate using Equation (9.7). Table 9.2 lists the values for some example QFP components.

When Equation (9.7) is considered, it becomes apparent that this calculation no longer contains any part that describes a dependency of Δx_R on x itself. Using the simplification $\cos\Delta\Theta = 1$, this part has been eliminated. It will be shown later that this simplification is appropriate.

Table 9.2. Maximum translatoral offset caused by the angular rotation by 0.5°

Component	Pitch in μm	L in mm	S in mm	Δx_R in μm
QFP 44	800	13.2	11.4	53.67
QFP 160	650	31.2	29.4	132.21
QFP 208	500	30.2	28.5	128.06
QFP 352	300	30.2	28.5	128.06
QFP 576	300	46.0	44.5	197.44

If the translatoral offset Δx_R in the x direction caused by the rotation is calculated for the two external pins in Figure 9.11 and the difference is formed, this produces:

$$\Delta(\Delta x_R) = \Delta x_{R0} - \Delta x_{R1} \tag{9.8}$$
$$= R_0 \cdot (\cos\alpha_0 \cdot (\cos\Delta\Theta - 1) + \sin\alpha_0 \cdot \sin\Delta\Theta) - \cdots$$
$$\cdots R_1 \cdot (\cos\alpha_1 \cdot (\cos\Delta\Theta - 1) + \sin\alpha_1 \cdot \sin\Delta)$$
$$= (R_0 \cdot \cos\alpha_0 - R_1 \cdot \cos\alpha_1) \cdot (\cos\Delta\Theta - 1) + \cdots$$
$$\cdots (R_0 \cdot \sin\alpha_0 - R_1 \cdot \sin\alpha_1) \cdot \sin\Delta\Theta$$

Because of $R_0 \cdot \sin\alpha_0 = R_1 \cdot \sin\alpha_1$ and with $r = R_0 \cdot \cos\alpha_0 - R_1 \cos\alpha_1$, Equation (9.8) becomes:

$$\Delta(\Delta x_R) = r \cdot (\cos\Delta\Theta - 1) \tag{9.9}$$

Namely, the difference between the Δx_R neighbouring pins depends on the pitch of the component and obviously on the angle of rotation. For a component with a pitch of 800 µm and for an angular misalignment of 0.5° (as in Table 9.2), this gives a $\Delta(\Delta x_R)$ of approximately 0.0305 µm. For a Δx_R of 53.67 µm, namely a change of 0.05%.

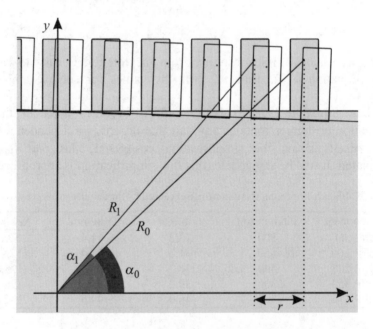

Figure 9.11. Pin sequence at a QFP with rotation

Thus, the change of the translatoral offset from pin to pin caused by the angular rotation lies in orders of magnitude below the true offset and so can be neglected. When the values for Δx_R (Δy_R analogue) in Table 9.2 are considered, it becomes firstly clear that Δx_R is only determined by the external geometrical dimensions of the component, and, secondly, depending on the components pitch, the effects can be quite different. The examples QFP 208 and QFP 352 result because of their identical dimensions in an equally large offset of approximately 128 μm, which, however, for a pitch of 500 μm for QFP 208, means in effect an offset by approximately 25%, for the 300-μm pitch of the QFP 352, however, already means more than 40% of the pitch. If, in addition, a translatoral offset Δx_T occurs, a successful assembly for the latter component becomes questionable.

Even more significant is the effect for the largest QFP contained in the table with 576 leads for a pitch of 300 μm. Although, this is admittedly an extremely demanding package form. An angular positioning error of 0.5° means here already an offset of more than 197 μm.

In a similar manner, values can be performed for the class of the so-called area-array components, namely packaging forms with connections (BGA, CSP, FC) arranged flat on the underside.

Considerations can also be made for the process of placing components in the surface mounted technology as to which individual function elements and function steps affect the machine accuracy. For a modern pick-and-place automatic machine, they are:

- the accuracy of the vision system that measures the position of the substrate in the machine
- the accuracy of the vision system that measures the position of the component at the placement head
- the accuracy of the positioning system with its individual function units for the various motion axes.

The effects of the tolerances and position deviations of the substrate and the components, together with the machine accuracy, also give the assembly accuracy for the placement.

9.4 Improvement of the Assembly Accuracy

The various effects on the assembly accuracy of the technological equipment for the solder paste printing and for the placment of surface-mounted components have been described in Sections 9.2 and 9.3. The position recognition and measuring of the involved partners (mask and substrate or substrate and component) are already measures that can be used to improve the achievable assembly accuracy. Positioning actions without such measuring, such as using purely mechanical stops or end position points, are unthinkable for the structure widths and dimensions used nowadays.

It is obvious that for the placement of components, a higher assembly accuracy can actually be achieved than for the solder paste printing, because the action of the measuring and the correction calculation of the specified setpoints is reperformed for each component, whereas for the solder paste printing, the positioning of the mask to the substrate occurs just once and the presence of a deviation of the dimension stability of both elements can only be compensated as an average.

For the user of technological equipment as considered here, it is initially of secondary importance which machine element caused the positioning error. Decisive is the result of the positioning action, namely, actually only the assembly action, that is, including the deviations caused by the involved assembly partners. The cause analysis for the resulting positioning errors is of interest for the designer and manufacturer of the equipment, and obviously for the user when he takes over the test to correct any determined deficiencies on the machine during running operations. Section 9.6 describes further considerations.

Every increase of the positioning accuracy that is already designed into the technological equipment must normally be bought expensively twice. Firstly, function elements with inherently higher accuracy parameters (*e.g.*, with linear drives instead of spindle-nut systems) increase the purchase costs of the equipment, and secondly, such realisations often cause a reduction of the productivity (*e.g.*, vision systems with higher resolution demand more computing time for the picture evaluation).

After all, with regard to the costs (also see Section 6.3) both for the design and for the selection of the technological equipment (in particular of the contained positioning equipment), the proven principle is true: "Not as accurate as possible, but rather as accurate as necessary." It is important before the selection, such as for an automatic placement machine, that both the requirements (current and future) and the capability limits of the existing equipment are known. Section 9.6 provides more details.

9.5 Assembly Accuracy as Quality Characteristic Variables

9.5.1 Introduction

Fundamental statements about the quality as a property of products and about the quality characteristic variables and their behaviour as random variables have already been discussed in detail in Chapter 5 of this book. Special aspects of the quality management in the electronics production are also shown in [12] (and also other sections).

The following summarising statements are significant:

- Whether an electronics (microtechnical) module was produced quality-conform, namely, a large number of factors influence whether this module performs its expected function under the relevant environmental conditions for its later use. These include both the properties of the initial products (components, substrate, adhesive or soldering materials) and also all assembly processes themselves. The complexity of some components, and obviously also those of the fully assembled module, scarcely allow the function proof for all subsequently possible operating and environmental conditions.

- Equally important as the proof of the produced quality (even when this can only be provided partially, *e.g.*, for static and/or low-dynamic operating states) is the proof of the reliability, namely, the proof of retaining the quality over the expected product life cycle.

The subsequent sections consider the assembly accuracy as one of many, but certainly very important, quality characteristic variable using the example of the placement. The formula apparatus and the general considerations can be transferred to the solder paste printing. As already discussed in Section 9.3, the relative positioning of the mask to the substrate for the solder paste printing is an analogous action for the positioning of a component on the substrate.

To allow the consideration of the assembly accuracy as quality characteristic variable, it must first be defined when a positioning is classified as being qualitatively good and qualitatively bad, respectively. This is only possible using some measurable quantity (quality characteristic) with defined limit values (see Sections 9.5.2 and 9.5.3).

Furthermore, we once again emphasise the difference between machine accuracy and assembly accuracy. The accuracy of the component and the structures on the substrate obviously have decisive importance for the question whether a component can be placed "qualitatively good" on a substrate. Namely, only the assembly of a real component on a real substrate and the evaluation of the result of this assembly provides a statement about the achieved quality. In other words: The assembly accuracy as quality characteristic variable can only be evaluated on the actual product.

It has already been determined in Sections 9.2 and 9.3 that a placed component will be positioned with an offset in the x and y directions and an angular misalignment ($\{\Delta x_S, \Delta y_S, \Delta\Theta\}$). The geometry of the component then can be used to calculate a resulting translatoral dispersion $\{\Delta x, \Delta y\}$ at any point of the component base surface.

The positioning deviations are random variables, namely, even a repeated positioning action with the same component at the same target position will always lead to new, slightly different results. This is caused by the individual function elements of the automatic placement machines, which themselves, like every other technical action, must work with random deviations or operate because of their constructive structure (e.g., the necessary clearance for the fit of mechanically moved parts). In general, it can be assumed that this is a continuous random variable, namely, the positioning deviations can assume theoretically infinitely many realisations within a finite interval. Even, for example, the associated discontinuous drives (stepping motors) do not change anything, although one could initially assume that these mean that only specific positioning values are possible. The large number of factors that effect such a system lead again to a continuous character of the random variable. Furthermore, the large number of influencing factors mentioned above means that a random variable with a normal distribution as the density function can be used as a good approximation.

9.5.2 Placement of Components with Peripheral Leads

The appropriate standards describe the acceptance criteria for electronic modules. For example, the IPC-A-610 Standard [1] contains sample diagrams and numerical values that can be used to evaluate the quality of the joints between the connections of electronic components (pins, leads, balls, *etc.*) and the connection surfaces on the substrate (pads). The defined criteria apply to the acceptance of the fully assembled module, namely, include the soldering, but not the intermediate event after the positioning of the components. A phenomenon often investigated, and frequently discussed at conferences or described in the literature, is the effect of the so-called "self-alignment" of components. It occurs as a result of the surface tension of the melting solder at the individual joints, which, in total over the complete component, can certainly "move" the component over short distances during the liquid phase of the soldering. Although this effect *can* occur, however, it should not be considered for a serious prediction of the expected process quality. Thus, it is legitimate to transfer the mentioned exception criteria already to the evaluation after the placement. A worsening of the positioning result is scarcely to be expected as a result of the soldering process (this ignores any other possible faults during soldering, such as the "tombstone effect") [1, 2].

The IPC-A-610 Standard assigns electronic modules to three product classes. These are:

- Class 1: general electronic products (consumer products; certain computer and computer peripherals; cosmetic deficiencies on the modules are acceptable)
- Class 2: electronic products with dedicated function (communications devices; sophisticated computer technology and other devices that demand high performance and long life; interruption-free function is desirable, but not critical; some cosmetic deficiencies are acceptable)
- Class 3: high-performance electronic products (covers products for which a continuous function or an on-demand function is critical; a failure cannot be tolerated, *e.g.*, life-support systems, air traffic control, *etc.*).

The aim of a placement process is obviously to position the pins of the component exactly at the associated pads on the substrate. Depending on the geometrical dimensions of these two elements, a positioning deviation leads to only a partial or possibly even a complete lack of matching between pin and pad. The Standard has introduced the "side-overhang" and "toe-overhang" quantities as quality characteristic.

Figure 9.12 shows schematically the geometrical quantities at a QFP. Where:

- SO side-overhang
- TO toe-overhang
- b_{Pad} width of the pad
- b_{Pin} width of the pin
- d_L distance of the pin to the next conductive line

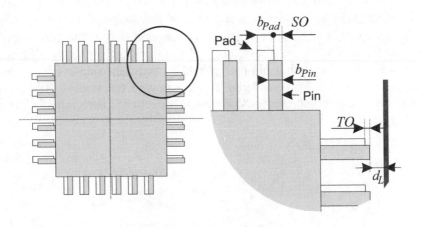

Figure 9.12. Side- and toe-overhang

The IPC-A-610 Standard defines the values shown in Table 9.3 for both quality characteristics that are still acceptable for a good solder joint.

Table 9.3. Specifications for acceptable values for SO and TO in accordance with [1]

	SO	TO
Class 1	$\leq \min\{0.5 \cdot b_{Pin}\,;\ 0.5\ \text{mm}\}$	(Determined by the electrically
Class 2		required separation to the next
Class 3	$\leq \min\{0.25 \cdot b_{Pin}\,;\ 0.5\ \text{mm}\}$	printed conductor)

Some conclusions can be made using the considerations in Section 9.3 and the geometric relationships. Under the prerequisite of one of the usual design rules for the following circuit-board layouts, it can, for example, be assumed that the toe-overhang at a pin row violates the acceptance criterion in accordance with Table 9.3 when the side-overhang at the associated orthogonal pin row already lies for quite some time in the critical range. Furthermore, the subsequent considerations on a pin row (here in the x direction) can be performed and transferred correspondingly to the other pin rows. As overlaying of the effects of the translatoral offset and of the rotary offset, the side-overhang is again the largest at the corner pin. Using Equation (9.7), the side-overhang for the upper pin row in the x direction can be calculated as:

$$SO_o = \begin{cases} \Delta x_T + \Delta x_R - \dfrac{1}{2} \cdot (b_{Pad} - b_{Pin}) & \text{for} \quad \begin{aligned}&(\Delta x_T + \Delta x_R)... \\ &...> \dfrac{1}{2} \cdot (b_{Pad} - b_{Pin}) \end{aligned} \\ 0 & \text{else} \end{cases} \tag{9.10}$$

Note that Δx_T and Δx_R are signed quantities. The translatoral offset Δx_T will be increased by the Δx_R part at a side of the component caused by an additional rotation $\Delta\Theta$ and reduced at the opposite side of the component. Because the quality-conform assembly of a component is already hindered by a single nonfunctioning pin-pad pair, the maximum side-overhang that occurs must always be considered at a positioned component. Thus, the maximum side-overhang for a "component direction", here initially the x direction, is:

$$SO_x = \max(SO_u\,;\,SO_o) \tag{9.11}$$

$$= \begin{cases} |\Delta x_T| + |\Delta x_R| - \dfrac{(b_{Pad} - b_{Pin})}{2} & \text{for} \quad \begin{aligned}&(|\Delta x_T| + |\Delta x_R|)... \\ &...> \dfrac{(b_{Pad} - b_{Pin})}{2} \end{aligned} \\ 0 & \text{else} \end{cases}$$

The side-overhang in the y direction is determined similarly. The maximum value for SO at all four component sides is used for the quality considerations.

The absolute value of the maximum permitted side-overhang is determined using Equations (9.10) and (9.11), and also using the geometry of the pins and the pads. Because the pads are usually wider than the pins, a reduced offset for the positioning initially does not cause any side-overhang. Namely, for example, the permitted total offset in the x direction that results from the addition of the translatoral and the proportional rotary offset can assume larger values than the permitted side-overhang. For the QFP components already used as examples in Section 9.3, the values for SO and Δx are listed in Table 9.4. The assumed pad widths b_{Pad} are based on the recommendations from [3].

Table 9.4. Example values for SO and Δx

Component	Pitch in μm	b_{Pin} in mm	b_{Pad} in mm	Maximum permitted SO in μm	Maximum permitted Δx in μm
QFP 44	800	0.37	0.5	92.5	157.5
QFP 160	650	0.3	0.4	75.0	125.0
QFP 208	500	0.2	0.3	50.0	100.0
QFP 352	300	0.1	0.17	25.0	60.0
QFP 576	300	0.1	0.17	25.0	60.0

When the values for the maximum permitted x offset in the last column of Table 9.4 are compared with the values for the 0.5° angle-positioning error caused by the proportional Δx_R x offset from Table 9.2, it is clear that only the QFP 44 component remains below the permitted value, the limit value, however, would already be exceeded for all other components in the table. Any additional purely translatoral offset is not considered here.

9.5.3 Placement of Area-array Components

In the previous section, the representation of the accuracy of the placement of components with peripheral connections was concentrated on one component side. This is possible, because the geometrical conditions mean that each orthogonal offset direction leads to a violation of the acceptance criteria only when the limit values have already been exceeded in the considered positioning direction.

This simplification cannot be made for area-array components (BGA, CGA, μBGA, CSP, FC) that have as a common characteristic a flat arrangement of the component connections on the underside of the component. The connections are present as spheres or columns, thus, they always have a circular cross section. The associated pads usually also have a circular form on the substrate.

Thus, the quality characteristic for the assembly of these components can be found in the covering of two circular areas, namely of the associated ball-pad combination. It is obvious that, for example, for considering the translatoral offset in the x direction while taking account of the corresponding part of a rotary offset, although a specific overlapping of these two circular areas results, the size of the overlapping area changes immediately with each offset in the orthogonal y direction (see Figure 9.13).

As for the components in Section 9.5.2, the position deviation for a ball-pad combination is always the largest at a "corner" of the component. This allows the calculations to be made for such a constellation.

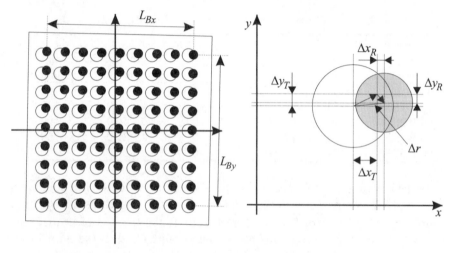

Figure 9.13. BGA with $\{\Delta x_S, \Delta y_S, \Delta\Theta\}$, right-hand ball-pad combination

Analogous to Equation (9.7), it is possible to calculate:

$$\Delta x_R = \Delta\Theta \cdot \frac{L_{By}}{2} \quad \text{and} \quad \Delta y_R = \Delta\Theta \cdot \frac{L_{Bx}}{2} \tag{9.12}$$

where L_{Bx} and L_{By} are the lengths of the BGA, each measured between the external balls. Equations (9.12) are approximate calculations with the same simplifications as discussed previously in the relationship with Equation (9.7).

Thus, the following equation is true for the overall deviation of the centres of the ball and pad circular area:

$$\Delta x = \Delta x_T + \Delta\Theta \cdot \frac{L_{By}}{2} \quad \text{and} \quad \Delta y = \Delta y_T + \Delta\Theta \cdot \frac{L_{Bx}}{2} \tag{9.13}$$

(Note that each of the deviations is signed.)

The evaluation of the overlapping of the ball area and the pad area as quality characteristic requires a quantity that can be calculated or measured. The degree of overlapping is used here, namely, the area part of the smaller of the two circles (normally the ball area) overlapped by the larger circular area.

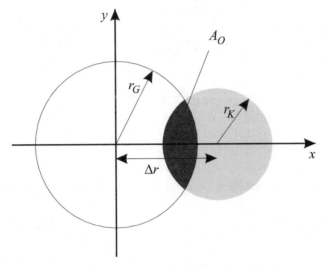

Figure 9.14. Overlapping of the ball-pad combination

The integration of the circle functions in the associated intervals produces the intersection area of the two circles. Related to the area of the smaller circle, this produces a percentage overlap factor for the value range $r_G-r_K < \Delta r < r_G+r_K$:

$$O(\Delta r) = -\frac{100}{\pi \cdot r_K^2} \cdot \left[\frac{1}{2} \cdot \sqrt{4 \cdot \Delta r^2 \cdot r_G^2 - \left(r_G^2 - r_K^2 + \Delta r^2 \right)^2} \cdots \right. \tag{9.14}$$
$$+ r_K^2 \cdot \arcsin\left(\frac{\Delta r^2 - r_G^2 + r_K^2}{2 \cdot \Delta r \cdot r_K} \right) \cdots$$
$$+ r_G^2 \cdot \arcsin\left(\frac{\Delta r^2 + r_G^2 - r_K^2}{2 \cdot \Delta r \cdot r_G} \right) \cdots$$
$$\left. -\frac{1}{2}\pi \cdot \left(r_K^2 + r_G^2 \right) \right]$$

The following relation is true outside of the mentioned interval:

$$O(\Delta r) = \begin{cases} 100\% & for \quad \Delta r \le r_G - r_K \\ 0 & for \quad \Delta r \ge r_G + r_K \end{cases}$$

Because Equation (9.14) is difficult to use in its entirety, a special case will be considered in more detail to give an impression of the general order of the overlap factor under various restrictions.

Special case: $r_G = r_K = r$, namely, the ball and the pad have the same diameter. Equation (9.14) then reduces to:

$$
\begin{aligned}
O(\Delta r) &= -\frac{100}{\pi \cdot r^2} \cdot \left[\begin{array}{l} \dfrac{1}{2}\sqrt{4 \cdot \Delta r^2 \cdot r^2 - \Delta r^4} \ldots \\[2mm] +2 \cdot r^2 \cdot \arcsin\left(\dfrac{\Delta r}{2 \cdot r}\right) - \pi \cdot r^2 \end{array} \right] \\[6mm]
&= -\frac{100}{\pi} \cdot \left[\begin{array}{l} \dfrac{1}{2}\sqrt{4 \cdot \left(\dfrac{\Delta r}{r}\right)^2 - \left(\dfrac{\Delta r}{r}\right)^4} \ldots \\[2mm] +2 \cdot \arcsin\left(\dfrac{1}{2} \cdot \dfrac{\Delta r}{r}\right) - \pi \end{array} \right]
\end{aligned}
$$

(9.15)

Equation (9.15) can then be used to calculate two significant threshold values:

$$
O = \begin{cases} \left(\dfrac{2}{3} - \dfrac{\sqrt{3}}{2 \cdot \pi}\right) \cdot 100 = 39.1\% & for \quad \Delta r = r_G = r_K = r \\[4mm] 50\% & for \quad \dfrac{\Delta r}{r} = 0.8079 \end{cases}
$$

Table 9.6 can be used as an aid for determining of the overlap factor of two circular areas. Thus, the overlap factor using Equation (9.14) depends on the two circle radiuses and on the distance between the centres of the two circles. The centre separation Δr can be calculated using Figure 9.13 and Equation (9.13):

$$\Delta r = \sqrt{\left(\Delta x_T + \Delta\Theta \cdot \frac{L_{By}}{2}\right)^2 + \left(\Delta y_T + \Delta\Theta \cdot \frac{L_{Bx}}{2}\right)^2}$$

(9.16)

Analogous to the example calculations for the effect of the angular misalignment for QFP components (see Table 9.2 and Table 9.4), the values for Δr and O for similarly large BGA components for an angular misalignment of 0.5° are listed below ($\Delta x_T = 0$ and $\Delta y_T = 0$).

In general, an overlap factor of at least 50% is demanded. The values in Table 9.5 show that the effect of the angular misalignment is much less than for the QFP components. This results from the significantly more favourable ratio between component size and the ball size. Thus, for example, the BGA 1089 with more than 40 mm edge length has a ball diameter of 750 μm, whereas the QFP 576, even with an edge length more than 46 mm, has a pin width of only 100 μm. Thus, it is clear that an equally large rotation of both components for the QFP causes a very much smaller overlapping for the pad.

Table 9.5. Effect of the angular misalignment on Δr and O

Component	L_{By} in mm	Δr in μm	$r_G = r_K$ in μm	O in %
PBGA 121	12.7	78.37	750	93.35
PBGA 225	17.8	109.84	750	90.68
PBGA 400	24.1	148.71	750	87.40
PBGA 1089	40.6	250.53	750	78.83
FBGA 144	8.8	54.30	200	82.77

Table 9.6. Overlap factor for different ratios for r_G/r_K and $\Delta r/r_K$

$\Delta r/r_K$ ↓ \ r_G/r_K →	1	1.1	1.2	1.3	1.4	1.5	1.6	1.7	1.8	1.9	2
2	0.00	1.36	3.91	7.26	11.26	15.83	20.89	26.37	32.20	38.32	44.66
1.9	1.33	3.83	7.13	11.09	15.62	20.65	26.10	31.91	38.03	44.38	50.89
1.8	3.74	6.98	10.89	15.38	20.37	25.80	31.60	37.71	44.06	50.58	57.21
1.7	6.81	10.67	15.10	20.06	25.46	31.24	37.35	43.70	50.24	56.90	63.57
1.6	10.41	14.79	19.70	25.07	30.84	36.94	43.30	49.86	56.54	63.26	69.91
1.5	14.43	19.29	24.63	30.37	36.47	42.84	49.43	56.14	62.90	69.60	76.16
1.4	18.81	24.11	29.84	35.93	42.32	48.93	55.68	62.48	69.25	75.87	82.21
1.3	23.51	29.22	35.31	41.71	48.35	55.15	62.01	68.85	75.54	81.95	87.93
1.2	28.48	34.57	41.00	47.68	54.53	61.46	68.38	75.15	81.66	87.72	93.12
1.1	33.68	40.14	46.87	53.80	60.81	67.82	74.70	81.32	87.49	92.99	97.45
1.0	39.10	45.90	52.91	60.04	67.16	74.17	80.91	87.21	92.83	97.39	100
0.9	44.70	51.83	59.09	66.36	73.52	80.42	86.88	92.64	97.32	100	100
0.8	50.46	57.90	65.36	72.72	79.82	86.47	92.40	97.23	100	100	100
0.7	56.36	64.09	71.71	79.06	85.95	92.11	97.13	100	100	100	100
0.6	62.38	70.36	78.07	85.29	91.74	96.99	100	100	100	100	100
0.5	68.50	76.71	84.38	91.23	96.81	100	100	100	100	100	100
0.4	74.71	83.07	90.51	96.55	100	100	100	100	100	100	100
0.3	80.97	89.38	96.15	100	100	100	100	100	100	100	100
0.2	87.29	95.44	100	100	100	100	100	100	100	100	100
0.1	93.64	100	100	100	100	100	100	100	100	100	100

9.5.4 Positioning Accuracy as Random Variable

It has already been indicated several times that the occurring positioning accuracies $\{\Delta x_S, \Delta y_S, \Delta\Theta\}$ or the resulting deviations for the pin-pad and ball-pad combinations of interest $\{\Delta x, \Delta y\}$ are random variables.

The following prerequisites are postulated for the considerations for an example of placing components:

- Several, mutually independent systems acting in the coordinate directions that do not affect each other are used for positioning of a component on the substrate surface. (*Note:* This prerequisite greatly simplifies the consideration of the multidimensional character of the distribution functions, *etc.* Obviously, depending on the actual constructive realisation of the positioning system, an absolute independence is not always present, however, compared with the actual random variables, the resulting effects can be neglected.)
- The positioning of a component is considered to be independent of the location of the target position on the substrate. (*Note:* Real placement equipment almost always exhibits a target position-dependent fault behaviour. Thus, for example, the deviations in the spindle lead result in a location-dependent change of the error quantity for a spindle-nut system used for the drive transmission. These effects can, however, only be taken into consideration for a specific case, in particular, for automatic placement machines using mapping files, *etc.*)
- The individual random variables have a normal distribution. (*Note:* The prerequisites for the presence of a normally distributed random variable are usually present for the positioning. Discussions for testing when the data obeys a normal distribution and for the procedure for the presence of a different distribution type are described in [7, 10, 12].
- The considerations are made using the pin-pad or ball-pad position deviation and, thus, using the Δx and Δy characteristic variables. The effect of the angular misalignment is included in these variables, as described in Section 9.3.

Using these prerequisites, the previous Δx and Δy variables produce two random variables:

$$\Delta X \quad \text{with} \quad f(\Delta x) = \frac{1}{\sqrt{2\cdot\pi}\cdot\sigma_{\Delta x}}\cdot\exp\left(-\frac{1}{2}\left(\frac{\Delta x - \mu_{\Delta x}}{\sigma_{\Delta x}}\right)^2\right) \tag{9.17}$$

$$\Delta Y \quad \text{with} \quad f(\Delta y) = \frac{1}{\sqrt{2\cdot\pi}\cdot\sigma_{\Delta y}}\cdot\exp\left(-\frac{1}{2}\left(\frac{\Delta y - \mu_{\Delta y}}{\sigma_{\Delta y}}\right)^2\right)$$

with the expected value $E\{\Delta X\} = \mu_{\Delta x}$ and the dispersion $D^2\{\Delta X\} = \sigma_{\Delta x}^2$ (ΔY similarly).

As before, the Δx and Δy positioning deviations actually consist of a translatoral and a rotary component (also see Section 9.3). Similar to the corresponding relations, assuming the mutual independency of the random variables, the following equations are also true:

$$\mu_{\Delta x} = \mu_{\Delta xT} + \lambda \cdot \mu_{\Delta\Theta} \quad \text{and} \quad \mu_{\Delta y} = \mu_{\Delta yT} + \lambda \cdot \mu_{\Delta\Theta}$$
$$\sigma_{\Delta x}^2 = \sigma_{\Delta xT}^2 + \lambda^2 \cdot \sigma_{\Delta\Theta}^2 \quad \text{and} \quad \sigma_{\Delta y}^2 = \sigma_{\Delta yT}^2 + \lambda^2 \cdot \sigma_{\Delta\Theta}^2$$

where λ is a conversion factor derived from the component geometry.

The mathematical handling of the normal distribution has been discussed in detail in previous chapters or can be read in [7, 10], *etc*.

For the considerations relevant here, the example of the placement of components with peripheral connections can again initially assume the one-dimensional case. The behaviour of the pin-pad combination has already been derived in Section 9.3. With regard to ΔX, this now gives the appearance shown in Figure 9.15.

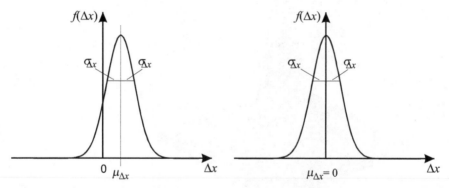

Figure 9.15. One-dimensional normal distribution (left – general case, right – special case with $\mu_{\Delta x} = 0$)

Namely, the translatoral offset varies with the dispersion σ^2 around the expected value μ. The right-hand part of the figure shows the special case $\mu = 0$. In this case, the actual position varies around the nominal position x_{nom} or the translatoral offset of zero.

The dispersion provides a measure of how well a nominal position is attained for the repeated positioning action. The corresponding machine brochures consequently often use the term "repeat accuracy". The expected value μ describes the systematic fault, in this case, the systematic offset. Such a systematic offset can, for example, result from an incorrectly calibrated vision system or for wrongly set correction values for the coordinate transformation. The same regularities apply to the y direction.

For an area-array component, the consideration cannot be restricted to a coordinate, because both the Δx and Δy components immediately affect the significant Δr centre separation parameter (see Section 9.5.3). Because, the definitions assumes both ΔX and ΔY as being mutually independent random variables, the multiplication of the density functions of the two single distributions can produce a two-dimensional normal distribution,

$$f(\Delta x, \Delta y) = \frac{1}{\sqrt{2\pi} \cdot \sigma_{\Delta x}} \cdot \exp\left(-\frac{(\Delta x - \mu_{\Delta x})^2}{2 \cdot \sigma_{\Delta x}^2}\right) \ldots \tag{9.18}$$

$$\cdot \frac{1}{\sqrt{2\pi} \cdot \sigma_{\Delta y}} \cdot \exp\left(-\frac{(\Delta y - \mu_{\Delta y})^2}{2 \cdot \sigma_{\Delta y}^2}\right)$$

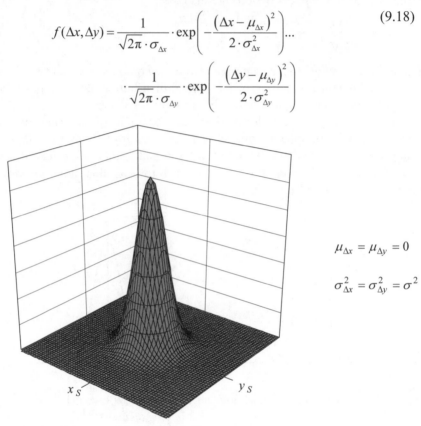

$$\mu_{\Delta x} = \mu_{\Delta y} = 0$$

$$\sigma_{\Delta x}^2 = \sigma_{\Delta y}^2 = \sigma^2$$

Figure 9.16. Two-dimensional normal distribution (parameters on the right-hand side of the figure)

Initially, a special case is assumed that no systematic offset is present in neither the x direction nor in the y direction and the dispersion is equally large in both directions. This gives a two-dimensional normal distribution whose vertex lies in the required nominal position (x_S, y_S) (see Figure 9.16).

Using these parameters, Equation (9.18) simplifies to:

$$f(\Delta x, \Delta y) = \frac{1}{2\pi \cdot \sigma^2} \cdot \exp\left(-\frac{\Delta x^2 + \Delta y^2}{2 \cdot \sigma^2}\right) \tag{9.19}$$

If the expected value for ΔX and/or ΔY now lies outside zero, the complete density function shifts without changing its form (see Figure 9.17, left). If, in addition, the two deviations differ, the normal distribution will also be compressed or stretched in a coordinate (see Figure 9.17, right) and so produces the generally valid form in accordance with Equation (9.18).

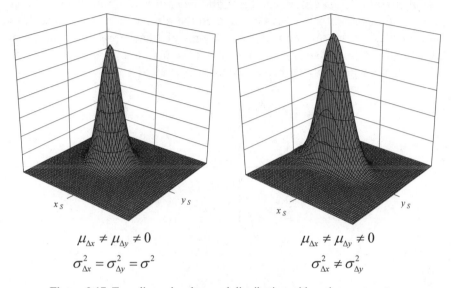

$$\mu_{\Delta x} \neq \mu_{\Delta y} \neq 0 \qquad\qquad \mu_{\Delta x} \neq \mu_{\Delta y} \neq 0$$
$$\sigma_{\Delta x}^2 = \sigma_{\Delta y}^2 = \sigma^2 \qquad\qquad \sigma_{\Delta x}^2 \neq \sigma_{\Delta y}^2$$

Figure 9.17. Two-dimensional normal distribution with various parameters

9.5.5 Defect Rate for Placement

Previously, the positioning deviations for the placement of components with peripheral connections and for area-array components was considered. The side-overhang and overlap factor characteristic variables introduced in Sections 9.5.2 and 9.5.3 can be used as quality characteristics for the positioning. If the character of the positioning deviations are now considered as random variables, this raises the question of the expected yield or the defect rate for the placement.

A random variable lies in the good domain when its realisations satisfy the quality characteristic; conversely, it lies in the bad domain when its realisations violate the quality characteristic. This requires initially to transfer the quality characteristics side-overhang (*SO*) and overlap factor (*O*) to the random variables ΔX and ΔY. It has already been shown that both *SO* and *O* depend on the geometric characteristics of the components. It follows that the defect rate or the yield can only be calculated for an actual component.

In general, initially for a one-dimensional normal distribution, namely, for the use for the placement of a component with peripheral connections, the random variable must lie within specific tolerance limits. For the placement of a component on a substrate with the goal of positioning each individual pin-pad combination as exact as possible to each other, a symmetrical tolerance field can generally be assumed. Using the previously described measure that always considered the pin-pad combination with the largest offset resulting from the overlaying of translatoral and rotary part (worst-case consideration), the reduced consideration in a dimension for components with peripheral connections is permitted.

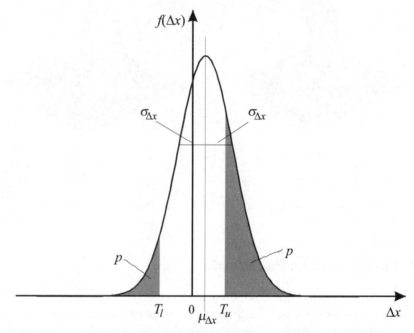

Figure 9.18. One-dimensional distributed random value with tolerance field

For a peripheral component, the maximum permitted side-overhang for modules of class 3 (here in the x direction) using Equation (9.11) and Table 9.3 is:

$$SO = |\Delta x| - \frac{1}{2} \cdot (b_{Pad} - b_{Pin}) \leq \frac{1}{4} \cdot b_{Pin}$$

This implies:

$$|\Delta x| \leq \frac{1}{2} \cdot b_{Pad} - \frac{1}{4} \cdot b_{Pin}$$

or for the tolerance limits:

$$|T_l| = |T_u| = \frac{1}{2} \cdot b_{Pad} - \frac{1}{4} \cdot b_{Pin} \tag{9.20}$$

The defect rate p is calculated with:

$$p = 1 - y \tag{9.21}$$

$$= 1 - \frac{1}{\sqrt{2\pi}\sigma_{\Delta x}} \cdot \int\limits_{T_l}^{T_u} \exp\left(-\frac{(\xi - \mu_{\Delta x})^2}{2 \cdot \sigma_{\Delta x}^2}\right) d\xi$$

Tables are available for calculating the distribution function ([8, 10]). The distribution function must be normalised to use these tables. For this purpose, Equation (9.21) is initially transformed into the following representation form:

$$p = P(\Delta X < T_l) + P(\Delta X > T_u)$$

Using the normalisation $Z = \dfrac{\Delta X - \mu_{\Delta x}}{\sigma_{\Delta x}}$, this yields:

$$p = P\left(\frac{\Delta X - \mu_{\Delta x}}{\sigma_{\Delta x}} < \frac{T_l - \mu_{\Delta x}}{\sigma_{\Delta x}}\right) + P\left(\frac{\Delta X - \mu_{\Delta x}}{\sigma_{\Delta x}} > \frac{T_u - \mu_{\Delta x}}{\sigma_{\Delta x}}\right) \tag{9.22}$$

$$= P\left(Z < \frac{T_l - \mu_{\Delta x}}{\sigma_{\Delta x}}\right) + P\left(Z > \frac{T_u - \mu_{\Delta x}}{\sigma_{\Delta x}}\right)$$

Example:
A QFP 352 must be set for an automatic placement machine for which it is known that the systematic offset in the x direction is 30 μm (expected value $\mu_{\Delta x} = 30$ μm), in the y direction is zero (expected value $\mu_{\Delta y} = 0$) and the standard deviation $\sigma_{\Delta x} = \sigma_{\Delta y} = 20$ μm. The values for the pin width and the pad width, and the permitted side-overhang can be taken from Table 9.4. Using Equation (9.20), this gives $T_l = T_u = 60$ μm as the calculated value. The given data allows us to conclude that the defect rate caused by the positioning in the x direction to be determined as

$$p = \Phi\left(\frac{T_l - \mu_{\Delta x}}{\sigma_{\Delta x}}\right) + 1 - \Phi\left(\frac{T_u - \mu_{\Delta x}}{\sigma_{\Delta x}}\right) = \Phi(-4.5) + 1 - \Phi(1.5) \approx 0.0668$$

The defect rate for this example is 6.68%.

As already mentioned in Section 9.5.4, the similar considerations for the placement of area-array components requires the use of a two-dimensional representation. Because the direction of the positional deviation in the plane has no significance for a ball-pad combination and the overlap factor as the quality criterion, other than for geometric conditions (ball and pad surface diameters), is determined only by the distance between the centres of both circles, the tolerance field obviously also lies in a circle around the nominal position. This leads to the term tolerance radius T_r, which is also derived from the overlap factor O.

For the distributions shown in Figure 9.16 and Figure 9.17, this means only each part of the realisations that lie in a circular field at (x_S, y_S) is also contained within the tolerance field and so belongs to the good domain.

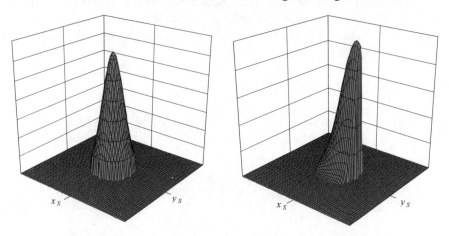

Figure 9.19. Yield component of the two-dimensional distributions

The position of the tolerance field and the distribution becomes clearer when considered as a top view

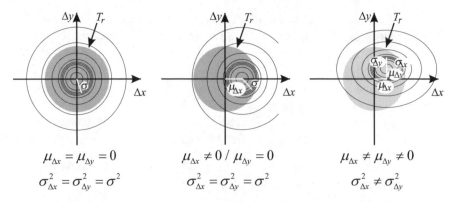

Figure 9.20. Position of the tolerance field and distributions in the top view

For simplification, the calculation of the defect rate uses the left-hand example in Figure 9.20, namely, for a central two-dimensional normal distribution with uniform dispersion. Initially, Equation (9.19) must be transformed into a representation using polar coordinates.

Using $\Delta x^2 + \Delta y^2 = \Delta r^2$, this yields:

$$f(\Delta r, \Delta \varphi) = \frac{1}{2\pi \cdot \sigma^2} \cdot \exp\left(-\frac{\Delta r^2}{2 \cdot \sigma^2}\right) \qquad (9.23)$$

Equation (9.23) is now transformed into a one-dimensional representation:

$$f(\Delta r) = \int_0^{2\pi} f(\Delta r, \Delta \varphi) \cdot \Delta r \cdot d\Delta \varphi \qquad (9.24)$$

$$= \frac{1}{2\pi \cdot \sigma^2} \cdot \int_0^{2\pi} \Delta r \cdot \exp\left(-\frac{\Delta r^2}{2 \cdot \sigma^2}\right) d\Delta \varphi$$

$$= \frac{1}{2\pi \cdot \sigma^2} \cdot \Delta r \cdot \exp\left(-\frac{\Delta r^2}{2 \cdot \sigma^2}\right) \cdot \int_0^{2\pi} d\Delta \varphi$$

$$f(\Delta r) = \frac{\Delta r}{\sigma^2} \cdot \exp\left(-\frac{\Delta r^2}{2 \cdot \sigma^2}\right)$$

This equation form corresponds to the density function of the Rayleigh distribution. For the calculation of the yield, this function must now be integrated over the range of the tolerance field [5, 6].

$$y = \int_0^{T_r} f(\Delta r) \, dr \qquad (9.25)$$

$$= \frac{1}{\sigma^2} \cdot \int_0^{T_r} \Delta r \cdot \exp\left(-\frac{\Delta r^2}{2 \cdot \sigma^2}\right) d\Delta r$$

Because the relationship $\dfrac{d}{d\Delta r}\left\{\exp\left(-\dfrac{\Delta r^2}{2 \cdot \sigma^2}\right)\right\} = -\dfrac{\Delta r}{\sigma^2} \cdot \exp\left(-\dfrac{\Delta r^2}{2 \cdot \sigma^2}\right)$ is true, this yields:

$$y = \left[\exp\left(-\frac{\Delta r^2}{2 \cdot \sigma^2}\right)\right]_0^{T_r} = 1 - \exp\left(-\frac{T_r^2}{2 \cdot \sigma^2}\right) \qquad (9.26)$$

The defect rate for the special case considered here, namely for $\mu_{\Delta x} = \mu_{\Delta y} = 0$ and $\sigma_{\Delta x} = \sigma_{\Delta y} = \sigma$, is thus:

$$p = \exp\left(-\frac{T_r^2}{2 \cdot \sigma^2}\right) \tag{9.27}$$

Figure 9.21 shows the defect rate p depending on the ratio T_r to σ.

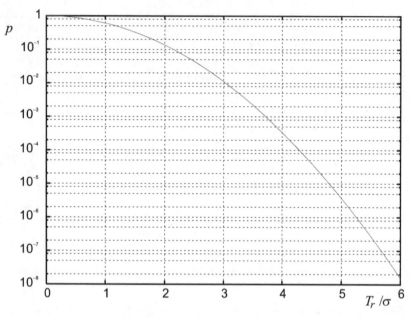

Figure 9.21. Defect rate as function of T_r/σ for $\mu = 0$

Example:
An automatic placement machine is to set an FBGA 144 component. It is known that the automatic machine has no systematic positioning error ($\mu_{\Delta x} = \mu_{\Delta y} = 0$) and the dispersion is $\sigma_{\Delta x} = \sigma_{\Delta y} = \sigma = 50$ μm. The geometric data (see Table 9.6) and a minimum required overlap factor of 50% allow the following calculation to be made:

$T_r = 0.8079 \cdot r = 161.58$ μm , and thus $p = 0.543\%$

In a similar manner as shown above, the yield or the defect rate for the other cases can also be derived from Figure 9.20. Obviously, the additional considered cause variables lead to complicated expressions. For example, the following equation can be used to calculate the defect rate for the case $\mu_{\Delta x} \neq 0$, $\mu_{\Delta y} = 0$, $\sigma_{\Delta x} = \sigma_{\Delta y} = \sigma$:

$$p = 1 - \frac{1}{2\pi \cdot \sigma^2} \cdot \exp\left(-\frac{\mu_{\Delta x}^2}{2\sigma^2}\right) \cdot \int_0^{T_r} \Delta r \cdot \exp\left(-\frac{\Delta r^2}{2\sigma^2}\right) \cdot \ldots$$

$$\left[\int_0^{2\pi} \exp\left(\frac{\mu_{\Delta x} \cdot \Delta r}{\sigma^2} \cdot \cos(\Delta\varphi)\right) d\Delta\varphi\right] d\Delta r$$

The contained integral can only be solved numerically.

9.6 Machine and Process Capability

In the previous section, the theoretically expected defect rates for the component placement were derived and calculated for different combinations. These defect rates always apply to a specific product and to exactly defined tolerance limits.

For several reasons, it is desirable to have a quasi-"product-neutral" quantity, which, for example, permits the capabilities of various equipment to be compared or predicted in advance and so determine whether a specific product can be produced on a specific piece of equipment.

The machine- and process-capability coefficients are such quantities. They originated from the automobile industry where they were introduced several decades ago to allow the qualification of component suppliers to be compared product-independent. In the meantime, these quantities have also been adopted in electronics production.

The correct determination and use of capability coefficients must be made very carefully, which, unfortunately, in particular with regard to their use as sales argument, is not always the case. Because Reference [12] provides an extremely comprehensive scientifically based discussion of this whole subject, an in-depth description does not need to be repeated here. The mathematical principles are also described in Section 5.8. Consequently, only some considerations with regard to the preceding sections are provided here. The classic calculation rule for the process capability coefficients for known tolerance limits is made using:

$$C_p = \frac{T_u - T_l}{6 \cdot \sigma} \tag{9.28}$$

Because, in general, the process does not need to be central in the tolerance field, the so-called *critical process capability coefficient* is calculated using:

$$C_{pk} = \min\left(C_{pl}, C_{pu}\right) = \min\left(\frac{\mu - T_l}{3 \cdot \sigma}, \frac{T_u - \mu}{3 \cdot \sigma}\right) \qquad (9.29)$$

The use of Equations (9.28) and (9.29) is subject to the following restrictions:

- A measurable quality characteristic must be specified (this allows it to be tested).
- Unique tolerance limits must be defined.
- The quality characteristic must have a normal distribution, and the expected value and the dispersion must be stable with regard to time.

If these restrictions are satisfied, a theoretical defect rate can be determined as an approximation:

$$p = 2 \cdot (1 - \Phi(3 \cdot C_p)) \quad \text{for } \mu = 0 \qquad (9.30)$$
$$p \approx 1 - \Phi(3 \cdot C_{pk}) \qquad \text{for } \mu \neq 0$$

Significant values for process-capability coefficients and defect rates are contained in Table 9.7 and Table 9.8.

Table 9.7. Process-capability coefficients and defect rates

Process-capability coefficient C_p	Theoretical defect rate in DPM (defects per million)	This is called a …
1.00	2,700	3-sigma process
1.33	63.34	4-sigma process
1.67	0.5734	5-sigma process
2.00	0.0019	6-sigma process

Table 9.8. Critical process-capability coefficients and defect rates

Critical process-capability coefficient C_{pk}	Theoretical defect rate in DPM (defects per million)	This is called a …
1.00	1,350	3-sigma process
1.33	31.67	4-sigma process
1.67	0.287	5-sigma process
2.00	0.0009	6-sigma process

The *process-capability coefficients* describe the capability of a complete process, including all influencing variables (material, human, machine, environment, *etc.*), to manufacture a specific product in good quality over an extended interval. The data for determining the coefficients should be obtained using an actual product, even when the coefficient will be reused later for other products.

The *machine-capability coefficients* C_m and C_{mk} consider only the pure capability of the machine with the largest possible elimination of other effects for their determination. It is obvious that the machine-capability coefficient must always be higher than the process-capability coefficient. The coefficients are calculated using the same relations as those for the process-capability coefficients. Machine-capability coefficients, because they do not consider the effects of the components actually involved on the process, however, are not suitable for determining a theoretical defect rate.

Detailed descriptions of the procedures for determining the coefficients, including the description of an appropriate measurement procedure, are contained in Reference [12].

References

[1]   *IPC-A-610, Revision C, Acceptability of Electronic Assemblies*, IPC – Association Connecting Electronics Industries, Northbrook, Illinois, USA, 2000

[2]   *IPC-CM-770 Revision D, Guidelines for Printed Board Component Mounting*, IPC – Association Connecting Electronics Industries, Northbrook, Illinois, USA, 1996.

[3]   *IPC-SM-782, Surface Mount Design and Land Pattern Standard*, IPC – Association Connecting Electronics Industries, Northbrook, Illinois, USA, 1999

[4]   KLEIN WASSINK, R. J.; VERGULD, M. M. F.: *Manufacturing Techniques for Surface Mounted Assemblies*. Electrochemical Publ. Ltd., Port Erin (Isle of Man) 1979.

[5]   LINß, G.: *Qualitätsmanagement für Ingenieure*. Fachbuchverlag Leipzig im Carl Hanser Verlag, Munich, Vienna 2002.

[6]   Linß, G.: *Qualitätsmanagement – Trainingsfragen – Praxisbeispiele – Multimediale Visualisierung* training manual. Fachbuchverlag Leipzig im Carl Hanser Verlag, Munich, Vienna 2003.

[7]   NOLLAU, V.; PARTZSCH, L.; STORM, R.: *Wahrscheinlichkeitsrechnung und Statistik in Beispielen und Aufgaben*, Teubner Verlagsgesellschaft Stuttgart, 1997

[8]   SAUER, W.; WOHLRABE, H.: *Qualitätssicherung* lecture script, Technische Universität Dresden, Fakultät Elektrotechnik und Informationstechnik, Institut für Elektronik-Technologie, 2002 edition

[9]   SCHEEL, W. (publisher): *Baugruppentechnologie der Elektronik – Montage*, Verlag Technik, Berlin 1997

[10]   Storm, R.: *Wahrscheinlichkeitsrechnung, mathematische Statistik und statistische Qualitätskontrolle*, 11th revised edition, Munich; Vienna: Fachbuchverlag Leipzig im Carl Hanser Verlag, 2001.

[11]   TUMMALA, RAO R.: *Fundamentals of Microsystems Packaging*, McGraw-Hill, New York, USA, 2001

[12]   WOHLRABE, H.: *Maschinen- und Prozessfähigkeit von Bestückausrüstungen der SMT*, Verlag Dr. Markus A. Detert, Templin 2000

[13]   WOLTER, K.-J.: *Elektroniktechnologie* lecture script. Technische Universität Dresden, Fakultät Elektrotechnik und Informationstechnik, Institut für Elektronik-Technologie, 2000 edition

Appendix – Tables

Table A.1. Distribution function $\phi(u)$ of the normal distribution

u	0.00	0.01	0.02	0.03	0.04	0.05	0.06	0.07	0.08	0.09
0.0	.500000	.503989	.507978	.511966	.515953	.519938	.523922	.527903	.531881	.535856
0.1	.539828	.543795	.547758	.551717	.555670	.559618	.563560	.567495	.571424	.575345
0.2	.579260	.583166	.587064	.590954	.594835	.598706	.602568	.606420	.610261	.614092
0.3	.617911	.621720	.625516	.629300	.633072	.636831	.640576	.644309	.648027	.651732
0.4	.655422	.659097	.662757	.666402	.670031	.673645	.677242	.680822	.684386	.687933
0.5	.691462	.694974	.698468	.701944	.705402	.708840	.712260	.715661	.719043	.722405
0.6	.725747	.729069	.732371	.735653	.738914	.742154	.745373	.748571	.751748	.754903
0.7	.758036	.761148	.764238	.767305	.770350	.773373	.776373	.779350	.782305	.785236
0.8	.788145	.791030	.793892	.796731	.799546	.802338	.805106	.807850	.810570	.813267
0.9	.815940	.818589	.821214	.823814	.826391	.828944	.831472	.833977	.836457	.838913
1.0	.841345	.843752	.846136	.848495	.850830	.853141	.855428	.857690	.859929	.862143
1.1	.864334	866500	.868643	.870762	.872857	.874928	.876976	.879000	.881000	.882977
1.2	.884930	.886861	.888768	.890651	.892512	.894350	.896165	.897958	.899727	.901475
1.3	.903200	.904902	.906582	.908241	.909877	.911492	.913085	.914656	.916207	.917736
1.4	.919243	.920730	.922196	.923642	.925066	.926471	.927855	.929219	.930563	.931889
1.5	.933193	.934478	.935744	.936992	.938220	.939429	.940620	.941792	.942947	.944083
1.6	.945201	.946301	.947384	.948449	.949497	.950528	.951543	.952540	.953521	.954486
1.7	.955434	.956367	.957284	.958185	.959070	.959941	.960796	.961636	.962462	.963273
1.8	.964070	.964852	.965620	.966375	.967116	.967843	.968557	.969258	.969946	.970621
1.9	.971283	.971933	.972571	.973197	.973810	.974412	.975002	.975581	.976148	.976704
2.0	.977250	.977784	.978308	.978822	.979325	.979818	.980301	.980774	.981237	.981691
2.1	.982136	.982571	.982997	.983414	.983823	.984222	.984614	.984997	.985371	.985738
2.2	.986097	.986447	.986791	.987126	.987454	.987776	.988089	.988396	.988696	.988989
2.3	.989276	.989556	.989830	.990097	.990358	.990613	.990862	.991106	.991344	.991576
2.4	.991802	.992024	.992240	.992451	.992656	.992857	.993053	.993244	.993431	.993613
2.5	.993790	.993963	.994132	.994297	.994457	.994614	.994766	.994915	.995060	.995201
2.6	.995339	.995473	.995604	.995731	.995855	.995975	.996093	.996207	.996319	.996427
2.7	.996533	.996636	.996736	.996833	.996928	.997020	.997110	.997197	.997282	.997365
2.8	.997445	.997523	.997599	.997673	.997744	.997814	.997882	.997948	.998012	.998074
2.9	.998134	.998193	.998250	.998305	.998359	.998411	.998462	.998511	.998559	.998605

	0.0	0.1	0.2	0.3	0.4	0.5	0.6	0.7	0.8	0.9
3.0	.998650	.999032	.999313	.999517	.999663	.999767	.999841	.999892	.999928	.999952

	4.0	4.5	5.0	6.0	7.0	8.0	10.0
	$1{-}3.167 \times 10^{-5}$	$1{-}3.398 \times 10^{-6}$	$1{-}2.867 \times 10^{-7}$	$1{-}9.866 \times 10^{-10}$	$1{-}1.280 \times 10^{-12}$	$1{-}6.221 \times 10^{-16}$	$1{-}7.62 \times 10^{-24}$

$$\Phi(-u) = 1 - \Phi(u)$$

$$\Phi(u) = 1 - \frac{1}{u\sqrt{2\pi}} \cdot e^{-\frac{u^2}{2}} \cdot \left(1 - \frac{1}{u^2} + \frac{3}{u^4} - \frac{15}{u^6} + \frac{105}{u^8}\right) \quad \text{(approximation for } u > 3.9\text{)}$$

Table A.2. $u_{1-\alpha}$ quantiles of the normal distribution

$1-\alpha$	0.8	0.9	0.95	0.975	0.990	0.995	0.9990	0.9995	0.9999	0.99995
$u_{1-\alpha}$	0.8416	1.2816	1.6449	1.9600	2.3263	2.5758	3.0902	3.2905	3.7190	3.8906

$$u_{\alpha} = -u_{1-\alpha}$$

Table A.3. $t_{m.q}$ quantiles of the t-distribution

m \ q	0.9	0.95	0.975	0.99	0.995	0.999	0.9995
1	3.08	6.31	12.7	31.8	63.7	318.3	636.6
2	1.89	2.92	4.30	6.96	9.92	22.3	31.6
3	1.64	2.35	3.18	4.54	5.84	10.2	12.9
4	1.53	2.13	2.78	3.75	4.6	7.17	8.61
5	1.48	2.02	2.57	3.37	4.03	5.89	6.87
6	1.44	1.94	2.45	3.14	3.71	5.21	5.96
7	1.41	1.89	2.36	3.00	3.50	4.79	5.41
8	1.40	1.86	2.31	2.90	3.36	4.50	5.04
9	1.38	1.83	2.26	2.82	3.25	4.30	4.78
10	1.37	1.81	2.23	2.76	3.17	4.14	4.59
11	1.36	1.80	2.20	2.72	3.11	4.02	4.44
12	1.36	1.78	2.18	2.68	3.05	3.93	4.32
13	1.35	1.77	2.16	2.65	3.01	3.85	4.22
14	1.35	1.76	2.14	2.62	2.98	3.79	4.14
15	1.34	1.75	2.13	2.60	2.95	3.73	4.07
16	1.34	1.75	2.12	2.58	2.92	3.69	4.01
17	1.33	1.74	2.11	2.57	2.90	3.65	3.97
18	1.33	1.73	2.10	2.55	2.88	3.61	3.92
19	1.33	1.73	2.09	2.54	2.86	3.58	3.88
20	1.33	1.72	2.09	2.53	2.85	3.55	3.85
21	1.32	1.72	2.08	2.52	2.83	3.53	3.82
22	1.32	1.72	2.07	2.51	2.82	3.51	3.79
23	1.32	1.71	2.07	2.50	2.81	3.49	3.77
24	1.32	1.71	2.06	2.49	2.80	3.47	3.75
25	1.32	1.71	2.06	2.49	2.79	3.45	3.73
26	1.31	1.71	2.06	2.48	2.78	3.44	3.71
27	1.31	1.70	2.05	2.47	2.77	3.42	3.69
28	1.31	1.70	2.05	2.47	2.76	3.41	3.67
29	1.31	1.70	2.05	2.46	2.76	3.40	3.66
30	1.31	1.70	2.04	2.46	2.75	3.39	3.65
40	1.30	1.68	2.02	2.42	2.70	3.31	3.55
60	1.30	1.67	2.00	2.39	2.66	3.23	3.46
120	1.29	1.66	1.98	2.36	2.62	3.16	3.37
∞	1.28	1.64	1.96	2.33	2.58	3.09	3.29

Approximation for large m $t_{\infty;q} = \Phi^{-1}(q)$

Table A.4. $\chi^2_{m;q}$ quantiles of the χ^2- distribution

q / m	0.005	0.01	0.025	0.05	0.1	0.9	0.95	0.975	0.99	0.995
1	0.00004	0.00016	0.00098	0.0039	0.016	2.71	3.84	5.02	6.63	7.88
2	0.01	0.0201	0.0506	0.103	0.211	4.61	5.99	7.38	9.21	10.6
3	0.0717	0.115	0.216	0.352	0.584	6.25	7.81	9.35	11.35	12.8
4	0.207	0.297	0.484	0.711	1.06	7.78	9.49	11.14	13.28	14.9
5	0.412	0.554	0.831	1.15	1.61	9.24	11.07	12.83	15.08	16.8
6	0.676	0.872	1.24	1.64	2.20	10.64	12.59	14.45	16.81	18.5
7	0.989	1.24	1.69	2.17	2.83	12.02	14.07	16.01	18.47	20.3
8	1.34	1.65	2.18	2.73	3.49	13.36	15.51	17.53	20.09	22.0
9	1.74	2.09	2.70	3.33	4.17	14.68	16.92	19.02	21.67	23.6
10	2.16	2.56	3.25	3.94	4.87	15.99	18.31	20.48	23.21	25.2
11	2.60	3.05	3.82	4.57	5.58	17.28	19.68	21.92	24.72	26.8
12	3.07	3.57	4.40	5.23	6.30	18.55	21.03	23.34	26.22	28.3
13	3.57	4.11	5.01	5.89	7.04	19.81	22.36	24.74	27.69	29.8
14	4.08	4.66	5.63	6.57	7.79	21.06	23.68	26.12	29.14	31.3
15	4.60	5.23	6.26	7.26	8.55	22.31	25.00	27.49	30.58	32.8
16	5.14	5.81	6.91	7.96	9.31	23.54	26.30	28.85	32.00	34.3
17	5.70	6.41	7.56	8.67	10.09	24.77	27.59	30.19	33.41	35.7
18	6.27	7.01	8.23	9.39	10.86	25.99	28.87	31.53	34.81	37.2
19	6.84	7.63	8.91	10.12	11.65	27.20	30.14	32.85	36.19	38.6
20	7.43	8.26	9.59	10.85	12.44	28.41	31.41	34.17	37.57	40.0
21	8.03	8.90	10.28	11.59	13.24	29.62	32.67	35.48	38.93	41.4
22	8.64	9.54	10.98	12.34	14.04	30.81	33.92	36.78	40.29	42.8
23	9.26	10.20	11.69	13.09	14.85	32.01	35.17	38.08	41.64	44.2
24	9.89	10.86	12.40	13.85	15.66	33.20	36.42	39.36	42.98	45.6
25	10.5	11.52	13.12	14.61	16.47	34.38	37.65	40.65	44.31	46.9
26	11.2	12.20	13.84	15.38	17.29	35.56	38.89	41.92	45.64	48.3
27	11.8	12.88	14.57	16.15	18.11	36.74	40.11	43.19	46.96	49.6
28	12.5	13.56	15.31	16.93	18.94	37.92	41.34	44.46	48.28	51.0
29	13.1	14.26	16.05	17.71	19.77	39.09	42.56	45.72	49.59	52.3
30	13.8	14.95	16.79	18.49	20.60	40.26	43.77	46.98	50.89	53.7
40	20.7	22.16	24.43	26.51	29.05	51.81	55.76	59.34	63.69	66.8
50	28.0	29.71	32.36	34.76	37.69	63.17	67.51	71.42	76.15	79.5
60	35.5	37.48	40.48	43.19	46.46	74.40	79.08	83.30	88.38	92.5
70	43.3	45.44	48.76	51.74	55.33	85.53	90.53	95.02	100.42	104.2
80	51.2	53.54	57.15	60.39	64.28	96.58	101.88	106.63	112.33	116.3
90	59.2	61.75	65.65	69.13	73.29	107.57	113.15	118.14	124.12	128.3
100	67.3	70.06	74.22	77.93	82.36	118.50	124.34	129.56	135.81	140.2

Approximation for large m

$$\chi^2_{m;q} \approx m + \sqrt{2m}\,\Phi^{-1}(q) \quad \text{or more exactly} \quad \chi^2_{m;q} \approx \frac{\left(\sqrt{2m-1}+\Phi^{-1}(q)\right)^2}{2}$$

Table A.5. $F_{m_1;m_2;q}$ quantiles of the F-distribution $q = 0.95$

m_2 \ m_1	1	2	3	4	5	6	7	8	10	12	14	16	20	30	50	75	100	500	∞
1	161	200	216	225	230	234	237	239	242	244	245	246	248	250	252	253	253	254	254
2	18.51	19.00	19.16	19.25	19.30	19.33	19.35	19.37	19.39	19.41	19.42	19.43	19.44	19.46	19.48	19.48	19.49	19.50	19.50
3	10.13	9.55	9.28	9.12	9.01	8.94	8.89	8.85	8.79	8.74	8.71	8.69	8.66	8.62	8.58	8.57	8.55	8.53	8.53
4	7.71	6.94	6.59	6.39	6.26	6.16	6.09	6.04	5.96	5.91	5.87	5.84	5.80	5.75	5.70	5.68	5.66	5.64	5.63
5	6.61	5.79	5.41	5.19	5.05	4.95	4.88	4.82	4.74	4.68	4.64	4.60	4.56	4.50	4.44	4.62	4.41	4.37	4.36
6	5.99	5.14	4.76	4.53	4.39	4.28	4.21	4.15	4.06	4.00	3.96	3.92	3.87	3.81	3.75	3.72	3.71	3.68	3.67
7	5.59	4.74	4.35	4.12	3.97	3.87	3.79	3.73	3.64	3.57	3.53	3.49	3.44	3.38	3.32	3.29	3.27	3.24	3.23
8	5.32	4.46	4.07	3.84	3.69	3.58	3.50	3.44	3.35	3.28	3.24	3.20	3.15	3.08	3.02	3.00	2.97	2.94	2.93
9	5.12	4.26	3.86	3.63	3.48	3.37	3.29	3.23	3.14	3.07	3.03	2.99	2.93	2.86	2.80	2.77	2.76	2.72	2.71
10	4.96	4.10	3.71	3.48	3.33	3.22	3.14	3.07	2.98	2.91	2.86	2.83	2.77	2.70	2.64	2.61	2.59	2.55	2.54
11	4.84	3.98	3.59	3.36	3.20	3.09	3.01	2.95	2.85	2.79	2.74	2.70	2.65	2.57	2.51	2.47	2.46	2.42	2.40
12	4.75	3.89	3.49	3.26	3.11	3.00	2.91	2.85	2.75	2.69	2.64	2.60	2.54	2.47	2.40	2.36	2.35	2.31	2.30
13	4.67	3.81	3.41	3.18	3.03	2.92	2.83	2.77	2.67	2.60	2.55	2.51	2.46	2.38	2.31	2.28	2.26	2.22	2.21
14	4.60	3.74	3.34	3.11	2.96	2.85	2.76	2.70	2.60	2.53	2.48	2.44	2.39	2.31	2.24	2.21	2.19	2.14	2.13
15	4.54	3.68	3.29	3.06	2.90	2.79	2.71	2.64	2.54	2.48	2.42	2.38	2.33	2.25	2.18	2.15	2.12	2.08	2.07
16	4.49	3.63	3.24	3.01	2.85	2.74	2.66	2.59	2.49	2.42	2.37	2.33	2.28	2.19	2.12	2.09	2.07	2.02	2.01
17	4.45	3.59	3.20	2.96	2.81	2.70	2.61	2.55	2.45	2.38	2.33	2.29	2.23	2.15	2.08	2.04	2.02	1.97	1.96
18	4.41	3.55	3.16	2.93	2.77	2.66	2.58	2.51	2.41	2.34	2.29	2.25	2.19	2.11	2.04	2.00	1.98	1.93	1.92
19	4.38	3.52	3.13	2.90	2.74	2.63	2.54	2.48	2.38	2.31	2.26	2.21	2.15	2.07	2.00	1.96	1.94	1.90	1.88
20	4.35	3.49	3.10	2.87	2.71	2.60	2.51	2.45	2.35	2.28	2.22	2.18	2.12	2.04	1.97	1.92	1.91	1.86	1.84
22	4.30	3.44	3.05	2.82	2.66	2.55	2.46	2.40	2.30	2.23	2.17	2.13	2.07	1.98	1.91	1.87	1.85	1.80	1.78
24	4.26	3.40	3.01	2.78	2.62	2.51	2.42	2.36	2.25	2.18	2.13	2.09	2.03	1.94	1.86	1.82	1.80	1.75	1.73
26	4.23	3.37	2.98	2.74	2.59	2.47	2.39	2.32	2.22	2.15	2.10	2.05	1.99	1.90	1.82	1.78	1.76	1.70	1.69
28	4.20	3.34	2.95	2.71	2.56	2.45	2.36	2.29	2.19	2.12	2.06	2.02	1.96	1.87	1.79	1.75	1.73	1.67	1.65
30	4.17	3.32	2.92	2.69	2.53	2.42	2.33	2.27	2.16	2.09	2.04	1.99	1.93	1.84	1.76	1.72	1.70	1.64	1.62
34	4.13	3.28	2.88	2.65	2.49	2.38	2.29	2.23	2.12	2.05	1.99	1.95	1.89	1.80	1.71	1.67	1.65	1.59	1.57
38	4.10	3.24	2.85	2.62	2.46	2.35	2.26	2.19	2.09	2.02	1.96	1.92	1.85	1.76	1.68	1.63	1.61	1.54	1.53
42	4.07	3.22	2.83	2.59	2.44	2.32	2.24	2.17	2.06	1.99	1.93	1.89	1.83	1.73	1.65	1.60	1.57	1.51	1.49
46	4.05	3.20	2.81	2.57	2.42	2.30	2.22	2.15	2.04	1.97	1.91	1.87	1.80	1.71	1.62	1.57	1.55	1.48	1.46
50	4.03	3.18	2.79	2.56	2.40	2.29	2.20	2.13	2.03	1.95	1.89	1.85	1.78	1.69	1.60	1.55	1.52	1.46	1.44
60	4.00	3.15	2.76	2.53	2.37	2.25	2.17	2.10	1.99	1.92	1.86	1.82	1.75	1.65	1.56	1.50	1.48	1.41	1.39
80	3.96	3.11	2.72	2.49	2.33	2.21	2.13	2.06	1.95	1.88	1.82	1.77	1.70	1.60	1.51	1.45	1.43	1.35	1.32
100	3.94	3.09	2.70	2.46	2.31	2.19	2.10	2.03	1.93	1.85	1.79	1.75	1.68	1.57	1.48	1.42	1.39	1.31	1.28
200	3.89	3.04	2.65	2.42	2.26	2.14	2.06	1.98	1.88	1.80	1.74	1.69	1.62	1.52	1.41	1.35	1.32	1.22	1.19
1000	3.85	3.00	2.61	2.38	2.22	2.11	2.02	1.95	1.84	1.76	1.70	1.65	1.58	1.47	1.36	1.30	1.26	1.13	1.08
∞	3.84	3.00	2.60	2.37	2.21	2.10	2.01	1.94	1.83	1.75	1.69	1.64	1.57	1.46	1.35	1.28	1.24	1.11	1.00

Table A.6. $F_{m_1,m_2;q}$ quantiles of the F-distribution $q = 0.99$

m_1 / m_2	1	2	3	4	5	6	8	10	12	14	16	20	30	50	75	100	500	∞
1	4052	4999	5403	5625	5764	5859	5981	6056	6106	6143	6169	6209	6261	6302	6323	6334	6361	6366
2	98.50	99.00	99.17	99.25	99.30	99.33	99.37	99.40	99.42	99.43	99.44	99.45	99.47	99.48	99.49	99.49	99.50	99.50
3	34.12	30.82	29.46	28.71	28.24	27.91	27.49	27.23	27.05	26.92	26.83	26.69	26.50	26.35	26.27	26.23	26.14	26.12
4	21.20	18.00	16.69	15.98	15.52	15.21	14.80	14.55	14.37	14.25	14.15	14.02	13.84	13.69	13.61	13.57	13.48	13.46
5	16.26	13.27	12.06	11.39	10.97	10.67	10.29	10.05	9.89	9.77	9.68	9.55	9.38	9.24	9.17	9.13	9.04	9.02
6	13.74	10.92	9.78	9.15	8.75	8.47	8.10	7.87	7.72	7.60	7.52	7.39	7.23	7.09	7.02	6.99	6.90	6.88
7	12.25	9.55	8.45	7.85	7.46	7.19	6.84	6.62	6.47	6.36	6.27	6.16	5.99	5.86	5.78	5.75	5.67	5.65
8	11.26	8.65	7.59	7.01	6.63	6.37	6.03	5.81	5.67	5.56	5.48	5.36	5.20	5.07	5.00	4.96	4.88	4.86
9	10.56	8.02	6.99	6.42	6.06	5.80	5.47	5.26	5.11	5.00	4.92	4.81	4.65	4.52	4.45	4.42	4.33	4.31
10	10.04	7.56	6.55	5.99	5.64	5.39	5.06	4.85	4.71	4.60	4.52	4.41	4.25	4.12	4.05	4.01	3.93	3.91
11	9.65	7.21	6.22	5.67	5.32	5.07	4.74	4.54	4.40	4.29	4.21	4.10	3.94	3.81	3.74	3.71	3.62	3.60
12	9.33	6.93	5.95	5.41	5.06	4.82	4.50	4.30	4.16	4.05	3.97	3.86	3.70	3.57	3.49	3.47	3.38	3.36
14	8.86	6.51	5.56	5.04	4.70	4.46	4.14	3.94	3.80	3.70	3.62	3.51	3.35	3.22	3.14	3.11	3.03	3.00
16	8.53	6.23	5.29	4.77	4.44	4.20	3.89	3.69	3.55	3.45	3.37	3.26	3.10	2.97	2.89	2.86	2.78	2.75
18	8.29	6.01	5.09	4.58	4.25	4.01	3.71	3.51	3.37	3.27	3.19	3.08	2.92	2.78	2.71	2.68	2.59	2.57
20	8.10	5.85	4.94	4.43	4.10	3.87	3.56	3.37	3.23	3.13	3.05	2.94	2.78	2.64	2.56	2.54	2.44	2.42
22	7.95	5.72	4.82	4.31	3.99	3.76	3.45	3.26	3.12	3.02	2.94	2.83	2.67	2.53	2.46	2.42	2.33	2.31
24	7.82	5.61	4.72	4.22	3.90	3.67	3.36	3.17	3.03	2.93	2.85	2.74	2.58	2.44	2.36	2.33	2.24	2.21
26	7.72	5.53	4.64	4.14	3.82	3.59	3.29	3.09	2.96	2.86	2.78	2.66	2.50	2.36	2.28	2.25	2.16	2.13
28	7.64	5.45	4.57	4.07	3.76	3.53	3.23	3.03	2.90	2.80	2.71	2.60	2.44	2.30	2.22	2.19	2.09	2.06
30	7.56	5.39	4.51	4.02	3.70	3.47	3.17	2.98	2.84	2.74	2.66	2.55	2.38	2.25	2.16	2.13	2.03	2.01
34	7.44	5.29	4.42	3.93	3.61	3.39	3.09	2.89	2.76	2.66	2.58	2.46	2.30	2.16	2.08	2.04	1.94	1.91
38	7.35	5.21	4.34	3.86	3.54	3.32	3.02	2.82	2.69	2.59	2.51	2.40	2.23	2.09	2.00	1.97	1.86	1.84
42	7.28	5.15	4.29	3.80	3.49	3.27	2.97	2.78	2.64	2.54	2.46	2.34	2.18	2.03	1.94	1.91	1.80	1.78
50	7.17	5.06	4.20	3.72	3.41	3.19	2.89	2.70	2.56	2.46	2.38	2.26	2.10	1.95	1.86	1.82	1.71	1.68
60	7.08	4.98	4.13	3.65	3.34	3.12	2.82	2.63	2.50	2.39	2.31	2.20	2.03	1.88	1.79	1.75	1.63	1.60
80	6.96	4.88	4.04	3.56	3.26	3.04	2.74	2.55	2.42	2.31	2.23	2.12	1.94	1.79	1.70	1.66	1.53	1.49
100	6.90	4.82	3.98	3.51	3.21	2.99	2.69	2.50	2.37	2.26	2.19	2.06	1.89	1.73	1.64	1.60	1.47	1.43
200	6.76	4.71	3.88	3.41	3.11	2.89	2.60	2.41	2.27	2.17	2.09	1.97	1.79	1.63	1.53	1.48	1.33	1.28
1000	6.66	4.63	3.80	3.34	3.04	2.82	2.53	2.34	2.20	2.09	2.02	1.89	1.71	1.54	1.44	1.38	1.19	1.11
∞	6.63	4.61	3.78	3.32	3.02	2.80	2.51	2.32	2.18	2.08	2.00	1.88	1.70	1.52	1.41	1.36	1.15	1.00

approximations:

1. $F_{m_1,\infty;q} = \dfrac{1}{m_1}\chi^2_{m_1;q}$

2. $F_{1,m_2;q} = t^2_{m_2,(1+q)/2}$

determination of the quantiles for $q < 0.5$

$$F_{m_1,m_2;q} = \frac{1}{F_{m_2,m_1;1-q}}$$

Table A.7. Correction factors for range R and empirical standard deviation s

n	a_n	c_n	d_n	e_n	e_n/d_n
2	0.798	1.000	1.128	0.853	0.756
3	0.886	1.160	1.693	0.888	0.525
4	0.921	1.092	2.059	0.880	0.427
5	0.940	1.197	2.326	0.864	0.371
6	0.952	1.135	2.564	0.848	0.335
7	0.959	1.214	2.704	0.833	0.308
8	0.965	1.160	2.847	0.820	0.288
9	0.969	1.223	2.970	0.808	0.272
10	0.973	1.176	3.078	0.797	0.259
11	0.975	1.228	3.173	0.787	0.248
12	0.978	1.187	3.258	0.778	0.239
13	0.979	1.232	3.336	0.770	0.231
14	0.981	1.196	3.407	0.762	0.224
15	0.982	1.235	3.472	0.755	0.217
16	0.983	1.202	3.532	0.749	0.212
17	0.985	1.237	3.588	0.743	0.207
18	0.985	1.207	3.640	0.738	0.203
19	0.986	1.239	3.689	0.733	0.199
20	0.987	1.212	3.735	0.729	0.195

Table A.8. Correction factors for quality-control charts (only for $\alpha = 1\%$)

n	Mean chart A_C	A_W	Median chart C_C	C_W	s-chart B_{UCL}	B_{UWL}	B_{LWL}	B_{LCL}	R-chart D_{UCL}	D_{UWL}	D_{LWL}	D_{LCL}
1	2.576	1.96	2.576	1.960								
2	1.821	1.386	1.821	1.386	2.807	2.241	0.031	0.006	3.970	3.170	0.044	0.009
3	1.487	1.132	1.725	1.313	2.302	1.921	0.159	0.071	4.424	3.682	0.303	0.135
4	1.288	0.980	1.406	1.070	2.069	1.765	0.268	0.155	4.694	3.984	0.595	0.343
5	1.152	0.877	1.380	1.050	1.927	1.669	0.348	0.227	4.886	4.197	0.850	0.555
6	1.052	0.800	1.194	0.908	1.830	1.602	0.408	0.287	5.033	4.361	1.066	0.749
7	0.974	0.741	1.182	0.899	1.758	1.552	0.454	0.336	5.154	4.494	1.251	0.922
8	0.911	0.693	1.056	0.804	1.702	1.512	0.491	0.376	5.255	4.605	1.410	1.075
9	0.859	0.653	1.050	0.799	1.657	1.480	0.522	0.410	5.341	4.700	1.550	1.212
10	0.815	0.620	0.958	0.729	1.619	1.454	0.548	0.439	5.418	4.784	1.674	1.335
11	0.777	0.591	0.954	0.726	1.587	1.431	0.570	0.464	5.485	4.858	1.784	1.446
12	0.744	0.566	0.883	0.672	1.560	1.412	0.589	0.486	5.546	4.925	1.884	1.547
13	0.714	0.544	0.880	0.670	1.536	1.395	0.606	0.506	5.602	4.985	1.976	1.639
14	0.688	0.524	0.823	0.626	1.515	1.379	0.621	0.524	5.652	5.041	2.059	1.724
15	0.665	0.506	0.821	0.625	1.496	1.366	0.634	0.540	5.699	5.092	2.136	1.803
16	0.644	0.490	0.774	0.589	1.479	1.354	0.646	0.554	5.742	5.139	2.207	1.876
17	0.625	0.475	0.773	0.588	1.463	1.343	0.657	0.567	5.783	5.183	2.274	1.944
18	0.607	0.462	0.733	0.558	1.450	1.333	0.667	0.579	5.820	5.224	2.336	2.008
19	0.591	0.450	0.732	0.557	1.437	1.323	0.676	0.590	5.856	5.262	2.394	2.068
20	0.576	0.438	0.698	0.531	1.425	1.315	0.685	0.600	5.889	5.299	2.449	2.125
21	0.562	0.428	0.697	0.530	1.414	1.307	0.692	0.610	5.921	5.333	2.500	2.178
22	0.549	0.418	0.668	0.508	1.404	1.300	0.700	0.619	5.951	5.365	2.549	2.229
23	0.537	0.409	0.667	0.507	1.395	1.293	0.707	0.627	5.979	5.396	2.596	2.277
24	0.526	0.400	0.640	0.487	1.386	1.287	0.713	0.635	6.006	5.425	2.640	2.323
25	0.515	0.392	0.640	0.487	1.378	1.281	0.719	0.642	6.032	5.453	2.682	2.366
26	0.505	0.384	0.617	0.469	1.370	1.275	0.724	0.649	6.057	5.480	2.722	2.408
27	0.496	0.377	0.616	0.469	1.363	1.270	0.730	0.655	6.080	5.506	2.761	2.448
28	0.487	0.370	0.595	0.453	1.356	1.265	0.735	0.661	6.103	5.530	2.798	2.487
29	0.478	0.364	0.595	0.453	1.350	1.260	0.739	0.667	6.125	5.554	2.833	2.524
30	0.470	0.358	0.576	0.438	1.343	1.256	0.744	0.673	6.146	5.577	2.867	2.559
31	0.463	0.352	0.576	0.438	1.338	1.251	0.748	0.678	6.166	5.599	2.900	2.593
33	0.448	0.341	0.558	0.425	1.327	1.243	0.756	0.688	6.205	5.640	2.962	2.657
35	0.435	0.331	0.543	0.413	1.317	1.236	0.763	0.697	6.241	5.679	3.019	2.718
37	0.423	0.322	0.528	0.401	1.308	1.230	0.770	0.705	6.275	5.716	3.072	2.773
39	0.412	0.314	0.514	0.391	1.300	1.224	0.776	0.712	6.307	5.750	3.123	2.826
41	0.402	0.306	0.502	0.382	1.292	1.218	0.782	0.719	6.337	5.782	3.172	2.876
43	0.393	0.299	0.490	0.373	1.285	1.213	0.787	0.726	6.365	5.813	3.217	2.923
45	0.384	0.292	0.479	0.364	1.278	1.208	0.792	0.732	6.392	5.842	3.260	2.967
47	0.376	0.286	0.469	0.357	1.272	1.203	0.796	0.738	6.418	5.870	3.300	3.010
49	0.368	0.280	0.459	0.349	1.266	1.199	0.800	0.743	6.442	5.896	3.339	3.050
50	0.364	0.277	0.450	0.343	1.264	1.197	0.802	0.746	6.454	5.909	3.357	3.070

Index

Printing: Krips bv, Meppel
Binding: Stürtz, Würzburg